# Progress in Mathematics

SADLIER-OXFORD

Rose Anita McDonnell

Catherine D. LeTourneau

Anne Veronica Burrows

Rita C. Smith

Theresa J. Talbot

Alice R. Martin

Francis H. Murphy

M. Winifred Kelly

*with*

Dr. Elinor R. Ford

## Series Consultants

Christine D. Thomas, Ph.D.
Assistant Professor of Mathematics Education
Georgia State University
Atlanta, GA

Margaret Mary Bell, S.H.C.J., Ph.D.
Director, Teacher Certification
Rosemont College
Rosemont, PA

Ana Maria Rodriguez
Math Specialist
St. Agatha School
Miami, FL

Re McClung, Ed.D.
Administrator of Instructional Services
and Staff Development
Centralia School District
Buena Park, CA

Alan Amundsen
Math Specialist
Los Angeles, CA

Dennis W. Nelson, Ed.D.
Director of Basic Skills
Mesa Public Schools
Mesa, AZ

Tim Mason
Math Specialist
Palm Beach County School District
West Palm Beach, FL

**Sadlier-Oxford**
A Division of William H. Sadlier, Inc.

**The publisher wishes to thank the following teachers and administrators, who read portions of the program prior to publication, for their comments and suggestions.**

Maria Bono
Whitestone, NY

Sr. Lynn Roebert
San Dimas, CA

Sr. Ruthanne Gypalo
East Rockaway, NY

Madonna Atwood
Creve Coeur, MO

Jennifer Fife
Yardley, PA

Anna Cano-Amato
Brooklyn, NY

Sr. Anita O'Dwyer
North Arlington, NJ

Marlene Kitrosser
Bronx, NY

Donna Violi
Melbourne, FL

Galen Chappelle
Los Angeles, CA

## Acknowledgments

Every good faith effort has been made to locate the owners of copyrighted material to arrange permission to reprint selections. In several cases this has proved impossible. The publisher will be pleased to consider necessary adjustments in future printings.

Thanks to the following for permission to reprint the copyrighted materials listed below.

"The Eraser Poem" (text only) © Louis Phillips.

"Five Cent Balloons" (text only) by Carl Sandburg from EARLY MOON by Carl Sandburg, copyright 1930 by Harcourt, Inc. and renewed 1958 by Carl Sandburg, reprinted by permission of the publisher.

"Little Pine" (text only) by Wang Jian from MAPLES IN THE MIST: Children's Poems of the Tang Dynasty. Translated by Minfong Ho. Copyright © 1996 by Minfong Ho. By permission of Lothrop, Lee & Shepard books, a division of William Morrow & Company, Inc.

"Marvelous Math" (text only) copyright © 1997 by Rebecca Kai Dotlich. First appeared in MARVELOUS MATH, published by Simon & Schuster, 1997. Reprinted by permission of Curtis Brown, Ltd.

"Nature Knows Its Math" (text only) by Joan Bransfield Graham from *Marvelous Math*, edited by Lee Bennett Hopkins, Simon & Schuster Books for Young Readers, 1997. Copyright © 1997 by Joan Bransfield Graham. Used by permission of the author, who controls all rights.

"Nine Mice" (text only) by Jack Prelutsky from THE NEW KID ON THE BLOCK by Jack Prelutsky. Copyright © 1984 by Jack Prelutsky. By permission of Greenwillow Books, a division of William Morrow & Company, Inc.

"Numbers, Numbers" (text only) by Lee Blair.

"Queue" (text only) by Sylvia Cassedy. Copyright © 1993 by Sylvia Cassedy. Used by permission of HarperCollins Publishers.

"School Bus" (text only) © 1996 by Lee Bennett Hopkins. Originally published in SCHOOL SUPPLIES. Published by Simon & Schuster. Reprinted by permission of Curtis Brown, Ltd.

"S.O.S." (text only) by Beverly McLoughland. "S.O.S." first appeared in INSTRUCTOR, February, 1988. Author controls all rights.

"Strategy for a Marathon" (text only) by Marnie Mueller. From *More Golden Apples*, edited by Sandra Haldeman Martz (Papier-Mache Press, 1986).

"Sunflakes" (text only) by Frank Asch from COUNTRY PIE by Frank Asch. © 1979 by Frank Asch. By permission of Greenwillow Books, a division of William Morrow & Company, Inc.

"Sweet Dreams" (text only) by Ogden Nash, from CUSTARD AND COMPANY by Ogden Nash. Copyright © 1961, 1962 by Ogden Nash. By permission Little, Brown and Company.

"Things You Don't Need to Know" (text only) by Kenn Nesbitt. Copyright © 1999 by Kenn Nesbitt. Reprinted by permission of the author.

Anastasia Suen, Literature Consultant

All manipulative products generously provided by ETA, Vernon Hills, IL.

## Photo Credits

Lori Berkowitz: 58, 137, 177, 221. Michael David Berkowitz: 57. Bettman Archive: 332 both. Myrleen Cate: 81, 82, 85, 167, 214, 224, 297, 333, 379. Neal Farris: 70. FPG/ Arthur Tilly: 9; Steven W. Jones:13; Telegraph Colour Library: 263; Jeffrey Meyers: 441. Richard Hutchings: 20-21, 40, 118, 134, 138, 228 top. Image Bank/ John P. Kelley: 44, 330, 360, 432. Ken Karp: 220, 275, 332. Clay Patrick McBride: 31, 67, 103, 133, 163, 170, 189, 209, 212 both, 216, 219, 253, 289, 292, 312, 321, 340, 347, 369, 375, 395, 397, 417. Photodisc: 67. H. Armstrong Roberts: 356. Stock Market/ Dimaggio Kalish: 406; Jon Feingresh: 426. Tony Stone Images/ Trevor Mein: 18; Connie Coleman: 36; Frank Cezus: 46; Rex Zlak: 96; Lori Adamski Peek: 108; Peter Cade: 191; Roger Tully: 270; Zigy Kaluzny: 280; Ryan Beyer:350; Alan Levenson: 384. Uniphoto: 122.

## Illustrators

Diane Ali
Batelman Illustration
Bob Berry
Don Bishop

Robert Burger
Ken Coffelt
Adam Gordon
Dave Jonason

Robin Kachantones
Bea Leute
Blaine Martin
Kathy O'Connell

John Quinn
Fernando Rangel
Sintora Vanderhorst
Dirk Wunderlich

# Contents

# Introduction to Problem Solving

* **Algebraic Reasoning**

# CHAPTER 1
# Place Value

# CHAPTER 2
# Addition

✱ Algebraic Reasoning

# CHAPTER 3
# Subtraction

# CHAPTER 4
# Multiplication Concepts and Facts

* Algebraic Reasoning

# CHAPTER 9
# Geometry

*Algebraic Reasoning

# CHAPTER 10
# Multiply by One Digit

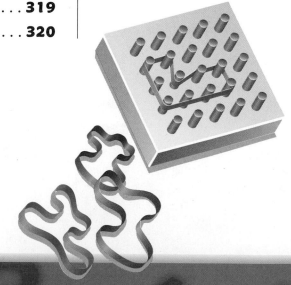

# CHAPTER 11
## Divide by One Digit

# CHAPTER 12
## Fractions

∗ **Algebraic Reasoning**

# End-of-Book Materials

$3 \times 6 - 5 = \underline{\ ?\ }$

# Progress in Mathematics

**W**hether you realize it or not, you see and use mathematics every day!

The lessons and activities in this textbook will help you enjoy mathematics *and* become a better mathematician.

This year you will build on the mathematical skills you already know, as you explore *new* ideas. Working in groups, you will solve problems using many different strategies. You will also have opportunities to make up your own problems and to keep a log of what you discover about math in your own personal Journal. You will learn more about money, time and measurement, and geometry—and even explore the world of fractions and decimals. Believe it or not, you will actually have fun learning about multiplication and division!

You will become a *Technowiz* by completing the computer lessons. These not only will teach you valuable skills, but also expose you to how technology is used in different situations by many people.

And you can use the Skills Update section at the beginning of this book throughout the year to sharpen and review any skills you need to brush up on.

We hope that as you work through this program you will become aware of how mathematics really is a *big* part of your life.

# An Introduction to Skills Update

*Progress in Mathematics* includes a "handbook" of essential skills, Skills Update, at the beginning of the text. These one-page lessons review skills you learned in previous years. It is important for you to know this content so that you can succeed in math this year.

Many lessons in your textbook refer to pages in the Skills Update "handbook". You can use the Skills Update lessons throughout the year, as needed, to help you understand skills you may be unclear about. Or, your teacher may choose to do the Skills Update lessons at the beginning of the year with you, so that you and your teacher can assess your understanding of these previously learned skills.

If you do need to review a skill in Skills Update, your teacher can work with you using ideas from the Teacher's Edition. You can practice the skill using manipulatives, which will help deepen your understanding of that skill.

You may even want to practice specific skills at home. If you need more practice than what is provided on the Skills Update page, you can use the exercises in the *Skills Update Practice Book*. It has an abundance of exercises for each one-page lesson.

## Ones and Tens

To write **numbers** use these ten **digits**.    0, 1, 2, 3, 4, 5, 6, 7, 8, 9

The **value** of the digit depends on its **place** in a number.

The number 4 has a value of 4 ones, or 4.

□ □ □ □
4 = 4 ones

The number 10 has a value of 10 ones, or 1 ten 0 ones.

□ □ □ □ □
□ □ □ □ □    ▭▭▭▭▭▭▭▭▭▭
10 = 10 ones = 1 ten

Standard Form: 44 ←

The digit 4 is in the *ones* place.
It has a value of 4 ones, or 4.

The digit 4 is in the *tens* place.
It has a value of 4 tens, or 40.

You can represent numbers in several ways.

| Standard Form | Place-Value Chart | Place-Value Model | Word Name |
|---|---|---|---|
| 4 | **ones** <br> 4 | □ □ □ □ | four |
| 44 | **tens** \| **ones** <br> 4 \| 4 | ▭▭▭▭▭ ▭▭▭▭▭ ▭▭▭▭▭ ▭▭▭▭▭ □ □ □ □ <br> 4 tens    4 ones | forty-four |

Write the number in standard form.

**1.** 5 ones

**2.**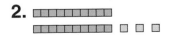

**3.** forty-seven

# Numeration II

## Skip Counting

You can count whole numbers: 0, 1, 2, . . .

▶ **Count by twos.**

Start at zero. Count by twos to 20.

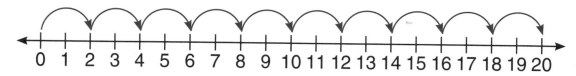

0, 2, 4, 6, 8, 10, 12, 14, 16, 18, 20

These numbers are called **even numbers**.
Even numbers end with the digits **0**, **2**, **4**, **6**, or **8**.

Start at one. Count by twos to 21.

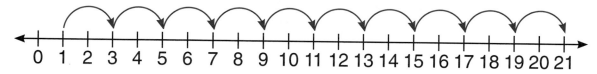

1, 3, 5, 7, 9, 11, 13, 15, 17, 19, 21

These numbers are called **odd numbers**.
Odd numbers end with the digits **1**, **3**, **5**, **7**, or **9**.

▶ **Count by fives to 50.**

0, 5, 10, 15, 20, 25, 30, 35, 40, 45, 50

▶ **Count by tens to 100.**

0, 10, 20, 30, 40, 50, 60, 70, 80, 90, 100

Skip count.

**1.** Count by twos.
Start at 12.
End at 26.

**2.** Count by twos.
Start at 21.
End at 35.

**3.** Count by fives.
Start at 15.
End at 40.

**4.** Count by tens.
Start at 40.
End at 80.

## Ordinal Numbers

Numbers that show order are called **ordinal numbers**.

Look at the calendar. Count to the second Tuesday. The second Tuesday is the fourteenth, or the 14th.

| October | | | | | | |
|---|---|---|---|---|---|---|
| S | M | T | W | TH | F | S |
| | | | 1 | 2 | 3 | 4 |
| 5 | 6 | 7 | 8 | 9 | 10 | 11 |
| 12 | 13 | 14 | 15 | 16 | 17 | 18 |
| 19 | 20 | 21 | 22 | 23 | 24 | 25 |
| 26 | 27 | 28 | 29 | 30 | 31 | |

| Ordinal Word Names | Ordinal Numbers |
|---|---|
| first | 1st |
| second | 2nd |
| third | 3rd |
| fourth | 4th |
| fifth | 5th |
| tenth | 10th |
| eleventh | 11th |
| twelfth | 12th |
| thirteenth | 13th |
| twenty-first | 21st |
| twenty-second | 22nd |
| twenty-third | 23rd |

Look at the ordinal word and number endings.

**Write the ordinal word name or number.**

1. 9th
2. seventh
3. twelfth
4. 17th
5. twentieth
6. 24th
7. eightieth
8. 31st
9. 22nd
10. thirtieth
11. twenty-fifth
12. 15th

### Problem Solving

13. Todd is seventeenth in line. Mika is twenty-first. How many people are between them?

14. There are 9 people ahead of you in line. What is your place?

# Whole Number Operations 1

## Addition Facts Through 18

Add: 9 + 4 = ___?___

```
9  ◄─── addend
+4  ◄─── addend
13  ◄─── sum
```

Think: Nine in the first addend.
　　　Count on: 10, 11, 12, 13.

Start at 0.
Go to 9.
Count on 4.

You can use the number line to add.

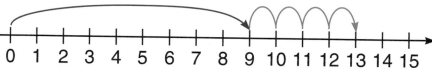

0  1  2  3  4  5  6  7  8  9  10 11 12 13 14 15

9 + 4 = 13
**number sentence**

```
  9
+ 4
 13
```

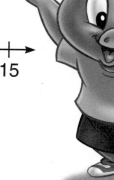

## Add. You may use a number line.

| | | | | | |
|---|---|---|---|---|---|
| **1.** $\begin{array}{r}1\\+2\\\hline\end{array}$ | **2.** $\begin{array}{r}5\\+3\\\hline\end{array}$ | **3.** $\begin{array}{r}4\\+5\\\hline\end{array}$ | **4.** $\begin{array}{r}3\\+3\\\hline\end{array}$ | **5.** $\begin{array}{r}2\\+7\\\hline\end{array}$ | **6.** $\begin{array}{r}2\\+4\\\hline\end{array}$ |
| **7.** $\begin{array}{r}7\\+8\\\hline\end{array}$ | **8.** $\begin{array}{r}3\\+7\\\hline\end{array}$ | **9.** $\begin{array}{r}6\\+5\\\hline\end{array}$ | **10.** $\begin{array}{r}9\\+4\\\hline\end{array}$ | **11.** $\begin{array}{r}4\\+8\\\hline\end{array}$ | **12.** $\begin{array}{r}6\\+6\\\hline\end{array}$ |
| **13.** $\begin{array}{r}6\\+1\\\hline\end{array}$ | **14.** $\begin{array}{r}0\\+9\\\hline\end{array}$ | **15.** $\begin{array}{r}5\\+2\\\hline\end{array}$ | **16.** $\begin{array}{r}8\\+6\\\hline\end{array}$ | **17.** $\begin{array}{r}7\\+9\\\hline\end{array}$ | **18.** $\begin{array}{r}5\\+6\\\hline\end{array}$ |

**19.** 8 + 3　　　　**20.** 9 + 9　　　　**21.** 6 + 4　　　　**22.** 8 + 2

**23.** 6 + 9　　　　**24.** 7 + 7　　　　**25.** 0 + 3　　　　**26.** 1 + 9

# Whole Number Operations II

## Order in Addition

You can use the Addition Table to find the sum of $3 + 6$ and $6 + 3$.

$3 + 6 = \underline{\ ?\ }$

▶ Find 3 along the side.
Find 6 along the top.
Trace across from 3 and down from 6 to meet at 9.

$3 + 6 = 9$ ◀——**sum**

**addends**

$6 + 3 = \underline{\ ?\ }$

▶ Now find 6 along the side.
Find 3 along the top.
Trace across from 6 and down from 3 to meet at 9.

$6 + 3 = 9$ ◀——**sum**

**addends**

▶ The sum of zero and a number is that number.

$6 + 0 = 6$         $0 + 6 = 6$

**Addition Table**

| + | 0 | 1 | 2 | 3 | 4 | 5 | 6 | 7 | 8 | 9 |
|---|---|---|---|---|---|---|---|---|---|---|
| 0 | 0 | 1 | 2 | 3 | 4 | 5 | 6 | 7 | 8 | 9 |
| 1 | 1 | 2 | 3 | 4 | 5 | 6 | 7 | 8 | 9 | 10 |
| 2 | 2 | 3 | 4 | 5 | 6 | 7 | 8 | 9 | 10 | 11 |
| 3 | 3 | 4 | 5 | 6 | 7 | 8 | 9 | 10 | 11 | 12 |
| 4 | 4 | 5 | 6 | 7 | 8 | 9 | 10 | 11 | 12 | 13 |
| 5 | 5 | 6 | 7 | 8 | 9 | 10 | 11 | 12 | 13 | 14 |
| 6 | 6 | 7 | 8 | 9 | 10 | 11 | 12 | 13 | 14 | 15 |
| 7 | 7 | 8 | 9 | 10 | 11 | 12 | 13 | 14 | 15 | 16 |
| 8 | 8 | 9 | 10 | 11 | 12 | 13 | 14 | 15 | 16 | 17 |
| 9 | 9 | 10 | 11 | 12 | 13 | 14 | 15 | 16 | 17 | 18 |

▶ Changing the order of the addends does *not* change the sum.

$3 + 6 = 9$         and         $6 + 3 = 9$

Find the sum. You may use the Addition Table.

**1.**  4      5     **2.**  2      7     **3.**  9      5     **4.**  6      7
     $+5$    $+4$          $+7$    $+2$          $+5$    $+9$          $+7$    $+6$

**5.** $3 + 9$        **6.** $7 + 0$        **7.** $6 + 9$        **8.** $3 + 8$
     $9 + 3$              $0 + 7$              $9 + 6$              $8 + 3$

# Whole number Operations III

## Subtraction Facts Through 18

Subtract: 13 − 4 = __?__

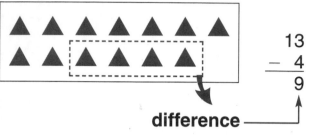

$$\begin{array}{r} 13 \\ -\ 4 \\ \hline 9 \end{array}$$

**difference**

Start at 0.
Go to 13.
Count back 4.

You can use the number line to subtract.

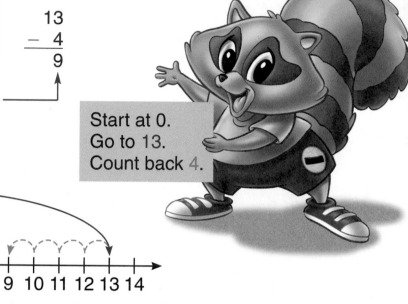

13 − 4 = 9          $$\begin{array}{r} 13 \\ -\ 4 \\ \hline 9 \end{array}$$
**number sentence**

| If you subtract zero from a number, the difference is the number. <br> 7 − 0 = 7 | If you subtract a number from itself, the difference is zero. <br> 7 − 7 = 0 |
|---|---|

Subtract. You may use a number line.

**1.**  $\begin{array}{r} 6 \\ -3 \\ \hline \end{array}$     **2.**  $\begin{array}{r} 10 \\ -4 \\ \hline \end{array}$     **3.**  $\begin{array}{r} 3 \\ -2 \\ \hline \end{array}$     **4.**  $\begin{array}{r} 12 \\ -8 \\ \hline \end{array}$     **5.**  $\begin{array}{r} 11 \\ -2 \\ \hline \end{array}$     **6.**  $\begin{array}{r} 18 \\ -9 \\ \hline \end{array}$

**7.**  $\begin{array}{r} 14 \\ -6 \\ \hline \end{array}$     **8.**  $\begin{array}{r} 8 \\ -0 \\ \hline \end{array}$     **9.**  $\begin{array}{r} 7 \\ -7 \\ \hline \end{array}$     **10.**  $\begin{array}{r} 12 \\ -3 \\ \hline \end{array}$     **11.**  $\begin{array}{r} 16 \\ -8 \\ \hline \end{array}$     **12.**  $\begin{array}{r} 10 \\ -5 \\ \hline \end{array}$

**13.** 9 − 5          **14.** 6 − 4          **15.** 16 − 7          **16.** 15 − 6

REVIEW OF GRADE 2 SKILLS

# Whole Number Operations IV

## Mental Math Strategies

▶ You can use fact families to subtract mentally.

Subtract: $11 - 6 =$ ?

> Think: $6 + 5 = 11$  So $11 - 6 = 5$.

| | Remember the fact family for 5, 6, and 11. | | |
|---|---|---|---|
| $\begin{array}{r} 6 \\ + 5 \\ \hline 11 \end{array}$ | $\begin{array}{r} 5 \\ + 6 \\ \hline 11 \end{array}$ | $\begin{array}{r} 11 \\ - 6 \\ \hline 5 \end{array}$ | $\begin{array}{r} 11 \\ - 5 \\ \hline 6 \end{array}$ |

▶ You can use doubles to add and subtract mentally.

### Doubles Plus 1

Add: $8 + 7 =$ ?
Think: $7 + 7 = 14$

1 more than $7 + 7$   $8 + 7 = 15$

Subtract: $15 - 7 =$ ?
Think: $14 - 7 = 7$

1 more than $14 - 7$   $15 - 7 = 8$

### Doubles Minus 1

Add: $4 + 5 =$ ?
Think: $5 + 5 = 10$

1 less than $5 + 5$   $4 + 5 = 9$

Subtract: $9 - 5 =$ ?
Think: $10 - 5 = 5$

1 less than $10 - 5$   $9 - 5 = 4$

▶ Look for doubles.
Add: $4 + 4 + 5 =$ ?

$4 + 4 + 5 = 13$
   $8$   Think: $8 + 5 = 13$

Look for sums of ten.
Add: $8 + 3 + 7 =$ ?

$8 + 3 + 7 = 18$
   $10$   Think: $8 + 10 = 18$

Add or subtract mentally.
Tell the strategy that helped you find the answer.

**1.** $15 - 8$    **2.** $6 + 7$    **3.** $4 + 3$    **4.** $11 - 6$

**5.** $3 + 7 + 9$    **6.** $9 + 1 + 8$    **7.** $6 + 4 + 6$

# Whole Number Operations V

## Patterns

Algebra ✓

Some patterns use numbers, shapes, and colors.

▶ Find the next number in the pattern.

18, 22, 21, 25, 24, 28, _?_

**Identify the rule:** $+4, -1$

**Apply the rule:** $24 + 4 = 28$, $28 - 1 = 27$

So the next number in the pattern is 27.

▶ Draw the next shape in the pattern.

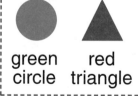

**Identify the rule:**

green circle  red triangle  green circle  red triangle  green circle  red triangle  ?

**Apply the rule:** add 1 red triangle, .

So the next shape in the pattern is .

Look for a pattern. Find the next number or shape.

**1.** 4, 8, 5, 9, 6, 10, _?_

**2.**

8

## Fractions: Parts of a Whole

▶ A **fraction** can name part of a whole.

Count the number of equal parts shaded blue: 3

Count the total number of equal parts: 4

To tell the part of the whole shaded blue, write the fraction $\frac{3}{4}$.

Read $\frac{3}{4}$ as: three fourths.

Word name: three fourths.

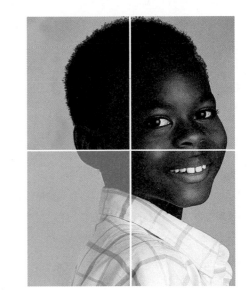

▶ Each figure is divided into equal parts. The number of equal parts names the fractional parts.

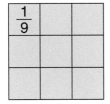

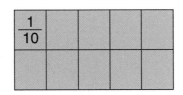

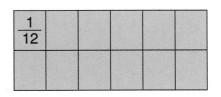

9 equal parts
**ninths**

10 equal parts
**tenths**

12 equal parts
**twelfths**

Write the fraction and word name for the shaded part of each figure.

**1.**

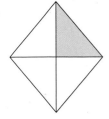

**2.**

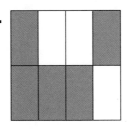

**3.**

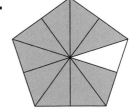

# Fractions II

## Fractions: Parts of a Set

**Fractions** can name parts of a set or groups of objects.

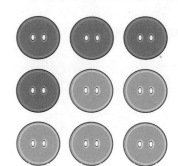

$\frac{4}{9}$ of the buttons are purple.

$\frac{5}{9}$ of the buttons are green.

Write the fraction for the part of each set that is shaded. Then write the fraction for the part that is *not* shaded.

1.

2.

3.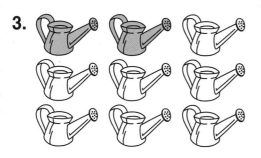

4.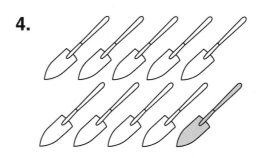

Write the fraction for each.

5. What fractional part of the flowers is purple?

6. What fractional part of the flowers is *not* purple?

7. What fractional part of the flowers is drooping?

8. What fractional part of the flowers is *not* drooping?

## Inch

The **inch** (**in.**) is a customary unit used to measure length.

The inch can be used to measure small objects.

The length of a small paper clip is *about* 1 inch (1 in.)

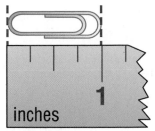

A small paper clip is a benchmark for an inch.

A **benchmark** is an object of known measure that can be used to estimate the measure of other objects.

Estimate how long each is.
Then measure with an inch ruler.

**1.**

**2.**

**3.**

**4.**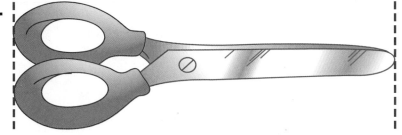

# Measurement II

## Centimeter

The **centimeter** (**cm**) is a metric unit used to measure length.

The centimeter can be used to measure small objects.

The width of your thumb is *about* 1 centimeter (1 cm). This is a benchmark for a centimeter.

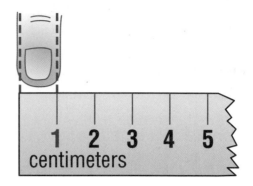

Use a centimeter ruler to measure an object to the nearest centimeter.

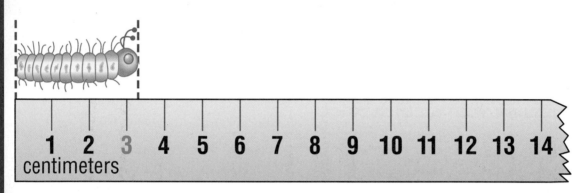

The caterpillar is *about* 3 cm long.

Draw a line for each length.

**1.** 3 cm     **2.** 7 cm     **3.** 12 cm     **4.** 10 cm

Use a centimeter ruler to measure each to the nearest centimeter.

**5.**

**6.**

# Perimeter

**Perimeter** is the distance around a figure.

To find the perimeter of the photo, add the length of its four sides.

Perimeter = 2 in. + 3 in. + 2 in. + 3 in.

Perimeter = 10 in.

The perimeter of the photo is 10 in.

3 in.

2 in.

**Find the perimeter of each figure.**

**1.**

4 in.

4 in.          4 in.

4 in.

**2.**

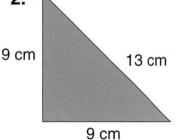

9 cm          13 cm

9 cm

**3.**

4 in.

9 in.          9 in.

4 in.

**4.**

5 cm          5 cm

8 cm          8 cm

5 cm          5 cm

**5.**

1 in.

1 in.          1 in.

2 in.

**6.**

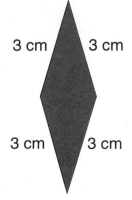

3 cm          3 cm

3 cm          3 cm

# Measurement IV

## Cup, Pint

The **cup (c)** and **pint (pt)** are customary units used to measure liquids.

1 cup                    1 pint

2 c = 1 pt

Do the pictures show the same amount?
Explain your answer.

**1.**

**2.**

**3.**

## Quart

The **quart (qt)** is a customary unit used to measure liquids.

1 quart

2 pt = 1 qt

Do the pictures show the same amount?
Explain your answer.

**1.**

**2.**

**3.**

## Hour

The short hand on a clock tells the **hour** (**h**). The long hand tells how many **minutes** (**min**).

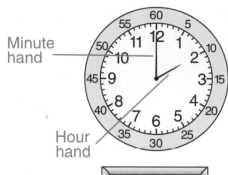

| 1 hour = 60 minutes |
| 1 h = 60 min |

Read this time as: two o'clock.

Write in standard form as: 2:00.

**Digital Time**

Write the time in standard form.

**1.**   **2.**   **3.**

**4.**   **5.**   **6.**

Draw the time. Show the hour and minute hands.

**7.**   **8.**   **9.**   **10.**

**11.** four o'clock  **12.** six o'clock  **13.** twelve o'clock  **14.** nine o'clock

## Half Hour

There are 30 minutes in one half hour.

Read this time as: two thirty,

or half past two,

or thirty minutes after two.

Write in standard form as: 2:30.

Write the time in standard form.

**1.**

**2.**

**3.**

**4.**

**5.**

**6.**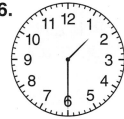

Draw the time. Show the hour and minute hands.

**7.**

**8.**

**9.**

**10.**

**11.** half past four

**12.** thirty minutes after 8

**13.** six thirty

REVIEW OF GRADE 2 SKILLS

# Geometry I

## Sides and Corners

A **side** is a straight **line segment** that bounds a closed figure.

The point where two sides meet is called a **corner**.

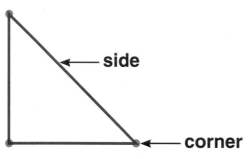

← **side**

← **corner**

A **triangle** has 3 sides and 3 corners.

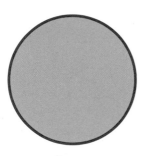

A **circle** is a closed flat figure with 0 sides and 0 corners.

Name each figure. Then write the number of straight sides and the number of corners each has.

**1.**

**2.**

**3.**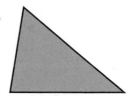

Draw each closed figure. You may use dot paper.

**4.** 5 sides          **5.** 6 corners          **6.** 4 sides

18

## Space Figures

**Space figures** are not flat.
They are sometimes called **solids**.

**cube**

**pyramid**

**rectangular prism**

curved surface

**cylinder**

**sphere**

**cone**

Write the name of each space figure pictured.

**1.**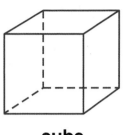

**2.**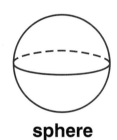

**3.**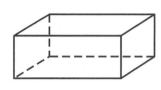

**4.**

**5.**

**6.**

**7.** Name some objects in the classroom, home, or outdoors that have the shape of space figures.

## Reading and Writing Tallies

| Student | Tally |
|---|---|
| Brianna | 卌 卌 IIII |
| Frederick | 卌 II |
| Jamie | IIII |
| Nicoletta | 卌 卌 卌 IIII |

A **tally chart** is a format used to organize information.

The tally chart to the left shows the number of votes each student received in the election for class representative.

A **tally mark** is a line used to count things. Write tally marks in groups of five.

| represents a count of 1

卌 represents a count of 5

> Every fifth tally mark is placed diagonally across the previous four marks.

Solve. Use the tally chart above.

1. How many votes did each student receive?

2. Who received the most votes? the least votes?

Write the number each represents.

3. 卌 |

4. 卌 卌

5. 卌 |||

6. 卌 卌 |

Represent each number using tally marks.

7. 15

8. 3

9. 13

10. 21

# Reading a Pictograph

A **pictograph** is a diagram that represents data in picture form.

The pictograph shows the number of books each student read during summer break.

The **key** in a pictograph tells how many each picture represents.

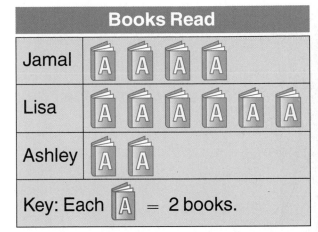

| Books Read | | | | | | |
|---|---|---|---|---|---|---|
| Jamal | A | A | A | A | | |
| Lisa | A | A | A | A | A | A |
| Ashley | A | A | | | | |
| Key: Each A = 2 books. | | | | | | |

Use the key to find how many books each student read.

**Skip count or add.**

Jamal:

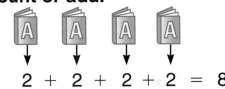

$$2 + 2 + 2 + 2 = 8$$

Jamal read 8 books during summer break.

Use the pictograph to solve each problem.

1. How many books did Lisa read? How many did Ashley read?

2. Who read the most books? Who read the least books?

3. How many more books did Jamal read than Ashley?

4. How many fewer books did Ashley read than Lisa?

# Statistics III

## Reading a Bar Graph

A **bar graph** is a graph that uses bars of different lengths to show data.

This bar graph shows the number of students who named each sport as their favorite to watch.

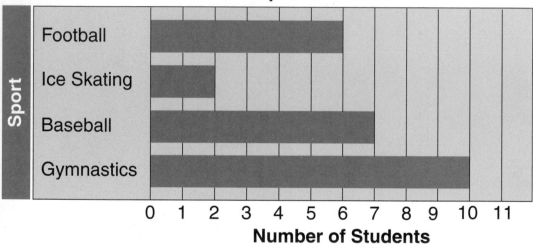

**Favorite Sport to Watch**

How many students prefer to watch ice skating?

Look at the end of the bar labeled *ice skating*. The bar ends at the 2 mark.

So 2 students named ice skating as their favorite sport to watch.

Use the bar graph to solve each problem.

1. What is the title of the bar graph?

2. What do the labels tell you about the graph?

3. How can you find which sport is preferred by the most number of students? preferred by the least number of students?

4. How many students prefer to watch football? baseball? gymnastics?

# Key Sequences

A calculator has 3 kinds of keys.

**Control Keys:**

**Number Keys:**

**Operation Keys:**

▶ Add: 45 + 36 = __?__

To add using a calculator,

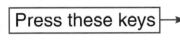

| Press these keys → | ON/AC | 4 | 5 | + | 3 | 6 | = |
|---|---|---|---|---|---|---|---|
| Display → | 0. | 4. | 45. | 45. | 3. | 36. | 81. |

So 45 + 36 = 81.

> The decimal point may not appear in the display.

▶ Subtract: 38 − 23 = __?__

To subtract using a calculator,

| Press these keys → | ON/AC | 3 | 8 | − | 2 | 3 | = |
|---|---|---|---|---|---|---|---|
| Display → | 0. | 3. | 38. | 38. | 2. | 23. | 15. |

So 38 − 23 = 15.

**Use a calculator to compute. Watch for + and −.**

**1.** 57 + 16      **2.** 98 − 42      **3.** 28 + 19

**4.** 86 − 28      **5.** 74 − 29      **6.** 33 + 47

# Introduction to Problem Solving

## Dear Student,

Problem solvers are super sleuths. We invite you to become a super sleuth by using these *five steps* when solving problems.

**1 IMAGINE**
Create a mental picture.

**2 NAME**
List the facts and the questions.

**3 THINK**
Choose and outline a plan.

**4 COMPUTE**
Work the plan.

**5 CHECK**
Test that the solution is reasonable.

Sleuths use clues to find a solution to a problem. When working together to solve a problem, you may choose to use one or more of these *strategies* as clues:

Write a Number Sentence
Multi-Step Problem
Use a Graph
Missing Information
Guess and Test
Make an Organized List

Make a Table
Draw a Picture
Use Simpler Numbers
Choose the Operation
Find a Pattern
Hidden Information

Logical Reasoning
Working Backwards
Interpret the Remainder
Use a Drawing or Model
Extra Information

## 1 ▶ IMAGINE

**Create a mental picture.**

As you read a problem, create a picture in your mind. Make believe you are there in the problem. This will help you think about:
- what facts you will need;
- what the problem is asking;
- how you will solve the problem.

After reading the problem, draw and label a picture of what you imagine the problem is all about.

## 2 ▶ NAME

**List the facts and the questions.**

Name or list all the facts given in the problem. Be aware of *extra* information not needed to solve the problem. Look for *hidden* information to help solve the problem. Name the question or questions the problem asks.

## 3 ▶ THINK

**Choose and outline a plan.**

Think about how to solve the problem by:
- looking at the picture you drew;
- thinking about what you did when you solved similar problems;
- choosing a strategy or strategies for solving the problem.

## 4 ▶ COMPUTE

**Work the plan.**

Work with the listed facts and the strategy to find the solution. Sometimes a problem will require you to add, subtract, multiply, or divide. Two-step problems require more than one choice of operation or strategy. It is good to estimate the answer before you compute.

## 5 ▶ CHECK

**Test that the solution is reasonable.**

Ask yourself:
- "Have you answered the question?"
- "Is the answer reasonable?"

Check the answer by comparing it to the estimate. If the answer is not reasonable, check your computation. You may use a calculator.

# Problem Solving

## Strategy: Guess and Test

 Algebra

**Problem:** Daryl read 16 mystery and science fiction books. He read three times as many mystery books as science fiction books. How many mystery books did he read?

 **1 IMAGINE** Create a picture in your mind.

 **2 NAME**

*Facts:*    16 books in all
3 times as many mystery books as science fiction

*Question:* How many mystery books did he read?

 **3 THINK** Make a guess. Test your guess by multiplying it by 3, then adding the product and your guess. Make a table to record your guesses.

**4 COMPUTE**

| Science Fiction | Mystery | Total | Test |
|---|---|---|---|
| 2 | $3 \times 2 = 6$ | $2 + 6 = 8$ | $8 < 16$ |
| 3 | $3 \times 3 = 9$ | $3 + 9 = 12$ | $9 < 16$ |
| 4 | $3 \times 4 = 12$ | $4 + 12 = 16$ | $16 = 16$ |

Daryl read 12 mystery books.

 **5 CHECK** Does the number of mystery books read equal 3 times the number of science fiction? Yes, because $12 = 3 \times 4$.

Does the total number of books equal 16? Yes, because $4 + 12 = 16$.

# Strategy: Hidden Information

**Problem:** Gerry has 50¢ in dimes. Jay has 3 dimes more than Gerry. Do they have enough money in all to buy a card for a dollar?

 **IMAGINE**   Act out the problem.

 **NAME**

*Facts:*   Gerry—50¢
Jay—3 dimes more than Gerry
card—costs a dollar

*Question:*   Do they have enough money to buy a card for a dollar?

 **THINK**

Is there enough information? Yes.
Is there hidden information? Yes.

one dollar = 100¢ or $1.00
Jay has 3 dimes, or 30¢, more than 50¢.

To find the amount Jay has,
add: 50¢ + 30¢ =  ?

To find if Gerry and Jay have enough money in all, first add the amount of money they have together. Then compare the total amount with the cost of the card.

 **COMPUTE**

|  |
|---|
| 50¢ |
| + 30¢ |
| 80¢  Jay's money |

|  |
|---|
| 50¢  Gerry's money |
| + 80¢  Jay's money |
| 130¢  in all |

Since 130¢ > 100¢, they have enough money.

 **CHECK**   Do the problem a different way. Since Jay has more than 50¢, they have enough money.
50¢ + 50¢ = 100¢ = $1.00

The answer
is
reasonable.

# Problem Solving

## Strategy: Find a Pattern

**Problem:** Raisa made up this number pattern. What are the missing eighth and ninth terms in her pattern?

1, 5, 4, 8, 7, 11, 10, _?_ , _?_ , 17, . . .

**1 ▶ IMAGINE**

As you look at the pattern, you notice that the second number is greater than the first, the third number is less than the second, and so on.

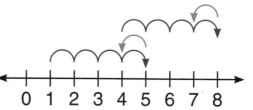

0 1 2 3 4 5 6 7 8

**2 ▶ NAME**

*Facts:* a number pattern of 10 terms
The numbers in the pattern increase, then decrease.

*Question:* What are the missing eighth and ninth terms?

**3 ▶ THINK**

Look at the numbers in the pattern.
Can you add to make the pattern? No.
Can you subtract to make the pattern? No.

Since the second number is greater and the third number is less, try adding first, subtracting next, and so on.

 **COMPUTE**

1 , 5 , 4 , 8 , 7 , 11 , 10
+ 4  − 1  + 4  − 1  + 4  − 1

The pattern is:
Add 4, subtract 1.

Use the pattern.
10 + 4 = 14 eighth term
14 − 1 = 13 ninth term

 **CHECK**

Check your computations.
Does 13 + 4 = 17? Yes.

28

## Strategy: Make a Table

**Problem:** The letter carrier left 3 letters at the first house. At the second house she left one more letter than at the first house. At the third house she left two more letters than at the first house, and so on. How many letters did she leave at the seventh house?

 **1 IMAGINE**

 **2 NAME**

*Facts:*

| | |
|---|---|
| 3 | 1st house |
| 3 + 1 | 2nd house |
| 3 + 2 | 3rd house |

*Question:* How many letters did she leave at the 7th house?

| House | Number of Letters |
|---|---|
| 1st | 3 |
| 2nd | (3 + 1) = 4 |
| 3rd | (3 + 2) = 5 |
| 4th | |
| 5th | |
| 6th | |
| 7th | |

 **3 THINK**

Make and complete a table.

Remember: Each house receives *one more* letter than the house before.

**4 COMPUTE**

(3 + 3) = 6    4th house
(3 + 4) = 7    5th house
(3 + 5) = 8    6th house
(3 + 6) = 9    7th house

9 letters are left at the 7th house.

**5 CHECK**

Is there one more letter at each house?
Look at your completed table to check.
      9 is 1 more than 8,
      8 is 1 more than 7, and so on.

# Problem Solving

## Applications

Choose a strategy from the list or use another strategy you know to solve each problem.

1. There are 4 white mice in a cage. There are double that number of black mice in another cage. How many mice are there in the two cages?

2. Murph has eight pet cages. She puts 3 pets in every other cage. Two of the pets have fur. How many pets does she put into cages?

3. Kara bought 3 pets. She spent exactly $12. Which pets did she buy if a fish cost $2, a hermit crab cost $3, a mouse cost $5, and a bird cost $6?

4. Every third person who came into the store bought 4 fish. Thirteen people came into the store. How many fish were sold?

5. Nichelle collects dinosaur and sports star pogs. She has 25 pogs altogether. If Nichelle has four times as many dinosaur pogs as sport star pogs, how many dinosaur pogs does she have?

6. There are 3 pet cages along the back wall of Murph's Pet Shop. On each of the other three sides of the shop there are 4 cages. How many pet cages are there altogether?

You can use these strategies:

Guess and Test
Find a Pattern
Extra Information
Make a Table
Hidden Information
Act It Out

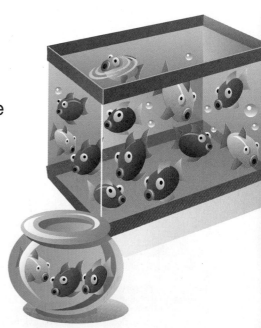

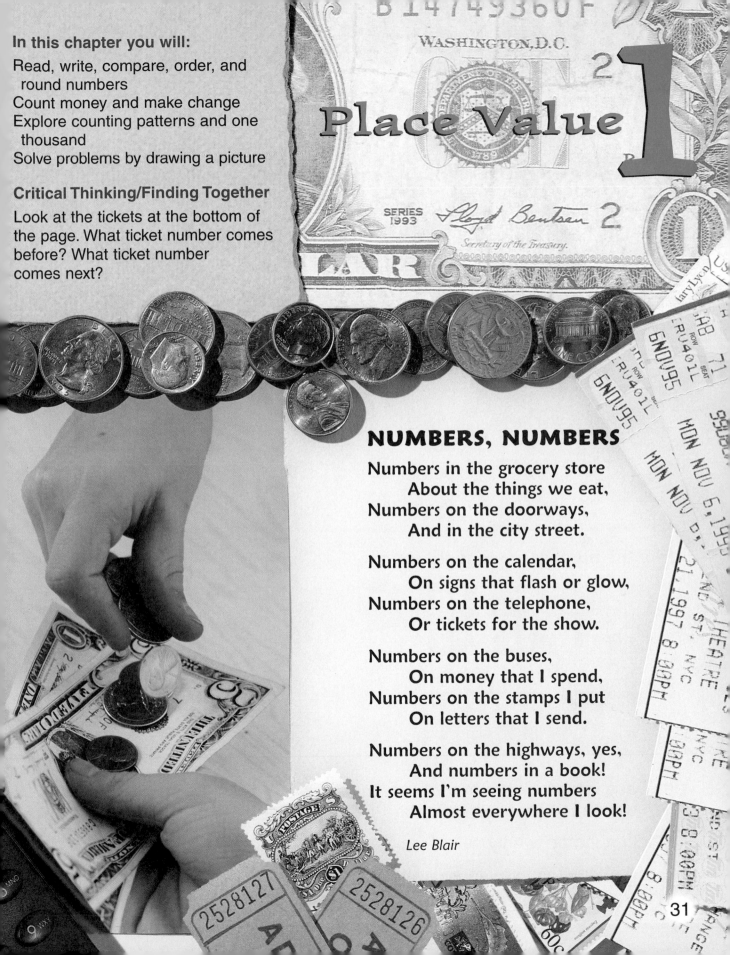

**In this chapter you will:**

Read, write, compare, order, and
  round numbers
Count money and make change
Explore counting patterns and one
  thousand
Solve problems by drawing a picture

**Critical Thinking/Finding Together**

Look at the tickets at the bottom of
the page. What ticket number comes
before? What ticket number
comes next?

# Place Value 1

## NUMBERS, NUMBERS

Numbers in the grocery store
    About the things we eat,
Numbers on the doorways,
    And in the city street.

Numbers on the calendar,
    On signs that flash or glow,
Numbers on the telephone,
    Or tickets for the show.

Numbers on the buses,
    On money that I spend,
Numbers on the stamps I put
    On letters that I send.

Numbers on the highways, yes,
    And numbers in a book!
It seems I'm seeing numbers
    Almost everywhere I look!

*Lee Blair*

31

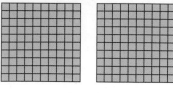

# 1-1 Hundreds

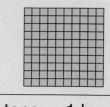

10 tens = 1 hundred
10 tens = 100

2 hundreds    3 tens    5 ones = 235

200    +    30    +    5    = 235

**Expanded Form**         **Standard Form**

Read 235 as: two hundred thirty-five.

## Study these examples.

| hundreds | tens | ones |
|----------|------|------|
| 9 | 5 | 3 |
| 8 | 0 | 4 |
| 6 | 9 | 0 |

900 + 50 + 3 = 953   nine hundred fifty-three
800 + 0 + 4 = 804   eight hundred four
600 + 90 + 0 = 690   six hundred ninety

## Write the number in standard form.

1.

2.

3.

4.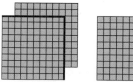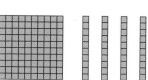

## Use base ten blocks to model each number.
## Then write the number in standard form.

5. four hundred twenty-seven

6. five hundred nineteen

7. three hundred thirty-three

8. one hundred ten

**Complete. Write each in expanded form.**

9. 576 = _5_ hundreds _7_ tens _6_ ones = _500_ + _70_ + _6_

10. 412 = _?_ hundreds _?_ ten _?_ ones = _?_ + _?_ + _?_

11. 890 = _?_ hundreds _?_ tens _?_ ones = _?_ + _?_ + _?_

12. 605 = _?_ hundreds _?_ tens _?_ ones = _?_ + _?_ + _?_

13. 500 = _?_ hundreds _?_ tens _?_ ones = _?_ + _?_ + _?_

**Write the number in standard form.**

14. 300 + 50 + 2          15. 200 + 40 + 8          16. 300 + 20 + 7

17. 700 + 70 + 7          18. 500 + 60 + 3          19. 800 + 70 + 0

20. 400 + 0 + 4           21. 600 + 8               22. 200 + 10

**In what place is the underlined digit? What is its value?**

23. 6<u>7</u>1          24. 85<u>7</u>          25. <u>7</u>91          26. 80<u>7</u>          27. <u>7</u>5

28. 96<u>4</u>          29. <u>4</u>05          30. <u>4</u>8           31. 7<u>4</u>1          32. <u>4</u>

33. 8<u>8</u>8          34. 80<u>8</u>          35. <u>8</u>00          36. 8<u>8</u>          37. <u>8</u>0

**Write the word name for each.**

| | hundreds | tens | ones |
|---|---|---|---|
| 38. | 6 | 5 | 4 |
| 40. | 1 | 2 | 9 |
| 42. | 1 | 1 | 0 |
| 44. | 8 | 7 | 1 |

| | hundreds | tens | ones |
|---|---|---|---|
| 39. | 3 | 7 | 8 |
| 41. | 4 | 2 | 6 |
| 43. | 9 | 0 | 2 |
| 45. | 6 | 0 | 0 |

 **Mental Math**

46. How much greater than 362 is: 462, 562, 662, 372, 382?

47. How much less than 957 is: 857, 757, 657, 947, 937?

# Comparing Numbers

Who has more votes, Jim or Terri?

To find who has more, compare the numbers.

| Election | Results |
|----------|---------|
| Person | Votes |
| Jim | 222 |
| Terri | 212 |

▶ Use base ten blocks to compare.

• Model each number.

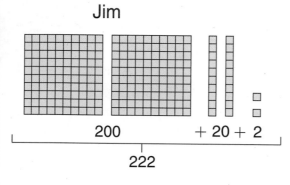

Jim

200    + 20 + 2

222

Terri

200    + 10 + 2

212

• Look at the models to compare the digits with the greatest place value.

2 hundreds = 2 hundreds
200 = 200

• Look at the models to compare the next digits.

2 tens > 1 ten
20 > 10

So 222 > 212.

Jim has more votes.

Remember: < is less than.
> is greater than.

▶ Use a number line to compare. 127 _?_ 136

125  126  127  128  129  130  131  132  133  134  135  136  137

Think:  127 comes *before* 136.

127 *is less than* 136.

So 127 < 136.

**Use base ten blocks or a number line to compare. Write $<$ or $>$.**

1. 26 _?_ 46

2. 35 _?_ 51

3. 44 _?_ 47

4. 90 _?_ 97

5. 74 _?_ 103

6. 318 _?_ 68

7. 148 _?_ 348

8. 719 _?_ 519

9. 212 _?_ 210

10. 384 _?_ 389

11. 949 _?_ 947

12. 333 _?_ 334

---

## Use a Place Value Chart to Compare

Compare: 847 _?_ 779

| h | t | o |
|---|---|---|
| 8 | 4 | 7 |
| 7 | 7 | 9 |

Look at the digits with the greatest place value.

Think: 800 $>$ 700
So      847 $>$ 779

Compare: 94 _?_ 104

| h | t | o |
|---|---|---|
|   | 9 | 4 |
| 1 | 0 | 4 |

There are no hundreds in 94. There is 1 hundred in 104.

Think:   0 $<$ 100
So       94 $<$ 104

Compare: 604 _?_ 610

| h | t | o |
|---|---|---|
| 6 | 0 | 4 |
| 6 | 1 | 0 |

Hundreds digits are the same. Look at the tens place.

Think:    0 $<$ 10
So       604 $<$ 610

Compare: 604 _?_ 601

| h | t | o |
|---|---|---|
| 6 | 0 | 4 |
| 6 | 0 | 1 |

Hundreds and tens digits are the same. Look at the ones place.

Think:    4 $>$ 1
So       604 $>$ 601

---

**Compare. Write $<$ or $>$.**

13. 53 _?_ 50

14. 87 _?_ 47

15. 398 _?_ 389

16. 105 _?_ 115

17. 451 _?_ 541

18. 209 _?_ 204

19. 241 _?_ 89

20. 692 _?_ 695

21. 712 _?_ 72

# 1-3 Ordering Numbers

List the waterfalls from highest to lowest.

| Three U.S. Waterfalls | |
|---|---|
| Name | Height |
| Ribbon | 491 meters |
| Widow's Tears | 357 meters |
| Upper Yosemite | 436 meters |

▶ Use a number line to order numbers.

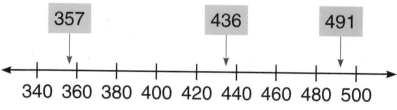

Think: 491 is to the right of 436.
491 *is greater than* 436.
436 is to the right of 357.
436 *is greater than* 357.

From greatest to least, the numbers are 491, 436, and 357.
From highest to lowest, the waterfalls are Ribbon, Upper Yosemite, and Widow's Tears.

▶ Use the value of the digits to order the numbers.

Write in order from least to greatest: 652, 424, 435.

| Compare hundreds. → | Compare tens. |
|---|---|
| 652 | 424 |
| 424 | 435 |
| 435 | |
| 600 > 400 | 20 < 30 |
| 652 is greatest. | 424 < 435 |

In order from least to greatest: 424, 435, 652.

Write in order from greatest to least: 635, 725, 88.

Compare hundreds.

635
725
 88 ◀ 88 has no hundreds. 88 is the least.

700 > 600
725 > 635    725 is the greatest.

In order from greatest to least: 725, 635, 88.

**Write in order from greatest to least.
Use the number line.**

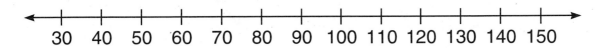

1. 80, 40, 60

2. 70, 50, 90

3. 35, 83, 69

4. 76, 52, 38

5. 42, 61, 56

6. 75, 136, 107

7. 121, 98, 108

8. 120, 112, 145

9. 145, 136, 109

**Write in order from least to greatest.**

10. 35, 27, 42

11. 88, 25, 63

12. 115, 92, 218

13. 157, 75, 213

14. 416, 747, 98

15. 224, 94, 386

16. 396, 385, 408

17. 566, 595, 481

18. 712, 716, 748

19. 523, 508, 509

20. 743, 790, 747

21. 225, 208, 221

**PROBLEM SOLVING**

22. Richard traveled 325 miles on vacation. Lorna traveled 875 miles. Jo traveled 850 miles. Who traveled the most miles?

23. Sarah read 325 pages. Robert read 97 pages. Jack read 147 pages. Who read the fewest pages?

**Finding Together**

24. Use the digits 4, 5, and 7 to write as many 3-digit numbers as you can, using each digit only once. Then write the numbers from least to greatest.

# 1-4 Skip Counting Patterns

Algebra

Lia skip counted by 3 to 100. Shyam skip counted by 4 to 100. Who said the most numbers?

## Hands-On Understanding

**You Will Need:** 2 hundred charts, 1 red and 1 blue crayon or colored pencil

| 1 | 2 | 3 | 4 | 5 | 6 | 7 | 8 | 9 | 10 |
|---|---|---|---|---|---|---|---|---|---|
| 11 | 12 | 13 | 14 | 15 | 16 | 17 | 18 | 19 | 20 |
| 21 | 22 | 23 | 24 | 25 | 26 | 27 | 28 | 29 | 30 |
| 31 | 32 | 33 | 34 | 35 | 36 | 37 | 38 | 39 | 40 |
| 41 | 42 | 43 | 44 | 45 | 46 | 47 | 48 | 49 | 50 |
| 51 | 52 | 53 | 54 | 55 | 56 | 57 | 58 | 59 | 60 |
| 61 | 62 | 63 | 64 | 65 | 66 | 67 | 68 | 69 | 70 |
| 71 | 72 | 73 | 74 | 75 | 76 | 77 | 78 | 79 | 80 |
| 81 | 82 | 83 | 84 | 85 | 86 | 87 | 88 | 89 | 90 |
| 91 | 92 | 93 | 94 | 95 | 96 | 97 | 98 | 99 | 100 |

**Step 1**
Use one of the hundred charts.
Start at 0.
Skip count by 3.
Color each count of 3 red.

**Step 2**
Use the second hundred chart.
Start at 0.
Skip count by 4.
Color each count of 4 blue.

| 1 | 2 | 3 | 4 | 5 | 6 | 7 | 8 | 9 | 10 |
|---|---|---|---|---|---|---|---|---|---|
| 11 | 12 | 13 | 14 | 15 | 16 | 17 | 18 | 19 | 20 |
| 21 | 22 | 23 | 24 | 25 | 26 | 27 | 28 | 29 | 30 |
| 31 | 32 | 33 | 34 | 35 | 36 | 37 | 38 | 39 | 40 |
| 41 | 42 | 43 | 44 | 45 | 46 | 47 | 48 | 49 | 50 |
| 51 | 52 | 53 | 54 | 55 | 56 | 57 | 58 | 59 | 60 |
| 61 | 62 | 63 | 64 | 65 | 66 | 67 | 68 | 69 | 70 |
| 71 | 72 | 73 | 74 | 75 | 76 | 77 | 78 | 79 | 80 |
| 81 | 82 | 83 | 84 | 85 | 86 | 87 | 88 | 89 | 90 |
| 91 | 92 | 93 | 94 | 95 | 96 | 97 | 98 | 99 | 100 |

**Step 3**
Look at each hundred chart.
Who said the most numbers?

## Use a different hundred chart for each exercise.

**1.** Skip count by 2.    **2.** Skip count by 5.    **3.** Skip count by 10.

## Communicate

**4.** For each hundred chart made, describe the pattern formed by the shading.

Math Journal

**Skip count by 3. You may use a hundred chart.**

**5.** Start at 33.
End at 48

**6.** Start at 42.
End at 27.

**7.** Start at 72.
End at 60.

**Skip count by 4. You may use a hundred chart.**

**8.** Start at 24.
End at 40.

**9.** Start at 37.
End at 21.

**10.** Start at 63.
End at 43.

**Write the missing numbers.**
**Explain how you skip count for each.**

Communicate ✓

**11.** 28, 30, 32, _?_ , 36, _?_ , 40

**12.** 40, 45, _?_ , 55, _?_ , _?_ , 70

**13.** 24, 27, _?_ , 33, _?_ , _?_ , 42

**14.** 70, 80, _?_ , 100, _?_ , 120

**15.** 49, 45, _?_ , 37, _?_ , 29, _?_ , 21

**16.** 86, 83, 80, _?_ , 74, _?_ , _?_ , 65

**Skip count to find the amount shown.**

**17.**

**18.**

**PROBLEM SOLVING**

**19.** I am an even number between 10 and 15.
You say me when you skip count by 3.
What number am I?

**20.** I have no ones. I am an even number
between 15 and 25. You say me when you
skip count by 4. What number am I?

39

# What Is One Thousand?

## Discover Together

**You Will Need:** 10 copies of a 10 × 10 grid, pencil

1,   2,   3,   4,   5,   6,   7,   8,   9,   10...

91,   92,   93,   94,   95,   96,   97,   98,   99,   100...

991, 992, 993, 994, 995, 996, 997, 998, 999, 1000

> One *thousand*, or 1000, is the next counting number after 999.

1. How is 1000 like 10 and like 100?

2. How is it different from both 10 and 100?

3. What patterns do you notice?

Find out just how large 1000 is.

Draw a dot in each square on 1 grid.

4. How many dots do you have altogether?

5. How many groups of 10 dots do you have?

Now draw a dot in each square on 4 more grids.

6. How many groups of 100 dots do you have?

7. How many groups of 10 dots do you have?

8. How many dots do you have altogether?

> 500 is halfway to 1000.

9. How many more groups of 10 dots do you think you need to show 1000 dots?

Draw the number of dots you need on more grids to show 1000.

10. How many groups of 10 dots are there?

11. How many groups of 100 dots do you have altogether?

# Communicate

12. How many tens is 1000 equal to?

13. How many hundreds is 1000 equal to?

## Project

14. Work with your classmates. Cut apart the 10 grids into strips of 10 dots each. Tape the strips of dots together to form a banner. Predict where in the classroom the banner of 1000 dots would end if it began at the edge of the board and extended around the room. Do you think the banner would wrap around the room? Explain. Then test your predictions.

# 1-6 Thousands

Lisa wrote the number she modeled in different ways.

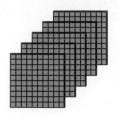 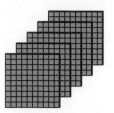 =

10 hundreds = 1 thousand
10 hundreds = 1000

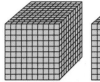 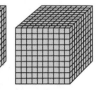  

| 2 thousands | | 3 hundreds | | 1 ten | | 4 ones | = | 2,314 |
|---|---|---|---|---|---|---|---|---|
| 2,000 | + | 300 | + | 10 | + | 4 | = | 2,314 |

**Expanded Form**                    **Standard Form**

Read 2,314 as: two thousand, three hundred fourteen.

## Study these examples.
Look at the place-value chart.

Four-digit numbers may be written with or without a comma.

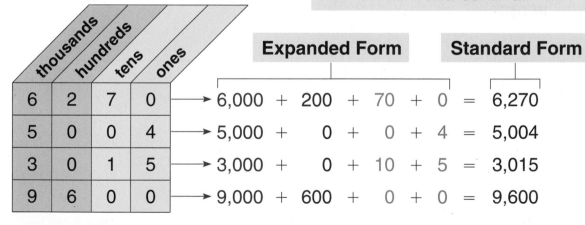

|  |  | | **Expanded Form** | | | | | | **Standard Form** |
|---|---|---|---|---|---|---|---|---|---|
| thousands | hundreds | tens | ones | | | | | | |
| 6 | 2 | 7 | 0 | → 6,000 + 200 + 70 + 0 | = | 6,270 |
| 5 | 0 | 0 | 4 | → 5,000 + 0 + 0 + 4 | = | 5,004 |
| 3 | 0 | 1 | 5 | → 3,000 + 0 + 10 + 5 | = | 3,015 |
| 9 | 6 | 0 | 0 | → 9,000 + 600 + 0 + 0 | = | 9,600 |

Read 9,600 as: nine thousand, six hundred.

## Write the number in standard form.

1.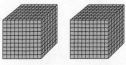

2.

3.

**Use base ten blocks to model each number. Then write the number in standard form.**

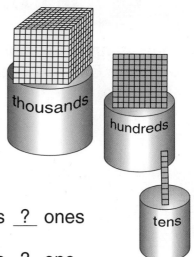

4. 8 thousands 0 hundreds 5 tens 7 ones

5. 4 thousands 8 hundreds 2 tens 6 ones

6. 3 thousands 5 hundreds 0 tens 0 ones

**Write the place value of the digits.**

7. $6{,}020 =$ ? thousands ? hundreds ? tens ? ones

8. $3{,}401 =$ ? thousands ? hundreds ? tens ? one

**Write the number in standard form.**

9. $8{,}000 + 400 + 20 + 6$

10. $3{,}000 + 900 + 70 + 8$

11. $7{,}000 + 800 + 0 + 3$

12. $9{,}000 + 0 + 40 + 0$

**Write the number in both standard and expanded forms.**

13. two thousand, four hundred ninety

14. one thousand, two hundred twelve

15. five thousand, fifty-six

16. three thousand, five

**In what place is the underlined digit? What is its value?**

17. 4,<u>1</u>81

18. 1,11<u>1</u>

19. 8,<u>1</u>81

20. <u>1</u>,818

21. 1,0<u>1</u>0

22. <u>5</u>,115

23. 1,<u>5</u>15

24. <u>5</u>,005

25. 5,0<u>5</u>0

26. 5,55<u>5</u>

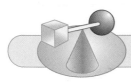

## Challenge

27. Write the largest 4-digit number.

28. Write the smallest 4-digit number that has no zeros.

# Ten Thousands and Hundred Thousands

At one summer Olympic event, there were 52,693 people. At 5 events, there were 204,372 people.

To show these numbers, extend the place-value chart.

| Thousands | | | Ones | | |
|---|---|---|---|---|---|
| hundred thousands (100,000) | ten thousands (10,000) | thousands (1000) | hundreds (100) | tens (10) | ones (1) |
| | 5 (50,000) | 2 (2000) | 6 (600) | 9 (90) | 3 (3) |
| **52 thousands** | | | **693 ones** | | |
| 2 (200,000) | 0 | 4 (4000) | 3 (300) | 7 (70) | 2 (2) |
| **204 thousands** | | | **372 ones** | | |

|  Expanded Form  |  Standard Form  |
|---|---|

50,000 + 2000 + 600 + 90 + 3 = 52,693

200,000 + 4000 + 300 + 70 + 2 = 204,372

Read 52,693 as: fifty-two thousand, six hundred ninety-three.

Read 204,372 as: two hundred four thousand, three hundred seventy-two.

**Write how many thousands.**

1. 40,000 = _?_ thousands

2. 57,000 = _?_ thousands

3. 749,000 = _?_ thousands

4. 926,000 = _?_ thousands

**Complete. Then write each number in expanded form.**

5. 82,346 = _?_ ten thousands _?_ thousands _?_ hundreds _?_ tens _?_ ones

6. 498,576 = _?_ hundred thousands _?_ ten thousands _?_ thousands _?_ hundreds _?_ tens _?_ ones

7. 310,430 = _?_ hundred thousands _?_ ten thousand _?_ thousands _?_ hundreds _?_ tens _?_ ones

8. 601,038 = _?_ hundred thousands _?_ ten thousands _?_ thousand _?_ hundreds _?_ tens _?_ ones

**Write the number in standard form.**

9. 2 ten thousands

10. 7 hundred thousands

11. 20,000 + 7000 + 40 + 3

12. 600,000 + 80,000 + 4000 + 90 + 1

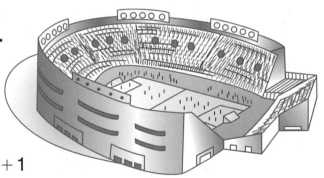

**In what place is the underlined digit? What is its value?**

13. 4̲7,896

14. 3̲16,000

15. 88̲1,720

16. 5̲73,128

**Write the number word in standard form.**

17. In 1959 ninety-two thousand, seven hundred six people went to a baseball game in Los Angeles.

18. In 1950 one hundred ninety-nine thousand, eight hundred fifty-four people went to a soccer match in Brazil.

45

# Comparing and Ordering Larger Numbers

The Sears Tower is 1454 feet tall. The Empire State Building is 1377 feet tall. Which building is taller?

Algebra ✓

▶ Compare: 1454 _?_ 1377

| th | h | t | o |
|----|---|---|---|
| 1  | 4 | 5 | 4 |
| 1  | 3 | 7 | 7 |

Compare thousands.
1000 = 1000

Compare hundreds.
400 > 300
So 1454 > 1377

The Sears Tower is taller.

▶ You can also order numbers by comparing the value of the digits.

Write in order from least to greatest:

5498, 4554, 5224

| Compare thousands. |  | Compare hundreds. |
|---|---|---|

| | |
|---|---|
| 5498 | 5498 |
| 4554 | 5224 |
| 5224 | |
| | 400 > 200 |
| 4000 < 5000 | 5498 > 5224 |
| 4554 is least. | 5498 is greatest. |

In order from least to greatest: 4554, 5224, 5498

## Which place is used to compare each?

**1.** 6000 < 8000      **2.** 5200 > 3200      **3.** 6380 < 6480

**4.** 9809 < 9990      **5.** 6305 < 6350      **6.** 3531 < 3587

**Compare. Write < or >.**

**7.** 2176 ? 1542     **8.** 3721 ? 2701     **9.** 5380 ? 5290

**10.** 8512 ? 8712     **11.** 5940 ? 5924     **12.** 4324 ? 4334

**13.** 530 ? 5320     **14.** 4852 ? 4859     **15.** 2708 ? 2709

**Write in order from least to greatest.**

**16.** 3510, 4510, 2510     **17.** 7208, 6208, 4878

**18.** 5177, 5140, 5320     **19.** 2730, 2708, 2834

**Write in order from greatest to least.**

**20.** 3280, 3260, 3400    3400, 3280, 3260

**21.** 4508, 5519, 3518     **22.** 7217, 9116, 8125

**23.** 3720, 3905, 3906     **24.** 8214, 8325, 8316

**PROBLEM SOLVING** Use the chart to solve each problem.

**25.** Which has a greater average depth, the Atlantic Ocean or the Indian Ocean? the Arctic Ocean or the Caribbean Sea?

**26.** Write the average depths of oceans and seas in order from greatest to least.

| Depths of Oceans and Seas | |
|---|---|
| Name | Average Depth |
| Pacific | 4028 meters |
| Atlantic | 3926 meters |
| Indian | 3963 meters |
| Arctic | 1205 meters |
| Caribbean | 2647 meters |

## Challenge

**Compare. Write < or >.**

**27.** 4731 ? 24,731     **28.** 16,320 ? 16,299

## 1-9 Rounding Numbers

Round numbers to tell *about* how many.

▶ **Round to the nearest ten.**

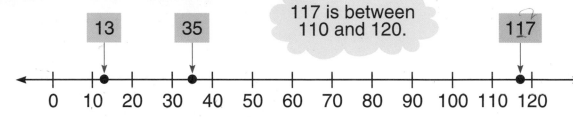

13 | 35 | 117 is between 110 and 120. | 117

0  10  20  30  40  50  60  70  80  90  100  110  120

13 is *nearer* to 10 than 20.     Round 13 *down* to 10.

35 is *halfway* between 30 and 40.     Round 35 *up* to 40.

117 is *nearer* to 120 than 110.     Round 117 *up* to 120.

▶ **Round to the nearest hundred.**

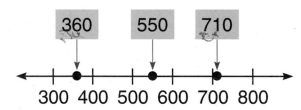

360 | 550 | 710

300  400  500  600  700  800

> Less than halfway round down.
> Halfway or more round up.

360 is *nearer* to 400 than 300.     Round 360 *up* to 400.

550 is *halfway* between 500 and 600.     Round 550 *up* to 600.

710 is *nearer* to 700 than 800.     Round 710 *down* to 700.

▶ **Round to the nearest thousand.**

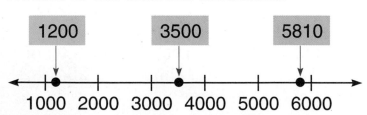

1200 | 3500 | 5810

1000  2000  3000  4000  5000  6000

1200 is *nearer* to 1000 than 2000.     Round 1200 *down* to 1000.

3500 is *halfway* between 3000 and 4000.     Round 3500 *up* to 4000.

5810 is *nearer* to 6000 than 5000.     Round 5810 *up* to 6000.

**Name the tens each is between. Round to the nearest ten.**

**1.** 86
80, 86, 90; 90

**2.** 54

**3.** 68

**4.** 25

**5.** 99

**6.** 245

**7.** 358

**8.** 171

**9.** 802

**10.** 555

**Name the hundreds each is between.
Round to the nearest hundred.**

**11.** 231

**12.** 459

**13.** 787

**14.** 486

**15.** 392

**16.** 110

**17.** 677

**18.** 543

**19.** 895

**20.** 550

**Name the thousands each is between.
Round to the nearest thousand.**

**21.** 1805

**22.** 3524

**23.** 8359

**24.** 6500

**25.** 7201

**26.** 4037

**27.** 2810

**28.** 6682

**29.** 1042

**30.** 5505

## Share Your Thinking

Communicate ✓

**31.** Think of how you rounded numbers to the nearest 10, 100, and 1000. Which place did you look at to round a number to the nearest 10? to the nearest 100? to the nearest 1000? Explain how you used these digits to round numbers.

**32.** Discuss the headline in the newspaper at the right. Do you think exactly 100,000 people attended the concert? Explain your reasoning to the class. Then talk about and identify situations when an exact answer is needed and those when a rounded answer is needed.

TODAY'S NEWS

100,000 People
Attend Jazz Concert
in Park

# Money Less Than $1.00

| penny | nickel | dime | quarter | half-dollar |
|---|---|---|---|---|
| 1¢ | 5¢ | 10¢ | 25¢ | 50¢ |
| $.01 | $.05 | $.10 | $.25 | $.50 |
| one cent | five cents | ten cents | twenty-five cents | fifty cents |

Donna has these coins. How much money does she have?

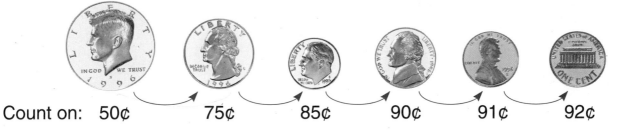

Count on:   50¢        75¢        85¢        90¢        91¢        92¢

Donna has 92¢, or $.92.

Arrange coins in order.

**Write the amount shown in two ways. First use the cent sign (¢).**
**Then use the dollar sign ($) and decimal point (.).**

1.

2.

3.

4.

# Match the coin amounts.

**5.**

**a.**

**6.**

**b.**

**7.**

**c.**

**8.**

**d.**

# Write the amount.

**9.** 1 quarter, 2 dimes, 2 nickels, 3 pennies

**10.** 1 quarter, 3 dimes, 4 pennies

**11.** 2 quarters, 4 dimes, 1 penny

**12.** 1 half-dollar, 5 nickels, 2 pennies

# Write the number of dimes for each amount.

**13.** $3.20　　　**14.** $.80　　　**15.** $2.00　　　**16.** $3.50

**17.** $1.10　　　**18.** $2.20　　　**19.** $.70　　　**20.** $1.90

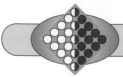 **Finding Together**

# Find as many ways as you can to make each amount.

**21.** 7¢　　　**22.** 12¢　　　**23.** 16¢　　　**24.** 20¢

## 1-11 Coins and Bills

one dollar
$1.00 or $1
100¢

five dollars
$5.00 or $5
500¢

ten dollars
$10.00 or $10
1000¢

Yolanda has 1 five-dollar bill, 1 one-dollar bill, 2 dimes, and 2 nickels. How much money does Yolanda have?

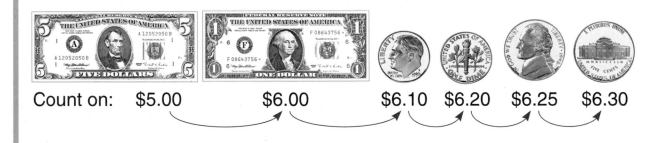

Count on:    $5.00        $6.00        $6.10   $6.20   $6.25   $6.30

Yolanda has $6.30.

**Write the amount. Use the dollar sign ($) and decimal point (.).**

**1.**

**2.**

**Write the amount. Use the dollar sign ($) and decimal point (.).**

3.

4.

**5.** 1 five-dollar bill, 2 nickels, 2 pennies

**6.** 3 one-dollar bills, 1 quarter, 3 dimes, 1 penny

**7.** 1 ten-dollar bill, 1 half-dollar, 4 dimes

**8.** 1 ten-dollar bill, 2 five-dollar bills, 3 quarters

**9.** 3 ten-dollar bills, 2 dimes, 6 pennies

## PROBLEM SOLVING

**10.** Robert has 1 quarter and 2 dimes. He earns one silver dollar. How much money does Robert have now?

**11.** Ty has 1 five-dollar bill, 2 one-dollar bills, one silver dollar, and 9 dimes. How much money does Ty have?

1 silver dollar =
$1.00 = 100 pennies

 **Skills to Remember**

**Complete each pattern.**

**12.** 5¢, 10¢, 15¢, 20¢, _?_ , _?_

**13.** 1¢, 3¢, 5¢, 7¢, _?_ , _?_

**14.** 25¢, 35¢, 45¢, 55¢, _?_ , _?_

**15.** 25¢, 50¢, 75¢, 100¢, _?_ , _?_

# Making and Counting Change

Paula buys a book for $8.79.
She gives the cashier $10.00.
How much change will Paula get?

To make change, start with the
cost of the item. Count up to the
amount paid.

Think: Start with $8.79.
Count up to $10.00.

$8.79      $8.80       $8.90       $9.00                    $10.00

└── +1¢ ──┘── +10¢ ──┘── +10¢ ──┘────────── +$1 ──┘

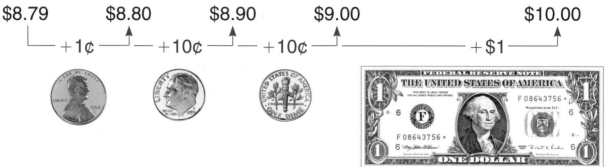

Arrange the money in order.
Count the change:

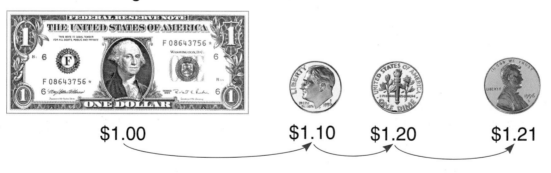

$1.00              $1.10    $1.20         $1.21

Paula will get $1.21.

Remember: There are many ways
to make change.

## Use play money to show each way.

**1.** Find another way Paula can get $1.21 change.

    **a.** use only coins

    **b.** use no dimes

    **c.** use only coins but no dimes

    **d.** use only nickels and pennies

**Choose the correct change.**

2. Robin buys a pen for $1.49. She gives the cashier $2.00.

a.     or b.

3. Bruce pays $3.19 for a notebook. He gives the cashier 3 dollars and 1 half-dollar.

a.    or b.

4. Noah buys a set of markers for $4.83. He gives the cashier $5.00.

a.    or b.

**PROBLEM SOLVING**
**Tell how the cashier can make change.**

5. Jane buys a ruler. She gives the cashier 3 quarters.

6. Morris buys a box of crayons. He gives the cashier $2.00.

7. Noriko buys a box of crayons. She gives the cashier 1 dollar and 2 quarters.

8. Cleo buys a set of paints. She gives the cashier $4.00.

9. Fred buys a notebook. He gives the cashier $10.00.

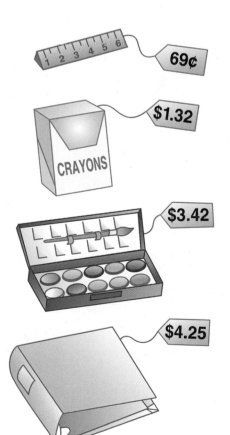

69¢

$1.32

CRAYONS

$3.42

$4.25

**Comparing and Rounding Money**  Algebra ✓

Nan spent $6.63 on her paperback book order. Cindy spent $7.52 on her order. Who spent more money?

Use the place-value chart to compare money as you do other numbers.

▶ Compare: $7.52 _?_ $6.63

| dollars | dimes | pennies |
|---------|-------|---------|
| 7 | 5 | 2 |
| 6 | 6 | 3 |

Look at the dollars.
  $7.00 > $6.00

So $7.52 > $6.63

Cindy spent more money than Nan.

▶ Compare: $4.84 _?_ $4.93

| dollars | dimes | pennies |
|---------|-------|---------|
| 4 | 8 | 4 |
| 4 | 9 | 3 |

Look at the dollars.
  $4.00 = $4.00
The dollars are the same.

Look at the dimes
  $.80 < $.90

So $4.84 < $4.93

▶ Compare: $5.60 _?_ $5.62

| dollars | dimes | pennies |
|---------|-------|---------|
| 5 | 6 | 0 |
| 5 | 6 | 2 |

Look at the dollars.
  $5.00 = $5.00

Look at the dimes
  $.60 = $.60

Look at the pennies
  $.00 < $.02

So $5.60 < $5.62

**Compare. Write < or >.**

**1.** $4.54 _?_ $3.55     **2.** $7.25 _?_ $8.25     **3.** $6.77 _?_ $6.67

**4.** $2.52 _?_ $2.67     **5.** $3.82 _?_ $3.85     **6.** $5.54 _?_ $5.56

---

### Rounding Money

**Round to the nearest dollar.**

$2.70        $5.50   $6.20

$2.00   $3.00   $4.00   $5.00   $6.00   $7.00

$2.70 is *nearer* to $3.00 than $2.00.    Round $2.70 *up* to $3.00.

$5.50 is *halfway* between          Round $5.50 *up* to $6.00.
$5.00 and $6.00.

$6.20 is *nearer* to $6.00 than $7.00.    Round $6.20 *down* to $6.00.

---

**Round to the nearest dollar.**

**7.** $4.65     **8.** $3.18     **9.** $7.50     **10.** $6.75     **11.** $5.15

**12.** $8.55     **13.** $2.45     **14.** $1.80     **15.** $4.50     **16.** $3.49

## Connections: Social Studies

Canadians use coins to represent two dollars, one dollar, twenty-five cents, ten cents, five cents, and one cent.

$2.00      $1.00      $.25      $.10      $.05      $.01

**17. Write the amount.**

# TECHNOLOGY

## The PRINT Statement

A **computer** is a machine that stores and processes information. One of the languages that a computer uses to process information is called BASIC.

A **keyboard** is used to **input**, or enter, information into the computer.

**A PRINT statement** in BASIC tells the computer what to display on the screen. The computer will print information exactly the way it appears in quotation marks.

Write a PRINT statement to display 4 + 3.

▶ Write a statement with quotation marks.

    Input: PRINT "4 + 3"
           END
    Press the [ENTER↵] key.

Remember: Press the [SHIFT] key and the [±] key for the plus sign.

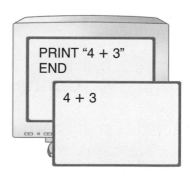

PRINT "4 + 3"
END

4 + 3

4 + 3 should appear on the screen.

4 + 3 is called the **output**.

Write a PRINT statement to display the sum of 4 and 3.

▶ Write a statement without quotation marks.

    Input: PRINT 4 + 3
           END
    Press the [ENTER↵] key.
    Output: 7

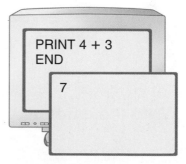

PRINT 4 + 3
END

7

## Write the output for each PRINT statement.

**1.**

```
PRINT "Good Morning"
END
```

**2.**

```
PRINT "$5.00 = $5"
END
```

**3.**
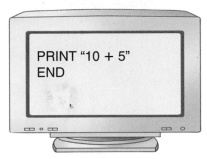

```
PRINT "10 + 5"
END
```

**4.**
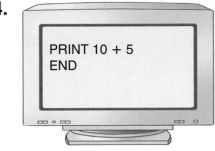

```
PRINT 10 + 5
END
```

**5.**
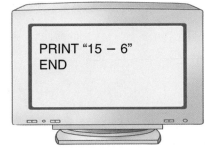

```
PRINT "15 − 6"
END
```

**6.**
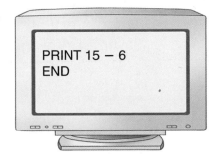

```
PRINT 15 − 6
END
```

**7.** PRINT "10 tens = 100"

**8.** PRINT "10 hundreds = 1000"

**9.** PRINT 16 − 8      **10.** PRINT 12 − 4      **11.** PRINT "$9 + $9"

**12.** PRINT "5 + 6 = 11"   **13.** PRINT 4 + 5 + 6   **14.** PRINT 10 − 2 − 4

## Match each PRINT statement that will give the same output.

**15.** PRINT 17 − 4

**16.** PRINT 3 + 5

**17.** PRINT 200 + 30 + 5

**18.** PRINT 18 − 15

**a.** PRINT 200 + 35

**b.** PRINT 9 − 6

**c.** PRINT 10 + 3

**d.** PRINT 18 − 3 − 7

# 1-15 | Problem Solving: Draw a Picture

**Problem:** Near the park there are 18 maple trees. Every 5th tree is to be removed. How many trees in all will be removed?

**1 IMAGINE** Create a mental picture.

**2 NAME**

*Facts:* 18 maple trees
remove every 5th tree

*Question:* How many trees will be removed?

**3 THINK** Look at the picture.
Mark every 5th tree.
Count the number of marked trees.

●●●●⊗●●●●
⊗●●●●⊗●●●

**4 COMPUTE** Count the number of trees removed.
Three trees will be removed.

**5 CHECK** Add the number of remaining trees and the number of trees removed.

$$15 \quad + \quad 3 \quad = \quad 18$$

| trees remaining | trees removed | total number of trees |

Does the sum equal 18?
Yes. Your answer is correct.

**Draw a picture to solve each problem.**

1. Shannon had $3.27 in her bank.
   Later she put 5 dimes and 9 pennies
   into her bank. How much money
   did she have in her bank then?

**IMAGINE**    Create a mental picture.

**NAME**    *Facts:*    She had $3.27.
                        She put 5 dimes, 9 pennies in.

            *Question:*  How much money
                         did she have then?

**THINK**    Count on from $3.27,
             beginning with the dimes.

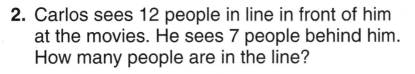

**COMPUTE** → **CHECK**

2. Carlos sees 12 people in line in front of him
   at the movies. He sees 7 people behind him.
   How many people are in the line?

3. The Anytown Train leaves Big City and goes
   100 miles to the 1st stop, 50 miles to the next,
   and 5 more miles to reach Anytown. The
   Clayville Express leaves Big City and goes
   20 miles to the 1st stop, 40 miles to the next,
   and 100 more miles to reach Clayville.
   Which city is closer to Big City by train?

4. If Di has one piece of art paper, how many cuts
   can she make to have just four equal strips
   of paper?

5. Jenny drew two straight lines in a rectangle.
   How many *triangles* did she make?

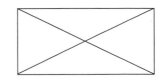

**Solve and explain the method you used.**

1. An aquarium ticket costs $6.00. Aileen has 5 one-dollar bills, 1 half-dollar, 3 quarters, and two dimes. Can Aileen buy a ticket?

2. Six hundred eighty children from Carlton School come to the aquarium. There are 659 children from London School at the aquarium. Which school has more children at the aquarium?

3. The keeper used 200 pounds of fish food in the morning, 80 pounds in the afternoon, and 7 pounds in the evening. How many pounds of fish food did the keeper use altogether?

4. Dall porpoises weigh up to 350 pounds. Harbor porpoises weigh up to 200 pounds. Which porpoises are heavier?

5. Whales have been known to dive to depths of three thousand two hundred feet. Write this depth as a standard numeral.

Imagine

Name

Think

Compute

Check

**Use the chart for problems 6–8.**

6. Which sea animals have a length of about 200 cm?

7. What whale is the shortest? the longest?

8. List the lengths of these animals from shortest to longest.

| Sea Animals | Length in Centimeters |
|---|---|
| Harbor porpoise | 180 cm |
| Sperm whale | 1800 cm |
| Dall porpoise | 210 cm |
| Blue whale | 3000 cm |
| Beluga whale | 500 cm |

**Choose a strategy from the list or use another strategy to solve each problem.**

9. Josh's library book on seals has a picture on every third page. There are 32 pages in the book. How many pictures are in the book?

10. Ted and Ana together have $1.40 in dimes. Ana has 4 fewer dimes than Ted. How much money does each have?

11. What even numbers between one thousand and two thousand have the same number of hundreds, tens, and ones?

12. Renée spent $2.25. Roberta spent 3 quarters more than Renée. How much did the two girls spend altogether?

13. There are 20 seats in the first row of the aquarium theater, 30 seats in the second row, 40 in the third row, and so on. How many seats are in the seventh row?

14. Tours of the aquarium last 1 hour. There were 140 people on the 10 o'clock tour. On each of the following tours there were 10 more people than on the one before. How many people were on the 2 o'clock tour?

Use these strategies:

Extra Information
Find a Pattern
Draw a Picture
Guess and Test
Hidden Information
Make a Table

**Share Your Thinking**

Math Journal

15. In your Math Journal, explain why you often have to use the strategies Find a Pattern and Make a Table together.

# Chapter Review and Practice

**Write the number in standard form.** *(See pp. 32–33, 40–45.)*

**1.** 5 hundreds 4 tens 3 ones

**2.** 6 hundreds 4 ones

**3.** 3000 + 500 + 40 + 1

**4.** 6 hundred thousands

**Write the number in expanded form.**

**5.** six hundred ten

**6.** 5025

**7.** two thousand, two

**Compare. Write < or >.** *(See pp. 34–35, 46–47, 56–57.)*

**8.** 46 _?_ 64

**9.** 917 _?_ 719

**10.** 5023 _?_ 5215

**11.** $3.21 _?_ $4.03

**12.** $2.61 _?_ $2.59

**13.** $7.59 _?_ $7.54

**Write in order from least to greatest.** *(See pp. 36–37, 46–47.)*

**14.** 782, 698, 739

**15.** 3998, 4012, 3897

**Write the missing numbers.** *(See pp. 38–39.)*

**16.** 66, 69, _?_ , _?_ , 78, _?_

**17.** 74, 78, _?_ , _?_ , 90, _?_

**Round to the nearest thousand or to the nearest dollar.** *(See pp. 48–49, 56–57.)*

**18.** 4023   **19.** 8521   **20.** 6666   **21.** $6.92   **22.** $2.39   **23.** $5.55

**Write the amount.** *(See pp. 50–53.)*

**24.** 1 quarter, 3 nickels

**25.** 5 one-dollar bills, 5 nickels, 3 pennies

## PROBLEM SOLVING *(See pp. 54–55, 60–63.)*

**26.** Mike buys a toy for $2.79. He gives the cashier $3.00. Name his change.

**27.** Kevin has 14 quarters. He puts every 4th quarter in his bank. How much money does Kevin put in his bank?

(See *Still More Practice*, p. 443.)

## WHAT IS A MILLION?

If you could travel 1 mile per second, you could get to places very quickly.

- You could travel from Maine to California in about 1 hour.
- You could travel to the moon in about 66 hours, or a little less than 3 days.
- But it would take you almost *3 years* to travel from Earth to the Sun.

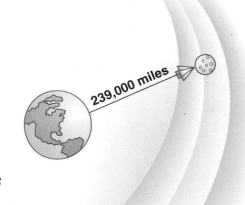

239,000 miles

Earth is about ninety-three million miles from the Sun. The other planets in our solar system are also millions of miles from the Sun.

**Millions** are used to name very large numbers.

| Planet | Distance from the Sun in Miles |
|---|---|
| Mercury | 36,000,000 or 36 million |
| Venus | 67,000,000 or 67 million |
| Earth | 93,000,000 or 93 million |
| Mars | 142,000,000 or 142 million |
| Jupiter | 484,000,000 or 484 million |
| Saturn | 885,000,000 or 885 million |

## Use the table above to solve these problems.

At 1 mile per second:

1. It would take about 1 year and 52 days to travel from Mercury to the Sun. How far is Mercury from the Sun?

2. It would take about four and a half years to travel from Mars to the Sun. How far is Mars from the Sun?

3. It would take about 15 years to travel from Jupiter to the Sun. How far is Jupiter from the Sun? How old will you be in 15 years?

4. It would take about 28 years to travel from Saturn to the Sun. How far is Saturn from the Sun? Is Jupiter or Saturn farther from the Sun?

# Check Your Mastery

## Performance Assessment

**Find each amount. Use < and >
to compare the amounts in two ways.**

**1.–2.**

| Jay | 1 five-dollar bill, 1 dime, 5 pennies |
|------|----------------------------------------|
| Drew | 4 one-dollar bills, 2 quarters, 3 dimes |
| Ana | 3 quarters, 1 half-dollar, 4 one-dollar bills |

**Write the number in standard form.**

**3.** 7 hundreds 6 tens 0 ones

**4.** 3 hundreds 5 tens 2 ones

**5.** 1000 + 400 + 50 + 9

**6.** 8 ten thousands

**Write the number in expanded form.**

**7.** 805

**8.** 3102

**9.** five thousand, forty-six

**Compare. Write < or >.**

**10.** 26 ? 62

**11.** $1.34 ? $1.38

**12.** 6330 ? 6319

**Write in order from least to greatest.**

**13.** 655, 695, 583

**14.** 5634, 5109, 4081

**Write the missing numbers.**

**15.** 52, 55, 58, ? , ? , ?

**16.** 71, 75, 79, ? , 87, ?

**Round to the nearest thousand or to the nearest dollar.**

**17.** 1704

**18.** 7493

**19.** 4535

**20.** $5.85

**21.** $1.61

**22.** $3.28

**PROBLEM SOLVING**  *Use a strategy you have learned.*

**23.** Sarah buys a doll for $3.99. She gives the
cashier $5.00. Name her change.

# Addition

# 2

## School Bus

This wide-awake
freshly-painted-yellow
school bus

readied for Fall

carries us all—

Sixteen boys—
Fourteen girls—

Thirty pairs of sleepy eyes

and

hundreds
upon
hundreds

of

school supplies.

*Lee Bennett Hopkins*

**In this chapter you will:**

Explore missing addends and
  regrouping in addition
Estimate and add whole
  numbers and money
Add money using a calculator
Solve problems using simpler
  numbers

**Critical Thinking/Finding Together**

Skip count to find how many sleepy
eyes there are on the bus.

## 2-1 More Than Two Addends

Jay and Darla added to find the distance around these figures.

They found the sum easily by using the strategies Looking for Tens and Doubles and by Counting On.

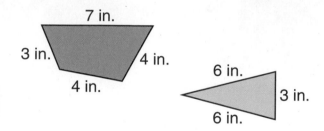

7 in.

3 in.      4 in.

4 in.

6 in.

6 in.

3 in.

$7 + 3 + 4 + 4 = \underline{\ ?\ }$          $3 + 6 + 6 = \underline{\ ?\ }$

Add down.

$\begin{array}{r} 7 \\ 3 \\ 4 \\ +\ 4 \\ \hline 18 \end{array}$

$7 + 3 = 10$

$4 + 4 = 8$

$10 + 8 = 18$

The distance around equals 18 inches.

Add up.

$\begin{array}{r} 3 \\ 6 \\ +\ 6 \\ \hline 15 \end{array}$

$12 + 3 = 15$

$6 + 6 = 12$

The distance around equals 15 inches.

**Find the sum.** Explain the strategy you use.

| 1. | 2. | 3. | 4. | 5. | 6. |
|---|---|---|---|---|---|
| 6 | 2 | 5 | 6 | 2 | 7 |
| 4 | 2 | 3 | 4 | 7 | 3 |
| +5 | +9 | +3 | +1 | +7 | +4 |

| 7. | 8. | 9. | 10. | 11. | 12. |
|---|---|---|---|---|---|
| 3 | 1 | 2 | 8 | 5 | 5 |
| 3 | 3 | 2 | 0 | 2 | 5 |
| 0 | 2 | 9 | 2 | 1 | 6 |
| +7 | +8 | +1 | +2 | +9 | +1 |

**13.** $2 + 8 + 5$          **14.** $7 + 9 + 1$          **15.** $4 + 4 + 3 + 7$

**16.** $4 + 6 + 1 + 1$      **17.** $5 + 5 + 0 + 3$      **18.** $9 + 0 + 7 + 3$

## Changing Order

Changing the order of adding the addends does *not* change the sum.

7¢ ⌐10¢
4¢ │
1¢ /        10¢ + 4¢ = 14¢
+3¢ /       14¢ + 1¢ = 15¢
15¢

3¢
6¢ ⌐12¢    12¢ + 3¢ = 15¢
2¢ │
+6¢ /      15¢ + 2¢ = 17¢
17¢

Remember: Look for tens and doubles.

**Find the sum.**

| **19.** | **20.** | **21.** | **22.** | **23.** | **24.** |
|---|---|---|---|---|---|
| 4¢ | 2¢ | 8¢ | 7¢ | 4¢ | 5¢ |
| 5¢ | 2¢ | 4¢ | 4¢ | 5¢ | 1¢ |
| 1¢ | 5¢ | 0¢ | 3¢ | 6¢ | 8¢ |
| +5¢ | +8¢ | +4¢ | +4¢ | +2¢ | +2¢ |

## PROBLEM SOLVING

**25.** In four play-off games, Duane scored 5, 3, 5, and 2 goals. How many goals did he score in all?

**26.** Laverne scored 4 goals in each of three games. How many goals did she score in all?

## Mental Math

**Find the answer.**

**27.** Start with 4 ⟶ Double it ⟶ Add 2 ⟶ Add 1 ⟶ Add 6.

**28.** Start with 3 ⟶ Double it ⟶ Subtract 4 ⟶ Add 6 ⟶ Add 4.

**29.** Start with 9 ⟶ Subtract 2 ⟶ Double it ⟶ Add 4 ⟶ Subtract 1.

# Missing Addends

You can find missing addends by using
a balance or number line.

$9¢ + \underline{\ ?\ } = 16¢$

## Hands-On Understanding

**You Will Need:** 40 pennies, balance, number line

**Step 1**   Place 9 pennies on the left pan and
16 on the right pan.

Are both sides equalized? Explain.

What can you do to the left pan to
make both sides the same?

**Step 2**   Add 1 penny at a time to the left
pan, counting on from 9¢ to 16¢.

Now are the sides equalized? Why?

How many more pennies were
needed to equalize the balance?

**Step 3**   Check your answer.     $9¢ + 7¢ = 16¢$

Use a number line to find the missing addend.     $8 + \underline{\ ?\ } = 14$

**Step 4**   Write the addend and sum on the number line

**Step 5**  Start with the addend and count on to the sum.

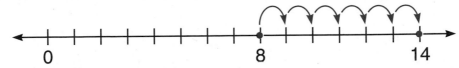

What numbers did you count?

How many numbers did you count to reach 14?

What is the missing addend?

**Step 6**  Check your answer.          $8 + 6 = 14$

## Communicate                                      Discuss

1. How are using a balance and a number line the same?

2. Explain why finding a missing addend is like finding the difference.

**Find the missing addend.** You may use a balance or a number line.

3. $3 + \underline{\ ?\ } = 7$          4. $9 + \underline{\ ?\ } = 10$          5. $2 + \underline{\ ?\ } = 5$

6. $\underline{\ ?\ } + 4 = 9$          7. $\underline{\ ?\ } + 3 = 6$          8. $5 + \underline{\ ?\ } = 7$

9. $12 = 4 + \underline{\ ?\ }$          10. $14 = \underline{\ ?\ } + 9$          11. $9 = \underline{\ ?\ } + 9$

12. $11 = 9 + \underline{\ ?\ }$          13. $17 = 8 + \underline{\ ?\ }$          14. $12 = 7 + \underline{\ ?\ }$

15. Explain how exercises 9–14 are different from exercises 3–8.

**Find the missing addends.**                        Math
Journal

16. $\underline{\ ?\ } + \underline{\ ?\ } + 6 = 16$          17. $10 = 2 + \underline{\ ?\ } + \underline{\ ?\ }$

18. Look at exercises 16 and 17. Can you find more solutions? Explain in your Math Journal why or why not.

# 2-3 | Adding: No Regrouping

If the third graders collect 500 cans for recycling, they will win a "Good Citizens" banner. The chart shows how many cans they have collected. Will they win the banner? How many cans did they collect in all?

| Cans Collected – 3rd Grade | |
|---|---|
| Boys | Girls |
| 224 | 312 |

▶ To find if they will win the banner, first estimate.
An estimate tells *about* how many.

| Add the front digits. | ⟶ | Write 0s for the other digits. |
|---|---|---|

$$\begin{array}{r} 224 \\ +\,312 \\ \hline 5 \end{array}$$

$$\begin{array}{r} 224 \\ +\,312 \\ \hline \text{about} \quad 500 \end{array}$$

The third graders will win the banner.

▶ To find how many cans in all,
add: 224 + 312 = __?__

| Add the ones. | ⟶ | Add the tens. | ⟶ | Add the hundreds. |
|---|---|---|---|---|

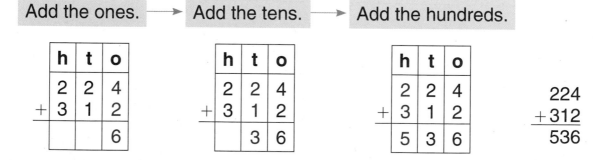

| h | t | o |
|---|---|---|
| 2 | 2 | 4 |
| + 3 | 1 | 2 |
| | | 6 |

| h | t | o |
|---|---|---|
| 2 | 2 | 4 |
| + 3 | 1 | 2 |
| | 3 | 6 |

| h | t | o |
|---|---|---|
| 2 | 2 | 4 |
| + 3 | 1 | 2 |
| 5 | 3 | 6 |

$$\begin{array}{r} 224 \\ +\,312 \\ \hline 536 \end{array}$$

The third graders collected 536 cans in all.

▶ Changing the order of the addends
does not change the sum.

224 + 312 = 536  and  312 + 224 = 536

**Estimate. Then add.**

| 1. | t | o |
|----|---|---|
|    | 3 | 3 |
| +  | 4 | 4 |
|    |   |   |

| 2. | t | o |
|----|---|---|
|    | 6 | 5 |
| +  | 1 | 1 |
|    |   |   |

| 3. | t | o |
|----|---|---|
|    | 2 | 6 |
| +  | 5 | 3 |
|    |   |   |

| 4. | h | t | o |
|----|---|---|---|
|    | 4 | 7 | 3 |
| +  | 2 | 1 | 6 |
|    |   |   |   |

| 5. | h | t | o |
|----|---|---|---|
|    | 8 | 0 | 3 |
| +  | 1 | 9 | 1 |
|    |   |   |   |

6.    753
    + 106

7.    615
    + 304

8.    824
    + 173

9.    564
    + 302

10.    656
     + 133

**Find the sum.**

11.     83         601
    + 601        + 83

12.    305          64
     + 64        + 305

13.    351         547
     + 547       + 351

**Align and add.**

**14.** 900 + 51     **15.** 752 + 146     **16.** 17 + 31     **17.** 52 + 6

**18.** 8 + 50     **19.** 20 + 617     **20.** 908 + 71     **21.** 47 + 340

**Add. Describe the pattern you see.** To 218 add:

*Communicate* ✓

**22.** 10     **23.** 20     **24.** 30     **25.** 40     **26.** 50

**27.** 700     **28.** 600     **29.** 500     **30.** 400     **31.** 300

**Number Sense**

*Algebra* ✓

**Find the missing addends.  Guess and test.**

**32.**    4 2
       +☐☐
        7 2

**33.**    ☐☐
        + 1 5
         9 6

**34.**    2 6
        +☐☐
         5 9

**35.**    3 4
        +☐☐
         6 8

**36.**    ☐☐
        + 2 2
         9 2

**37.** ☐☐ + 21 = 72     **38.** ☐☐ + 47 = 69     **39.** 63 + ☐☐ = 99

**Estimating Sums**

Mickey bought a pennant, a program, and a baseball cap at the baseball game. About how much did he spend?

An estimate can also tell *about* how much. You can estimate sums by rounding.

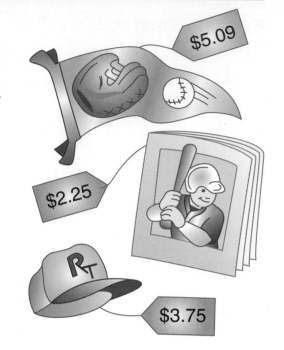

Estimate: $5.09 + $2.25 + $3.75

• Round each amount to the nearest dollar.

• Then add the rounded amounts.

$$
\begin{array}{rcr}
\$5.09 & \longrightarrow & \$5.00 \\
2.25 & \longrightarrow & 2.00 \\
+\ 3.75 & \longrightarrow & +\ 4.00 \\
& \text{about} & \$11.00
\end{array}
$$

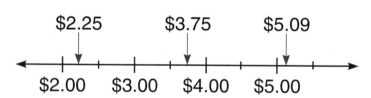

Mickey spent about $11.00.

**Study these examples.**

| Round to the nearest ten. | Round to the nearest ten cents. | Round to the nearest hundred. |
|---|---|---|

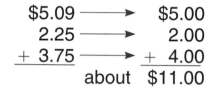

$$
\begin{array}{rcr}
82 & \longrightarrow & 80 \\
12 & \longrightarrow & 10 \\
+\ 5 & \longrightarrow & +10 \\
& \text{about} & 100
\end{array}
$$

$$
\begin{array}{rcr}
\$1.45 & \longrightarrow & \$1.50 \\
+\ .26 & \longrightarrow & +\ .30 \\
& \text{about} & \$1.80
\end{array}
$$

$$
\begin{array}{rcr}
624 & \longrightarrow & 600 \\
+317 & \longrightarrow & +300 \\
& \text{about} & 900
\end{array}
$$

Remember: Write the dollar sign and the decimal point.

74

**Estimate by rounding to the nearest ten or ten cents.**

1.  24
    13
    +47

2.  49
    64
    +31

3.  72
    51
    +25

4.  $.12
    .61
    + .14

5.  $.14
    .28
    + .33

6.  92
    +41

7.  37
    +52

8.  74
    +83

9.  $.47
    + .19

10. $.29
    + .29

11. 37 + 78 + 24

12. 93 + 64 + 6

13. $.05 + $.16 + $.44

**Estimate by rounding to the nearest hundred or dollar.**

14. 133
    +288

15. 917
    +245

16. 822
    +413

17. $1.95
    + 1.77

18. $5.79
    + 9.04

19. $1.36
    + 1.03

20. $7.18
    + 4.30

21. $2.49
    + 3.16

22. $5.25
    + 4.12

23. $6.81
    + 2.09

24. $1.17 + $1.92

25. $2.81 + $.81

26. 776 + 52

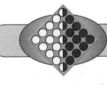

## Finding Together

Danielle has $7.00 to spend at the baseball game. Estimate. Does she have enough money to buy:

27. 1 hot dog and 1 soda?

28. 1 pennant and 1 bag of peanuts?

29. 2 hot dogs and 2 sodas?

30. 1 bucket of popcorn and 1 baseball cap?

31. 1 calendar and 1 pennant?

32. 1 hot dog, 1 soda, and 1 pennant?

........... $2.25
........... $1.60
........... $1.25
........... $1.10

........... $4.99

........... $2.80

........... $3.75

75

**Adding Money**

Jason bought a lunch box and a set of markers. How much money did he spend?

▶ First estimate    $4.25 ⟶ $4.00
  the sum:        + 1.30 ⟶ + 1.00
                        about   $5.00

▶ Then to find out how much Jason spent,
  add: $4.25 + $1.30 = __?__

| Line up the dollars and cents. | Add as usual. | Write $ and . in the sum. |
|---|---|---|
| $4.25<br>+ 1.30 | $4.25<br>+ 1.30<br>5 55 | $4.25<br>+ 1.30<br>$5.55 |

Jason spent $5.55.

Think: $5.55 is close to $5.00.
The answer is reasonable.

**Study these examples.**

| $.25 | 45¢ | $2.24 |
|---|---|---|
| + .31 | +23¢ | + .65 |
| $.56 | 68¢ | $2.89 |

**Estimate. Then add.**

1.  $7.50
   + 1.25

2.  $4.25
   + 2.61

3.  $1.05
   + 2.10

4.  $2.30
   + 4.19

5.  $3.62
   + 2.17

6.  33¢
   +54¢

7.  16¢
   +12¢

8.  $.72
   + .16

9.  $.44
   + .34

10.  $.27
    + .31

Sometimes you need to regroup ones as tens.

13 ones = _?_

## Hands-On Understanding

**You Will Need:** base ten blocks

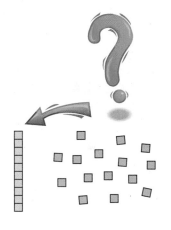

**Step 1** | Place 1 tens rod on your worktable.

How many ones units fit on top of the tens rod?

What do you notice about 1 ten and 10 ones?

**Step 2** | Model 13 ones.

What number does the model show?

**Step 3** | Now model the same number, 13, using fewer blocks.

What blocks did you use?

So 13 ones can be regrouped as _?_ ten _?_ ones.

## Communicate

Math Journal

**1.** Explain when you can regroup ones as tens.

**Model each number using only ones units.
Then model each number using the fewest blocks.
Describe the regrouped number.**

**2.** 18     **3.** 15     **4.** 28     **5.** 11     **6.** 30

## 2-7 Adding with Regrouping

How many toy cars do Ned and Jay have in all?

▶ First estimate the sum.

$$24 \longrightarrow 20$$
$$\underline{+19} \longrightarrow \underline{+20}$$
$$\text{about} \quad 40$$

| Toy Car Collections | |
|---|---|
| Ned | 24 |
| Jay | 19 |
| Nessa | 24 |
| Jaclyn | 59 |

▶ Then to find how many toy cars in all, add: $24 + 19 = \underline{\ ?\ }$

| Add the ones. Regroup. | → | Add the tens. |
|---|---|---|

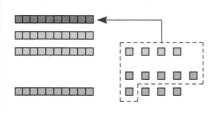

| tens | ones |
|---|---|
| →1 | |
| 2 | 4 |
| + 1 | 9 |
| | 3 |

| tens | ones |
|---|---|
| 1 | |
| 2 | 4 |
| + 1 | 9 |
| 4 | 3 |

$$\begin{array}{r} 24 \\ +19 \\ \hline 43 \end{array}$$

13 ones = 1 ten 3 ones

43 is close to 40. The answer is reasonable.

Ned and Jay have 43 toy cars in all.

**Study these examples.**

| tens | ones |
|---|---|
| →1 | |
| 5 | 1 |
| + 2 | 9 |
| 8 | 0 |

$$\begin{array}{r} 51 \\ +29 \\ \hline 80 \end{array}$$

10 ones = 1 ten 0 ones

| dimes | pennies |
|---|---|
| →1 | |
| 2 | 3 |
| + 4 | 9 |
| 7 | 2 |

$$\begin{array}{r} 23¢ \\ +49¢ \\ \hline 72¢ \end{array}$$

12 pennies = 1 dime 2 pennies

**Add.**

| 1. | t | o |
|---|---|---|
|   | 2 | 5 |
| + | 5 | 9 |
|   |   |   |

| 2. | t | o |
|---|---|---|
|   | 1 | 3 |
| + | 3 | 9 |
|   |   |   |

| 3. | t | o |
|---|---|---|
|   | 4 | 7 |
| + |   | 8 |
|   |   |   |

| 4. | d | p |
|---|---|---|
|   | 6 | 1 |
| + | 1 | 9 |
|   |   |   |

| 5. | d | p |
|---|---|---|
|   | 2 | 8 |
| + | 4 | 2 |
|   |   |   |

**Estimate. Then find the sum.**

6.  54
   +26

7.  38
   +29

8.  58
   +31

9.  33
   +29

10.  49
    +24

11.  78
    +18

12.  17
    +58

13.  24
    +66

14.  34
    +49

15.  45
    +19

16.  23
    +37

17.  62
    +24

**Align and add.**

18. 58 + 25

19. 29 + 25

20. 69 + 27

21. 80 + 17

22. 19 + 3

23. 47¢ + 33¢

24. 45¢ + 45¢

25. 6¢ + 92¢

26. 5¢ + 86¢

**PROBLEM SOLVING  Use the chart on page 78.**

Algebra

27. Alvin has double the number of toy cars that Nessa has. How many toy cars do the two have altogether?

28. Rosita has 35 more toy cars than Jaclyn. How many toy cars does Rosita have?

 **Connections: Geography**

29. Use a map of the United States and locate San Francisco, Palo Alto, and San Jose, California. It is 29 miles from San Francisco to Palo Alto and 15 miles from Palo Alto to San Jose. Using the same route, how many miles is it from San Francisco to San Jose?

## 2-8 | Regrouping Tens

Sometimes you need to regroup tens as hundreds.

12 tens = _?_ hundreds

 **Hands-On Understanding**

**You Will Need:** base ten blocks

 **Step 1** Place 1 hundred flat on your worktable.

How many tens rods fit on top of the hundred flat?

What do you notice about 1 hundred and 10 tens?

**Step 2** Model 12 tens.

What number does the model show?

**Step 3** Now model the same number, 120, using fewer blocks.

What blocks did you use?

So 12 tens can be regrouped as _?_ hundred _?_ tens.

## Communicate

**Math Journal** ✓

1. Explain when you can regroup tens as hundreds.

**Model each number using only tens rods.
Then model each number using the fewest blocks.
Describe the regrouped number.**

**2.** 18 tens    **3.** 15 tens    **4.** 13 tens    **5.** 16 tens    **6.** 19 tens

**7.** 20 tens    **8.** 31 tens    **9.** 24 tens    **10.** 29 tens    **11.** 10 tens

# Adding: Regrouping Tens

Tasha sold 56 hot dogs and 81 hamburgers. How many hot dogs and hamburgers did she sell altogether?

To find how many she sold altogether, add: 56 + 81 = _?_

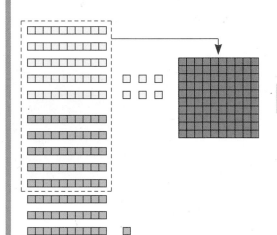

Add the ones. → Add the tens. Regroup.

| h | t | o |
|---|---|---|
|   | 5 | 6 |
| + | 8 | 1 |
|   |   | 7 |

| h | t | o |
|---|---|---|
|   | 5 | 6 |
| + | 8 | 1 |
| 1 | 3 | 7 |

```
  56
+ 81
 137
```

13 tens =
1 hundred 3 tens

Tasha sold 137 hot dogs and hamburgers.

**Complete.**

| 1. | 2. | 3. | 4. | 5. |
|---|---|---|---|---|
| 64 | 51 | 84 | 90 | 34 |
| +73 | +96 | +75 | +87 | +74 |
| ??7 | 1?? | ?5? | ??? | ??? |

**Add.**

| 6. | 7. | 8. | 9. | 10. | 11. |
|---|---|---|---|---|---|
| 43 | 86 | 72 | 32 | 27 | 21 |
| +72 | +83 | +71 | +97 | +81 | +95 |

| 12. | 13. | 14. | 15. | 16. | 17. |
|---|---|---|---|---|---|
| $.64 | $.76 | $.82 | $.55 | $.91 | $.63 |
| + .72 | + .92 | + .45 | + .71 | + .63 | + .84 |
| $1.36 | | | | | |

# Adding: Regrouping Twice

Last week April collected 58 cans for recycling. This week she collected 76 cans. How many cans did she collect in the two weeks?

▶ First estimate the sum.
$58 + 76 \longrightarrow 60 + 80 = 140$

▶ Then to find the number of cans, add: $58 + 76 = \underline{\ ?\ }$

Use base ten blocks to help you add 58 and 76.

 ## Hands-On Understanding

**You Will Need:** base ten blocks

| Step 1 | Model the first addend, 58, using the fewest number of blocks. |

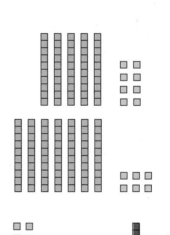

| Step 2 | Model the second addend, 76, using the fewest number of blocks. |

| Step 3 | Combine the ones for both addends. Regroup ones as tens. How many ones do you have? |

| Step 4 | Combine the tens for both addends. Regroup tens as hundreds. How many tens do you have? How many hundreds? What is sum of 58 and 76? |

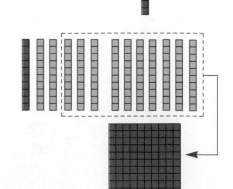

**Use base ten blocks to find each sum. Record your results.**

| 1. | 65 | 2. | 49 | 3. | 56 | 4. | 38 | 5. | 99 | 6. | 15 |
|---|---|---|---|---|---|---|---|---|---|---|---|
| | +99 | | +97 | | +44 | | +83 | | +86 | | +95 |

7. 16 + 99          8. 28 + 77          9. 43 + 88

▶ You can also use paper and pencil to find the sum of 58 and 76.

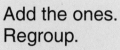
Add the ones. Regroup.  →  Add the tens. Regroup.

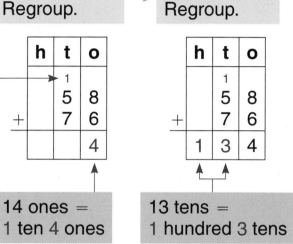

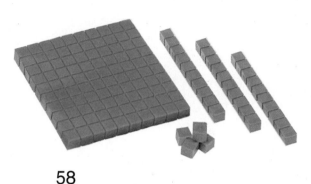

58
+76
134

14 ones = 1 ten 4 ones

13 tens = 1 hundred 3 tens

134 is close to 140. The answer is reasonable.

April collected 134 cans in the two weeks.

**Find the sum.**

| 10. | 39 | 11. | 54 | 12. | 79 | 13. | 76 | 14. | 27 | 15. | 99 |
|---|---|---|---|---|---|---|---|---|---|---|---|
| | +82 | | +64 | | +99 | | +45 | | +57 | | +33 |

| 16. | 63 | 17. | 86 | 18. | 67 | 19. | 47 | 20. | 4 | 21. | 8 |
|---|---|---|---|---|---|---|---|---|---|---|---|
| | +89 | | +87 | | +56 | | +62 | | +96 | | +98 |

## Communicate

22. If you regroup twice when adding 2 two-digit numbers, how many digits would be in the sum? Explain.

## 2-11 Three-Digit Addition

This year 236 girls and 192 boys attend a school on Hill Street. How many students attend the school in all?

▶ First estimate the sum.

200 + 200 = 400

▶ Then to find how many students in all, add: 236 + 192 = ?

| Add the ones. | Add the tens. Regroup. | Add the hundreds. |
|---|---|---|

| h | t | o |
|---|---|---|
| 2 | 3 | 6 |
| +1 | 9 | 2 |
|   |   | 8 |

|   | h | t | o |
|---|---|---|---|
| 1 |   |   |   |
|   | 2 | 3 | 6 |
| + | 1 | 9 | 2 |
|   |   | 2 | 8 |

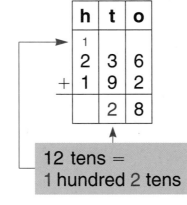

12 tens = 1 hundred 2 tens

|   | h | t | o |
|---|---|---|---|
| 1 |   |   |   |
|   | 2 | 3 | 6 |
| + | 1 | 9 | 2 |
|   | 4 | 2 | 8 |

```
  236
+ 192
  428
```

This year 428 students attend the school.

428 is close to the estimate of 400.

### Study this example.

First estimate.

```
  $2.63  ────▶    $3.00
+ 4.82  ────▶  + 5.00
            about   $8.00
```

Then add.
```
     1
  $2.63
+ 4.82
  $7.45
```

**Estimate. Then add.**

| 1. | 2. | 3. | 4. | 5. |
|---|---|---|---|---|
| 264<br>+127 | 365<br>+418 | 490<br>+333 | 271<br>+452 | 609<br>+156 |

| 6. | 7. | 8. | 9. | 10. |
|---|---|---|---|---|
| $4.17<br>+ 3.19 | $2.93<br>+ 3.86 | $1.36<br>+ 2.83 | $6.25<br>+ 1.09 | $3.72<br>+ 2.55 |

**Add and check.**

11.
```
    1
   271  | Add
  +352  ↓ down.
   623
```
```
    1
   271  ↑ Check:
  +352  | add up.
   623
```

| 12. | 13. |
|---|---|
| 925<br>+ 39 | 192<br>+ 42 |

| 14. | 15. | 16. | 17. | 18. |
|---|---|---|---|---|
| 509<br>+231 | 246<br>+183 | 126<br>+126 | $4.36<br>+ 4.93 | $1.84<br>+ 7.25 |

**Align and add.**

**19.** 677 + 151      **20.** 408 + 129      **21.** 281 + 181

**22.** 526 + 318      **23.** 776 + 52      **24.** 293 + 635

**25.** 117 + 192      **26.** 682 + 222      **27.** 24 + 866

**PROBLEM SOLVING**

**28.** At lunchtime the students in grades 1–3 need 126 places in the lunchroom. The students in grades 4–6 need 281 places. If the six grades eat lunch at the same time, how many places are needed in the lunchroom?

**29.** On Tuesday 86 students ordered pizza and 193 students ordered hot dogs for lunch. How many students ordered these foods?

## 2-12  More Regrouping in Addition

The Lincoln Garden Club planted 568 tulips and 285 daisies in the park. How many flowers did the club plant?

▶ First estimate the sum.

600 + 300 = 900

▶ Then to find how many flowers the club planted, add: 568 + 285 = ?

| Add the ones. Regroup. | Add the tens. Regroup. | Add the hundreds. |
|---|---|---|

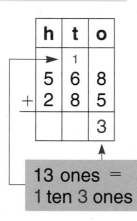

| h | t | o |
|---|---|---|
|   | 1 |   |
| 5 | 6 | 8 |
| + 2 | 8 | 5 |
|   |   | 3 |

**13 ones = 1 ten 3 ones**

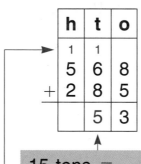

| h | t | o |
|---|---|---|
| 1 | 1 |   |
| 5 | 6 | 8 |
| + 2 | 8 | 5 |
|   | 5 | 3 |

**15 tens = 1 hundred 5 tens**

| h | t | o |
|---|---|---|
| 1 | 1 |   |
| 5 | 6 | 8 |
| + 2 | 8 | 5 |
| 8 | 5 | 3 |

853 is close to the estimate of 900.

The Lincoln Garden Club planted 853 flowers.

**Estimate. Then add.**

| 1. | 524 <br> + 387 | 2. | 429 <br> + 295 | 3. | 324 <br> + 578 | 4. | $6.05 <br> + 1.97 | 5. | $3.29 <br> + 4.89 |
|---|---|---|---|---|---|---|---|---|---|
| 6. | 759 <br> + 189 | 7. | 873 <br> + 68 | 8. | 427 <br> + 96 | 9. | $3.54 <br> + .56 | 10. | $8.09 <br> + .99 |
| 11. | 495 <br> + 15 | 12. | 326 <br> + 83 | 13. | 507 <br> + 97 | 14. | $1.53 <br> + .47 | 15. | $5.61 <br> + .69 |

**Add and check.**

| 16. | 178<br>+342 | 17. | 676<br>+285 | 18. | 382<br>+317 | 19. | 857<br>+ 99 | 20. | 386<br>+479 |
|---|---|---|---|---|---|---|---|---|---|
| 21. | 753<br>+177 | 22. | 429<br>+245 | 23. | 328<br>+295 | 24. | 191<br>+609 | 25. | 337<br>+333 |
| 26. | $2.98<br>+ 5.18 | 27. | $3.75<br>+ 5.46 | 28. | $1.24<br>+ 7.55 | 29. | $6.38<br>+ 2.93 | 30. | $1.96<br>+ 1.06 |

**31.** 388 + 226          **32.** 686 + 224          **33.** 438 + 369

**34.** 884 + 47          **35.** 461 + 472          **36.** 729 + 181

## PROBLEM SOLVING

**37.** The Lincoln Garden Club chose 165 red tulips and 158 yellow tulips to plant around the park fountain. How many tulips in all did the club plant?

**38.** Smithtown Garden Center sold 568 tulips to the Lincoln Garden Club and 219 tulips and 49 daisies to the Roeder Garden Club. How many tulips in all were sold to the two garden clubs?

 **Skills to Remember**

**Add.**      $7 + 7 + 4 =$
              $\underbrace{14}\ \ + 4 = 18$

**39.** 8 + 3 + 6          **40.** 6 + 9 + 5          **41.** 7 + 8 + 9

**42.** 2 + 8 + 7          **43.** 5 + 5 + 8          **44.** 9 + 3 + 3

## 2-13 Mental Math

Algebra

Here are two ways to find sums mentally.

▶ **Look for patterns.**

You know 6 + 3 = 9.           You know 8 + 6 = 14.
   So 16 + 3 = 19                 So 18 + 6 = 24
      26 + 3 = 29                    28 + 6 = 34
      36 + 3 = 39                    38 + 6 = 44

▶ **Break apart numbers to find tens.**

$$28 + 4 = \underline{\ ?\ }$$
Think: 20 + 8 + 4
$$20 + \quad 12 \quad = 32$$

$$67 \quad + \quad 25 \quad = \underline{\ ?\ }$$
Think: (60 + 7)  + (20 + 5)
$$(60 + 20) + \quad (7 + 5)$$
$$80 \quad + \quad 12 \quad = 92$$

Changing the **order** and the **grouping** of the addends does not change the sum.

**Add mentally.**

**1.** 19 + 9        **2.** 29 + 9        **3.** 39 + 9        **4.** 49 + 9

**5.** 5 + 56        **6.** 5 + 66        **7.** 5 + 76        **8.** 5 + 86

**9.** 23 + 7        **10.** 33 + 7        **11.** 43 + 7        **12.** 53 + 7

**13.**  39        **14.**  16        **15.**  45        **16.**  26        **17.**  83        **18.**  34
     + 7              + 9              + 6              + 8              + 9              + 6

**19.**  64        **20.**  19        **21.**  37        **22.**  83        **23.**  75        **24.**  54
     +22              +78              +44              +17              +58              +29

# Regrouping Hundreds

Sometimes you need to regroup
hundreds as thousands.

1 thousand 12 hundreds = _?_

## Hands-On Understanding

**You Will Need:** base ten blocks

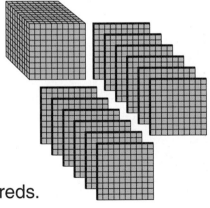

**Step 1**  Place 1 thousand cube and 1 hundred
flat on your worktable.

How many hundred flats can you stack
on top of each other to equal 1 thousand cube?

What do you notice about 1 thousand and 10 hundreds?

**Step 2**  Model 1 thousand 12 hundreds.

What number did you model?

**Step 3**  Now model the same number,
using fewer blocks.

Which blocks did you use?

So 1 thousand 12 hundreds can be
regrouped as _?_ thousands _?_ hundreds.

## Communicate

1. Explain when you can regroup hundreds as thousands.

**Model each number using only hundred flats.
Then model each using the fewest blocks.
Describe the regrouped number.**

**2.** 19 hundreds

**3.** 11 hundreds

**4.** 20 hundreds

## 2-15 | Three or More Addends

How many pounds of newspaper were collected by grades 2–4 of Cameron School in the 1st week?

**1st Week**

| Grade | Pounds Collected |
|-------|------------------|
| 2nd   | 335              |
| 3rd   | 834              |
| 4th   | 155              |

▶ First estimate the sum.
300 + 800 + 200 = 1300

▶ Then to find how many,
add: 335 + 834 + 155 = ?

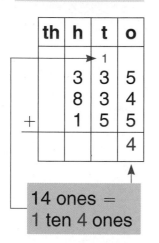

| Add the ones. Regroup. | Add the tens. Regroup. | Add the hundreds. Regroup. |
|---|---|---|

| th | h | t | o |
|----|---|---|---|
|    |   | 1 |   |
|    | 3 | 3 | 5 |
|    | 8 | 3 | 4 |
| +  | 1 | 5 | 5 |
|    |   |   | 4 |

| th | h | t | o |
|----|---|---|---|
|    | 1 | 1 |   |
|    | 3 | 3 | 5 |
|    | 8 | 3 | 4 |
| +  | 1 | 5 | 5 |
|    |   | 2 | 4 |

| th | h | t | o |
|----|---|---|---|
|    | 1 | 1 |   |
|    | 3 | 3 | 5 |
|    | 8 | 3 | 4 |
| +  | 1 | 5 | 5 |
| 1  | 3 | 2 | 4 |

| 14 ones = 1 ten 4 ones | 12 tens = 1 hundred 2 tens | 13 hundreds = 1 thousand 3 tens |
|---|---|---|

1324 pounds of newspaper were collected.

1324 is close to the estimate of 1300.

## Estimate. Then add.

| 1. | 382 | 2. | 145 | 3. | 173 | 4. | $1.41 | 5. | $1.17 |
|----|-----|----|-----|----|-----|----|-------|----|-------|
|    | 754 |    | 362 |    | 460 |    | 8.07  |    | 4.17  |
|    | +127 |   | +656 |   | +625 |   | + 3.12 |   | + 5.34 |

| 6. | 284 | 7. | 416 | 8. | 335 | 9. | $9.33 | 10. | $1.82 |
|----|-----|----|-----|----|-----|----|-------|-----|-------|
|    | 800 |    | 377 |    | 115 |    | 2.33  |     | 1.42  |
|    | +491 |   | +186 |   | +206 |   | + 1.55 |    | + 1.25 |

**Add and check.**

| 11. | 143 | 12. | 134 | 13. | 325 | 14. | $2.13 | 15. | $1.75 |
|---|---|---|---|---|---|---|---|---|---|
| | 287 | | 718 | | 275 | | 9.04 | | .48 |
| | +651 | | + 79 | | +350 | | + 4.92 | | + 2.00 |

| 16. | 82 | 17. | 530 | 18. | 415 | 19. | $3.85 | 20. | $4.63 |
|---|---|---|---|---|---|---|---|---|---|
| | 763 | | 607 | | 514 | | 3.27 | | 5.38 |
| | +385 | | + 73 | | +145 | | + 3.58 | | + .07 |

---

**Four Three-Digit Addends**                          Algebra

Add: 216 + 301 + 570 + 132 = _?_

| | Add. Start with |
| Align. | the ones. |

You can estimate the sum.
200 + ⌐300 + 600 + 100⌐ = _?_
200 + 1000 = 1200

| Align. | Add. Start with the ones. |
|---|---|
| | 1 |
| 216 | 216 |
| 301 | 301 |
| 570 | 570 |
| +132 | +132 |
| | 1219 |

216 + 301 + 570 + 132 = 1219

---

**Align and add.**

**21.** 326 + 101 + 480 + 229          **22.** 350 + 406 + 232 + 581

**23.** 122 + 334 + 445 + 551          **24.** 177 + 201 + 418 + 352

**Challenge**

Algebra

**Complete the patterns.**

**25.** 0, 1, 3, 6, 10, 15, 21, _?_ , _?_ , _?_

**26.** 1, 2, 4, 8, 16, 32, 64, 128, _?_ , _?_ , _?_

**27.** 1000, 1300, 1900, 2200, 2800, _?_ , _?_ , _?_

# Adding Four-Digit Numbers

For two weeks, Oakhill School held a food drive to help stock a soup kitchen. How many canned goods were collected in all during the food drive?

| Week | Number of Cans |
|------|----------------|
| 1st  | 5458 |
| 2nd  | 2797 |

▶ Estimate the sum. $5000 + 3000 = 8000$

▶ To find how many, add: $5458 + 2797 = \underline{?}$

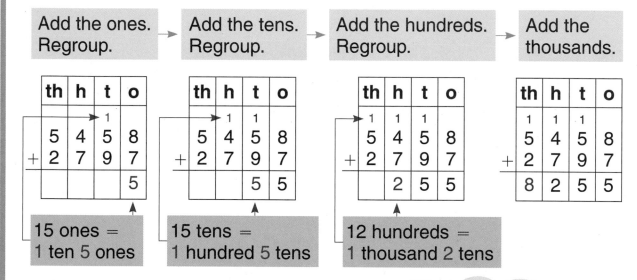

| Add the ones. Regroup. | Add the tens. Regroup. | Add the hundreds. Regroup. | Add the thousands. |

15 ones = 1 ten 5 ones

15 tens = 1 hundred 5 tens

12 hundreds = 1 thousand 2 tens

During the food drive, 8255 canned goods were collected.

8255 is close to 8000. The answer is reasonable.

## Estimate. Then add.

| 1. | 2935 <br> + 4508 | 2. | 4853 <br> + 3649 | 3. | 8147 <br> + 1269 | 4. | 1007 <br> + 6397 | 5. | 3333 <br> + 1777 |

## Add. Use the $ and . when needed.

| 6. | 6551 <br> + 2709 | 7. | 1173 <br> + 5416 | 8. | 848 <br> + 8152 | 9. | 2126 <br> + 3929 | 10. | $55.55 <br> + 25.55 |

| 11. | 3784 <br> + 5247 | 12. | $23.99 <br> + 6.06 | 13. | 1647 <br> + 1843 | 14. | $40.62 <br> + 20.76 | 15. | 3784 <br> + 5247 |

**Find the sum.**

| | | | | |
|---|---|---|---|---|
| **16.** 5783<br>+1811 | **17.** 4623<br>+ 899 | **18.** 7053<br>+1008 | **19.** 4716<br>+ 604 | **20.** 7777<br>+1223 |
| **21.** $60.40<br>+ 23.82 | **22.** $30.86<br>+ .90 | **23.** $ 5.19<br>+ 49.86 | **24.** $60.81<br>+ 20.19 | **25.** $80.80<br>+ 10.30 |
| **26.** 6379<br>+2669 | **27.** 4056<br>+4157 | **28.** 295<br>+5786 | **29.** 6049<br>+ 92 | **30.** 8999<br>+ 111 |

**31.** 2633 + 199          **32.** 87 + 1604          **33.** $4.01 + $31.95

**34.** 5999 + 19          **35.** $46.51 + $21.23          **36.** $61.53 + $10.87

**37.** $43.31 + $24.22          **38.** $34.52 + $18.30          **39.** $56.72 + $33.33

**40.** 66 + 6666          **41.** 2222 + 99          **42.** 1111 + 899

## PROBLEM SOLVING

**43.** The children at Oakhill School raised $31.48 at their craft fair and $46.16 at a bake sale. How much money did they raise?

**44.** The third graders collected 3784 labels for gym equipment in April and 4730 in May. How many labels did they collect altogether?

 **Mental Math**

**Add mentally.**

| | | | | |
|---|---|---|---|---|
| **45.** 5600<br>+ 39 | **46.** 5600<br>+ 139 | **47.** 5600<br>+ 239 | **48.** 5600<br>+ 339 | **49.** 5600<br>+ 439 |
| **50.** 4020<br>+1100 | **51.** 3020<br>+1100 | **52.** 2020<br>+1100 | **53.** 1020<br>+1100 | **54.** 920<br>+1100 |

# TECHNOLOGY

## Calculating Money

You can add money on a **calculator**
by using the decimal point key,
to separate dollars and cents.

Add:　$8.07
　　　+ 5.94

▶ First estimate the sum:

| | |
|---|---|
| $8.07 ⟶ | $8.00 |
| + 5.94 ⟶ | + 6.00 |
| | about $14.00 |

▶ Then use a calculator to find the sum.

To add $8.07 + $5.94 on a calculator,

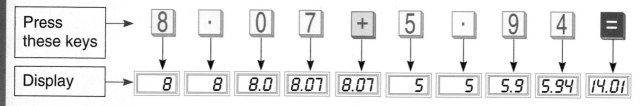

| Press these keys | 8 | · | 0 | 7 | + | 5 | · | 9 | 4 | = |
|---|---|---|---|---|---|---|---|---|---|---|
| Display | 8 | 8 | 8.0 | 8.07 | 8.07 | 5 | 5 | 5.9 | 5.94 | 14.01 |

$8.07
+ 5.94
$14.01

Think: $14.01 is close to $14.
The answer is reasonable.

Remember: Write the dollar sign
and the decimal point.

---

**Estimate. Tell if the sum is greater or less than $10.
Then use a calculator to find the sum.**

**1.** $6.36 + $1.95

**2.** $7.70 + $1.01

**3.** $5.62 + $4.80

**4.** $3.29 + $3.29

**5.** $3.94 + $8.15

**6.** $6.02 + $9.47

**7.** $3.87 + $8.98

**8.** $6.30 + $2.75

**9.** $8.05 + $1.99

**Estimate. Then use a calculator to find the sum.**

| | | | | |
|---|---|---|---|---|
| **10.** $3.32<br>+ 6.27 | **11.** $5.43<br>+ 3.26 | **12.** $7.18<br>+ 2.54 | **13.** $6.39<br>+ 1.43 | **14.** $4.46<br>+ 5.16 |
| **15.** $7.28<br>+ 1.63 | **16.** $5.93<br>+ 1.59 | **17.** $4.16<br>+ 2.88 | **18.** $6.07<br>+ 7.98 | **19.** $3.36<br>+ 8.79 |
| **20.** $15.25<br>+ 10.40 | **21.** $70.40<br>+ 21.05 | **22.** $45.27<br>+ 16.01 | **23.** $33.33<br>+ 47.33 | **24.** $ 7.52<br>+ 12.03 |
| **25.** $18.06<br>+ 4.91 | **26.** $3.10<br>1.52<br>+ 4.85 | **27.** $6.15<br>3.76<br>+ 9.00 | **28.** $ 2.67<br>12.81<br>+ .55 | **29.** $40.07<br>.20<br>+ 5.19 |

**Estimate. Tell if the given sum is close to your estimate.
If not, use a calculator to find the correct sum.**

**30.** $5.16
3.92
+ 8.40    | 9.92 |

**31.** $6.89
4.75
+ 9.81    | 21.45 |

**32.** $2.19
.72
+ 6.45    | 9.36 |

**33.** $3.80
9.12
+ .64    | 18.56 |

**34.** $5.22
9.70
6.41
+ 3.09    | 36.42 |

**35.** $4.10
7.22
5.78
+ 9.46    | 26.56 |

## Critical Thinking

**36.** Bobby added $3.28 + $2.42 using a calculator. The calculator displayed | 5.7 |. How would you write the sum using a dollar sign and decimal point?

**37.** Maria added $1.30 + $1.32 using a calculator. The calculator displayed | 2.622 |. What mistake did Maria make when entering the addends?

95

## 2-18 | Problem Solving: Use Simpler Numbers

**Problem:** In the Great Pine Forest 2184 seedlings were planted after a fire. The next year 1076 more seedlings were planted. How many seedlings were planted in the two years?

**1 IMAGINE** Think about the problem.

**2 NAME**

*Facts:*   2184 seedlings
1076 more seedlings

*Question:*   How many seedlings were planted in two years?

**3 THINK** Use simpler numbers like 20 and 10. Use 20 for 2184 and 10 for 1076. Reread the problem, using the simpler numbers.

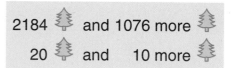

2184 🌲 and 1076 more 🌲
20 🌲 and    10 more 🌲

*Think:*   If 20 seedlings were planted one year and 10 more were planted the next year, how many seedlings were planted in all?   30

How did you get your answer? Add:   $20 + 10 = 30$

So to find the answer, add 2184 and 1076.

$2184 + 1076 = \underline{\ ?\ }$

**4 COMPUTE**

```
  11
  2184
+ 1076
  3260
```
There were 3260 seedlings planted.

**5 CHECK** Is 3260 greater than 2184?  Yes.
Your answer makes sense.

**Use simpler numbers to solve each problem.**

1. For Saturday's game 2563 tickets were sold on Thursday, and 998 tickets were sold on Friday. What is the total number of tickets sold on the two days?

| Soccer | |
|---|---|
| Wednesday | SOLD OUT |
| Thursday | 2563 |
| Friday | 998 |

**IMAGINE**   Look at the table.
Think about the problem.

**NAME**   *Facts:*   Thursday   2563 tickets sold
Friday      998 tickets sold

*Question:*   What is the total number sold?

**THINK**   Reread the problem, using 20 as 2563 and 10 as 998.

*Think:*   What do you need to do to find the total number of tickets? Add.

Now add the numbers given in the problem.

**COMPUTE** ⟶ **CHECK**

2. At Landis School 8453 students have goldfish. Another 1396 have tropical fish. How many students have fish as pets?

3. Roberta flew 425 miles to a layover in Texas. Then she flew 215 miles farther. How far did she fly in all?

4. A truck delivered 210 pounds of potatoes, 195 pounds of meat, and 230 pounds of fruit. What was the total weight of the items delivered?

**Problem-Solving Applications**

**Solve each problem and explain the method you used.**

1. A sports shop sold 32 red caps and 28 blue caps. Each cap costs $6.99. How many caps did the shop sell altogether?

2. Juanita had 175 baseball cards in her collection. She bought a dozen more. How many baseball cards does she have in her collection now?

3. On Thursday 5144 fans went to the swim meet. On Friday 4288 fans went to the track meet. What is the total number of fans who went to the two meets?

4. The game warden stocked the lake with 419 bass and 539 trout. How many fish were put into the lake?

5. The distance from Sportville to the stadium is double the distance from Sportville to the lake. How far is it from Sportville to the stadium?

6. Ed spent $19.20 at the sports shop. He spent $24.90 at the lake. About how many dollars did Ed spend?

**Use the table to solve problems 7–9.**

7. Marty has $15. What three different items can she buy?

8. Carlos spent between $15 and $20. What two items did he buy?

9. Laurie bought 3 jump ropes. How much did she spend?

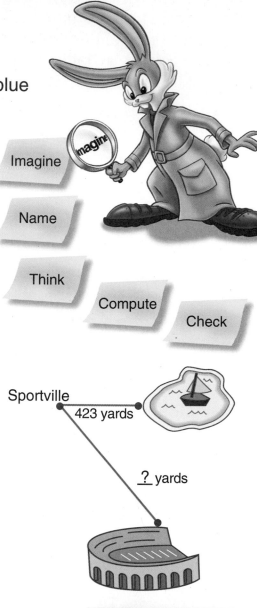

Imagine

Name

Think

Compute

Check

Sportville

423 yards

? yards

| ★ SALE ★ | |
|---|---|
| baseball caps | $6.99 |
| jump ropes | $4.89 |
| weights | $16.99 |
| sweat-bands | $2.49 |

**Choose a strategy from the list or use another strategy you know to solve each problem.**

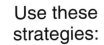

Use these strategies:

Draw a Picture
Hidden Information
Guess and Test
Make a Table
Find a Pattern
Extra Information
Use Simpler Numbers

10. There are 18 sailboats at the dock. Every fourth sailboat is red. How many sailboats at the dock are *not* red?

11. On Monday Raul did 5 sit-ups. On Tuesday he did 10 sit-ups. Each day after, he doubled the number of sit-ups. How many sit-ups did Raul do on Friday?

12. Elena exercised 35 minutes on Monday. Then she read for 45 minutes. The next day she exercised for 25 minutes in the morning and again at noon for 40 minutes. How long did she exercise in all?

13. Thea's soccer team scored 12 goals in the first three games. The team scored the same number of goals in each of the first two games. The team scored 3 fewer goals in the third game than in the second. How many goals were scored in each game?

14. The jersey numbers of the runners in a race form a pattern: 6, 11, 9, 14, 12, and so on. What is the number of the eighth runner?

**Use the graph to solve problems 15–17.**

15. Ms. Myers needs 35 mats. How many more mats does she need to buy?

16. Fifteen students in the morning gym class each used a jump rope and a scooter. How many pieces of equipment did the class use?

17. How many pieces of equipment are there in all?

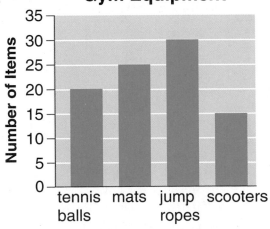

**Gym Equipment**

Number of Items

tennis balls  mats  jump ropes  scooters

**Type of Equipment**

# Chapter 2

# Chapter Review and Practice

**Find the sum.** *(See pp. 68–69.)*

**1.** $4 + 3 + 7$       **2.** $5 + 5 + 7$       **3.** $6 + 4 + 3 + 3$

**Find the missing addend.** *(See pp. 70–71.)*

**4.** $7 + \underline{\ ?\ } = 15$       **5.** $3 + \underline{\ ?\ } = 12$       **6.** $18 = 9 + \underline{\ ?\ }$

**Regroup.** *(See pp. 77, 80, 89.)*

**7.** 7 tens 13 ones =
   <u>8</u> tens <u>3</u> ones

**8** 2 tens 14 ones =
   <u>?</u> tens <u>?</u> ones

**9.** 29 tens =
   <u>?</u> hundreds <u>?</u> tens

**10.** 1 hundred 15 tens =
   <u>?</u> hundreds <u>?</u> tens

**11.** 13 hundreds =
   <u>?</u> thousand <u>?</u> hundreds

**12.** 3 thousands 15 hundreds =
   <u>?</u> thousands <u>?</u> hundreds

**13.** 15 dimes =
   <u>?</u> dollar <u>?</u> dimes

**14.** 4 dollars 16 dimes =
   <u>?</u> dollars <u>?</u> dimes

**Estimate the sum. Then add and check.** *(See pp. 72–76, 78–79, 81–87, 90–93.)*

| 15. | 16. | 17. | 18. | 19. |
|---|---|---|---|---|
| 37 | 49 | 73 | 28 | $.65 |
| +16 | +13 | +27 | +31 | + .89 |

| 20. | 21. | 22. | 23. | 24. |
|---|---|---|---|---|
| 185 | $2.27 | $7.34 | 214 | 256 |
| +136 | + 6.84 | + 2.95 | 331 | 572 |
|  |  |  | +568 | + 193 |

| 25. | 26. | 27. | 28. | 29. |
|---|---|---|---|---|
| $3.49 | $7.19 | $8.93 | 2563 | 4375 |
| + 8.73 | + 3.29 | + 1.37 | +6374 | +2856 |

## PROBLEM SOLVING

*(See pp. 96–99.)*

**30.** Linh had 1048 stickers in one pile and 2161 in another pile. How many stickers were there altogether?

*(See Still More Practice, p. 444.)*

## COIN COMBINATIONS

Tom has $4.00 to buy a football for $3.75.

**PROBLEM SOLVING** Use the table.

1. Name 5 other ways Tom can receive 25¢ in change.

2. Which way uses the least number of coins?

3. Which way uses the greatest number of coins?

| Change for a Quarter | | |
|:---:|:---:|:---:|
|  | | |
| 2 | 1 | |
| 1 | 3 | |
| | 5 | |
| | | 25 |
| | 1 | 20 |
| 1 | | 15 |
| 1 | 1 | 10 |

**Name two other ways to show each amount.**

4.

5.

6. What is the least number of coins that shows 35¢?

7. What is the least number of coins that shows 65¢?

8. Jan has 18¢. She has 6 coins. Name the coins.

9. Fran has 96¢. She has 6 coins. Name two possible coin combinations.

101

# Check Your Mastery

## Performance Assessment

Predict which exercise involves two regroupings. Use base ten blocks to model each and explain the regrouping.

**1.** 4326 + 946        **2.** 6253 + 476

**Find the sum.**

**3.** 3¢ + 2¢ + 7¢ + 5¢      **4.** 1¢ + 5¢ + 6¢ + 5¢

**Find the missing addend.**

**5.** $9 + \underline{\ ?\ } = 15$      **6.** $6 + \underline{\ ?\ } = 13$      **7.** $17 = \underline{\ ?\ } + 8$

**Regroup.**

**8.** 25 dimes =                **9.** 9 dollars 10 dimes =
     $\underline{\ ?\ }$ dollars $\underline{\ ?\ }$ dimes         $\underline{\ ?\ }$ dollars $\underline{\ ?\ }$ dimes

**Add and check.**

| **10.** | **11.** | **12.** | **13.** | **14.** |
|---|---|---|---|---|
| 63 | 19¢ | $.68 | 249 | $8.32 |
| +17 | +27¢ | +.48 | +173 | + .42 |

| **15.** | **16.** | **17.** | **18.** | **19.** |
|---|---|---|---|---|
| 650 | 361 | $7.84 | 1469 | 3526 |
| 542 | 27 | + 3.28 | + 376 | + 724 |
| +183 | +447 | | | |

**Estimate the sum.**

| **20.** | **21.** | **22.** | **23.** | **24.** |
|---|---|---|---|---|
| 39 | 559 | $9.05 | $74.80 | 5328 |
| +23 | +361 | + 2.75 | + 8.62 | +1009 |

**PROBLEM SOLVING**    *Use a strategy you have learned.*

**25.** There were 327 students who attended the show on Monday. On Tuesday 276 attended. On Wednesday 112 attended. How many students attended altogether?

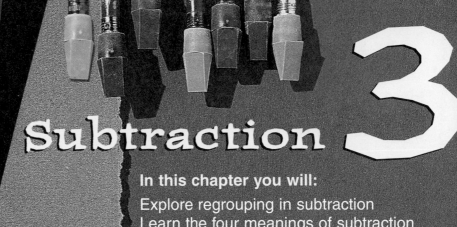

# Subtraction 3

**In this chapter you will:**

Explore regrouping in subtraction
Learn the four meanings of subtraction
Estimate and subtract whole numbers
and money
Solve problems by choosing the operation

**Critical Thinking/Finding Together**

Write a poem using your full name,
following the pattern in "The Eraser Poem."
Is your poem shorter or longer than "The
Eraser Poem"? By how many lines?

## THE ERASER POEM

The eraser poem.
The eraser poem
The eraser poe
The eraser po
The eraser p
The eraser
The erase
The eras
The era
The er
The e
The
Th
T

*Louis Phillips*

# 3-1 Subtraction Concepts

*Algebra* ✓

Here are four meanings of subtraction.

▶ **Take Away**

Anita had 26 baseball caps.
She sold 13 of them.
How many caps are left?

26 − 13 = 13
Thirteen caps are left.

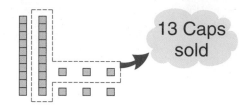

13 Caps sold

▶ **Find Part of a Whole Set**

A sporting goods store has
37 football helmets.
Fifteen have stripes.
How many do *not* have stripes?

37 − 15 = 22
Twenty-two helmets have no stripes.

15 Striped helmets

▶ **Compare**

Marika has 48 football cards. Pablo
has 32. How many more football
cards does Marika have than Pablo?

48 − 32 = 16
Marika has 16 more football cards than Pablo.

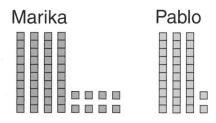

Marika       Pablo

▶ **Find How Many More Are Needed**

Lara has 13 dollars. She needs
18 dollars for a baseball pennant.
How many more dollars does
she need?

$18 − $13 = $5
Lara needs $5 more.

Dollars needed

**Solve. Tell which meaning of subtraction is shown.**

Communicate ✓

1. Juan has 75 baseball cards from Team A and 72 baseball cards from Team B. How many more cards from Team A than from Team B does he have?

2. Lori kicked the ball 28 yards. Mona kicked the ball 16 yards. How many fewer yards did Mona kick the ball than Lori?

3. A baseball team had 12 bats. Two of the bats were broken in a game. How many baseball bats were left?

4. Hank had 57 toy cars. He gave 7 cars to his brother. How many toy cars does Hank have now?

5. There are 18 players on a football team. Nine of them are in the locker room. The rest are on the field at practice. How many players are on the field?

6. The library got 89 new books. Eighteen of them were science books. How many were other kinds of books?

7. The football team has 22 yards to go to the goal line. They gain 20 yards. How many more yards does the team need to make a touchdown?

8. Marco has thirty-two dollars. He needs 45 dollars to buy a baseball glove. How many more dollars does he need?

9. Peggy is 54 years old. Fred is 51 years old. How much younger is Fred than Peggy?

10. A video store sold 66 videotapes. Twenty of them were about sports. How many videotapes sold were not about sports?

## Connections: Language Arts

**Write a subtraction story for each meaning of subtraction. You may use the picture.**

11. Take Away

12. Find Part of a Whole Set

13. Compare

14. Find How Many More Are Needed

| Team | Number of Awards |
|------|------------------|
| A | 🎀🎀🎀🎀 |
| B | 🎀🎀🎀🎀🎀 |
| C | 🎀🎀🎀 |
| D | 🎀🎀🎀🎀🎀🎀🎀🎀 |

## 3-2 Subtracting: No Regrouping

At the fair a clown had 279 balloons to sell.
In the morning he sold 157 balloons.
How many does he have left to sell?

▶ First estimate the difference.

| Subtract the front digits. | → | Write 0s for the other digits. |

$$
\begin{array}{r} 279 \\ -157 \\ \hline 1 \end{array}
$$

$$
\begin{array}{r} 279 \\ -157 \\ \hline \text{about} \quad 100 \end{array}
$$

About 100 balloons are left.

▶ Then to find how many left,
subtract: 279 − 157 = ?

| Subtract the ones. | → | Subtract the tens. | → | Subtract the hundreds. |

| h | t | o |
|---|---|---|
| 2 | 7 | 9 |
| − 1 | 5 | 7 |
|   |   | 2 |

| h | t | o |
|---|---|---|
| 2 | 7 | 9 |
| − 1 | 5 | 7 |
|   | 2 | 2 |

| h | t | o |
|---|---|---|
| 2 | 7 | 9 |
| − 1 | 5 | 7 |
| 1 | 2 | 2 |

$$
\begin{array}{r} 279 \\ -157 \\ \hline 122 \end{array}
$$

The clown has 122 balloons left to sell.

## Estimate. Then find the difference.

**1.** 
$$\begin{array}{r} 424 \\ -412 \\ \hline \end{array}$$

**2.** 
$$\begin{array}{r} 475 \\ -304 \\ \hline \end{array}$$

**3.** 
$$\begin{array}{r} 296 \\ -240 \\ \hline \end{array}$$

**4.** 
$$\begin{array}{r} 263 \\ -\ 42 \\ \hline \end{array}$$

**5.** 
$$\begin{array}{r} 917 \\ -\ 12 \\ \hline \end{array}$$

**6.** 
$$\begin{array}{r} 257 \\ -\ 43 \\ \hline \end{array}$$

**7.** 
$$\begin{array}{r} 999 \\ -909 \\ \hline \end{array}$$

**8.** 
$$\begin{array}{r} 551 \\ -\ 51 \\ \hline \end{array}$$

**9.** 
$$\begin{array}{r} 675 \\ -\ 55 \\ \hline \end{array}$$

**10.** 
$$\begin{array}{r} 988 \\ -\ 83 \\ \hline \end{array}$$

**11.** 476 − 35

**12.** 217 − 12

**13.** 869 − 59

# 3-3 Subtracting Money

Vera lent $2.61 to her brother, Vinnie. He now has $4.84. How much money did Vinnie already have?

▶ First estimate the difference.

$$\begin{array}{r} \$4.84 \longrightarrow \$4.00 \\ -\ 2.61 \longrightarrow -\ 2.00 \\ \hline \text{about} \quad \$2.00 \end{array}$$

▶ To find how much Vinnie had, subtract: $4.84 − $2.61 = __?__

| Line up the dollars and cents. | Subtract as usual. | Write $ and . in the difference. |
|---|---|---|
| $4.84<br>− 2.61 | $4.84<br>− 2.61<br>2 23 | $4.84<br>− 2.61<br>$2.23 |

Vinnie already had $2.23.

> Think: $2.23 is close to $2.00. The answer is reasonable.

**Study these examples.**

| $7.50 | 27¢ | $.99 |
|---|---|---|
| − 4.50 | −17¢ | − .26 |
| $3.00 | 10¢ | $.73 |

---

**Estimate. Then subtract.**

| | | | | |
|---|---|---|---|---|
| **1.** 54¢<br> −33¢ | **2.** 76¢<br> −25¢ | **3.** $3.69<br> − 1.37 | **4.** $9.95<br> − 7.64 | **5.** $5.06<br> − 2.05 |
| **6.** $.16<br> − .12 | **7.** $.86<br> − .41 | **8.** $6.55<br> − 4.50 | **9.** $7.47<br> − 6.35 | **10.** $6.78<br> − 4.32 |

Ice skates are on sale for $76.45. Julie has saved $64.90. About how much more does she need to buy the skates on sale?

An estimate tells *about* how much or *about* how many. You can estimate differences by rounding.

Estimate: $76.45 − $64.90

- Round the amounts to the nearest dollar.

- Then subtract the rounded amounts.

$$\begin{array}{r} \$76.45 \longrightarrow \$76.00 \\ -\ 64.90 \longrightarrow -\ 65.00 \\ \hline \text{about}\quad \$11.00 \end{array}$$

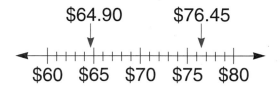

$64.90          $76.45

$60  $65  $70  $75  $80

Julie needs about $11.00 more.

**Study these examples.**

| Round to the nearest ten. | Round to the nearest ten cents. | Round to the nearest hundred. |
|---|---|---|
| $\begin{array}{r}63 \longrightarrow 60 \\ -28 \longrightarrow -30 \\ \hline \text{about}\ \ 30\end{array}$ | $\begin{array}{r}\$1.57 \longrightarrow \$1.60 \\ -\ .39 \longrightarrow -\ .40 \\ \hline \text{about}\ \ \$1.20\end{array}$ | $\begin{array}{r}859 \longrightarrow 900 \\ -623 \longrightarrow -600 \\ \hline \text{about}\ \ 300\end{array}$ |

Remember: Write the dollar sign and the decimal point.

**Estimate by rounding to the nearest ten or ten cents.**

| | | | | |
|---|---|---|---|---|
| **1.** 46 <br> −24 | **2.** 53 <br> −41 | **3.** 85 <br> −16 | **4.** $.17 <br> − .12 | **5.** $.37 <br> − .22 |

| | | | | |
|---|---|---|---|---|
| **6.** 35 <br> −13 | **7.** 77 <br> −36 | **8.** $.94 <br> − .43 | **9.** $.74 <br> − .36 | **10.** $.88 <br> − .29 |

**11.** 61 − 34          **12.** 58 − 23          **13.** $.38 − $.29

**14.** $.92 − $.75          **15.** $1.49 − $.26          **16.** $2.76 − $.15

**Estimate by rounding to the nearest hundred or dollar.**

| | | | | |
|---|---|---|---|---|
| **17.** 674 <br> −211 | **18.** 883 <br> −502 | **19.** 949 <br> −631 | **20.** $6.79 <br> − 5.51 | **21.** $34.29 <br> − 11.09 |

| | | | | |
|---|---|---|---|---|
| **22.** 794 <br> −364 | **23.** 888 <br> −444 | **24.** 574 <br> −262 | **25.** $9.95 <br> − 7.61 | **26.** $86.43 <br> − 25.02 |

**27.** 893 − 763          **28.** $66.82 − $54.21          **29.** $28.16 − $13.05

 **Finding Together**

**30.** Andrew has $25 to buy world map games.
About how much does he spend
if the cashier gives him:

a) $3.96 change?          b) $4.27 change?

c) $2.72 change?          d) $1.88 change?

e) $2.17 change?          f) $4.51 change?

g) $5.45 change?          h) $5.89 change?

## 3-5 Subtracting with Regrouping

How many more points does Jeff need to reach 35 points?

| Name | Points Scored |
|---|---|
| Jeff | 18 |
| Kyle | 26 |
| Lara | 29 |
| Sandra | 17 |

▶ First estimate the difference.

$$35 \longrightarrow 40$$
$$-18 \longrightarrow -20$$
$$\text{about} \quad 20$$

▶ Then to find how many more points, subtract: 35 − 18 = __?__

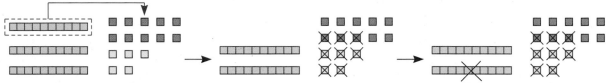

| More ones needed. Regroup tens. | → | Subtract the ones. | → | Subtract the tens. |
|---|---|---|---|---|

| tens | ones |
|---|---|
| 2 | 15 |
| 3̸ | 5̸ |
| − 1 | 8 |
| | |

3 tens 5 ones = 2 tens 15 ones

| tens | ones |
|---|---|
| 2 | 15 |
| 3̸ | 5̸ |
| − 1 | 8 |
| | 7 |

| tens | ones |
|---|---|
| 2 | 15 |
| 3̸ | 5̸ |
| − 1 | 8 |
| 1 | 7 |

$$35$$
$$-18$$
$$17$$

Jeff needs 17 more points.

> 17 is close to 20. The answer is reasonable.

**Study these examples.**

| tens | ones |
|---|---|
| 2 | 10 |
| 3̸ | 0̸ |
| − | 9 |
| 2 | 1 |

$$30$$
$$-9$$
$$21$$

3 tens 0 ones = 2 tens 10 ones

| dimes | pennies |
|---|---|
| 8 | 12 |
| 9̸ | 2̸ |
| − 4 | 6 |
| 4 | 6 |

$$92¢$$
$$-46¢$$
$$46¢$$

9 dimes 2 pennies = 8 dimes 12 pennies

110

## Subtract. Use base ten blocks to help.

**1.**

| t | o |
|---|---|
| 5 | 4 |
| − 3 | 5 |
|   |   |

**2.**

| t | o |
|---|---|
| 3 | 3 |
| − 1 | 9 |
|   |   |

**3.**

| t | o |
|---|---|
| 2 | 0 |
| − 1 | 2 |
|   |   |

**4.**

| d | p |
|---|---|
| 4 | 7 |
| − 2 | 8 |
|   |   |

**5.**

| d | p |
|---|---|
| 6 | 1 |
| − 1 | 9 |
|   |   |

## Estimate. Then find the difference.

**6.**  54
     − 27

**7.**  38
     − 29

**8.**  57
     − 31

**9.**  44
     − 29

**10.**  73
      − 18

**11.**  61
      − 43

**12.**  24
      − 16

**13.**  82
      − 65

**14.**  76
      − 48

**15.**  51
      − 33

**16.**  46
      − 17

**17.**  62
      − 54

## Align and subtract.

**18.** 55 − 28

**19.** 42 − 35

**20.** 94 − 5

**21.** 81¢ − 17¢

**22.** 67¢ − 29¢

**23.** 25¢ − 19¢

**24.** 43¢ − 37¢

**25.** 92¢ − 6¢

**26.** $.76 − $.28

**27.** $.52 − $.28

**28.** $.36 − $.09

**29.** $.84 − $.55

## Share Your Thinking

**30.** Compare your estimates from exercises 6–17 with the exact differences. What do you notice?

**31.** How does estimating help you determine if an exact answer is reasonable?

**32.** Show and explain in your Math Journal the steps that you would use to subtract 8 from 97. Explain why it is important to align the digits.

Math Journal

## 3-6 Regrouping Hundreds and Dollars

You can regroup hundreds as tens and dollars as dimes.

3 hundreds 2 tens = __?__
3 dollars 4 dimes = __?__

## Hands-On Understanding

**You Will Need:** base ten blocks, play money

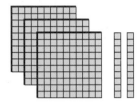

**Step 1** Model 320, using the fewest number of blocks.

How did you model 320?

**Step 2** Now model the same number. This time regroup one of the hundreds as tens.

How many hundreds do you have?

How many tens do you have?

So 3 hundreds 2 tens can be regrouped as __?__ hundreds __?__ tens.

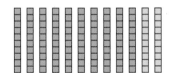

**Model each number using the fewest blocks. Then model each number regrouping one of the hundreds as tens. Describe the regrouped number.**

Discuss

**1.** 680     **2.** 450     **3.** 700     **4.** 990     **5.** 440

You can also regroup money.

**Step 3** Model $3.40 using dollars and dimes only.

How did you model $3.40?

112

Now model the same amount. This time regroup 1 dollar as dimes.

How many dollars do you have?

How many dimes do you have?

So 3 dollars 4 dimes can be regrouped as _?_ dollars _?_ dimes.

**Model each amount using dollars and dimes. Then model each amount regrouping 1 dollar as dimes. Describe the regrouped amount.**

Discuss

**6.** $4.30 **7.** $6.50 **8.** $9.10 **9.** $5.00 **10.** $1.00

**Regroup. Use base ten blocks and play money to help.**

**11.** 7 hundreds 7 tens =
_?_ hundreds _?_ tens

**12.** 5 hundreds 1 ten =
_?_ hundreds _?_ tens

**13.** 2 hundreds 8 tens =
_?_ hundreds _?_ tens

**14.** 3 hundreds 0 tens =
_?_ hundreds _?_ tens

**15.** 4 dollars 1 dime =
_?_ dollars _?_ dimes

**16.** 6 dollars 0 dimes =
_?_ dollars _?_ dimes

# Communicate

**17.** Explain what happens to the hundreds and the tens when you regroup 1 hundred as tens.

**18.** Explain what happens to the dollars and the dimes when you regroup 1 dollar as dimes.

**19.** How is regrouping whole numbers the same as regrouping money?

# Regrouping Once in Subtraction

An express monorail can seat 326 passengers.
A local monorail holds 145. How many more
people can ride the express than the local?

▶ First estimate: 300 − 100 = 200

▶ Then to find how many more
people, subtract: 326 − 145 = __?__

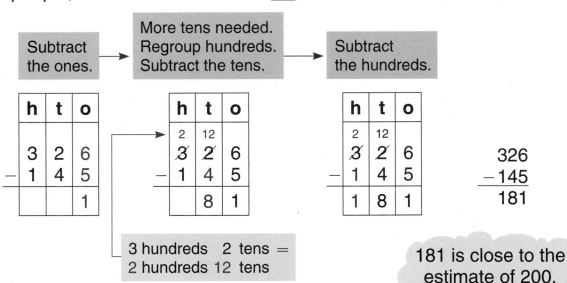

Subtract
the ones.

More tens needed.
Regroup hundreds.
Subtract the tens.

Subtract
the hundreds.

| h | t | o |
|---|---|---|
| 3 | 2 | 6 |
| −1 | 4 | 5 |
| | | 1 |

| h | t | o |
|---|---|---|
| | 2 | 12 |
| 3̸ | 2̸ | 6 |
| −1 | 4 | 5 |
| | 8 | 1 |

| h | t | o |
|---|---|---|
| | 2 | 12 |
| 3̸ | 2̸ | 6 |
| −1 | 4 | 5 |
| 1 | 8 | 1 |

$$326$$
$$-145$$
$$\overline{181}$$

3 hundreds  2 tens =
2 hundreds 12 tens

181 is close to the
estimate of 200.

181 more people can ride the express monorail.

**Study this example.**

Line up the
dollars and cents.

Subtract
as usual.

Write $ and .
in the difference.

$7.57 − $2.83 = __?__

$$\$7.57$$
$$-\ 2.83$$

6 15
$$\$7.\cancel{5}7$$
$$-\ 2.83$$
$$\overline{4\ 7\ 4}$$

6 15
$$\$7.\cancel{5}7$$
$$-\ 2.83$$
$$\overline{\$4.74}$$

**Complete.**

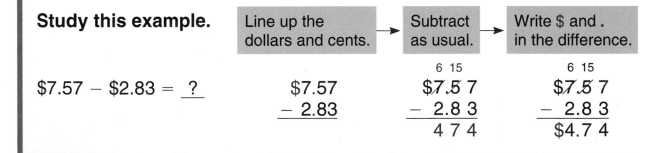

1.    2 12
   3̸ 2̸ 9
  −1 6 5
   ? ? 4

2.    5 14
   6̸ 4̸ 7
  −1 8 3
   ? ? ?

3.      6 10
   2 7̸ 0̸
  −1 3 7
   ? ? ?

4.    ? ?
   6̸ 5̸ 8
  −  9 7
   ? ? ?

5.    ? ?
   7 8̸ 6̸
  −3 2 9
   ? ? ?

**Estimate. Then subtract.**

6.  446
    − 256

7.  555
    − 473

8.  839
    − 449

9.  214
    − 122

10. 920
    − 709

11. 692
    − 288

12. 186
    −  78

13. 139
    −  69

14. 319
    − 175

15. 659
    − 289

16. $6.78
    − 4.84

17. $3.12
    − 1.61

18. $2.90
    − 1.25

19. $7.74
    −  .26

20. $9.12
    −  .07

**Align and subtract.**

21. 379 − 188

22. 869 − 697

23. 555 − 464

24. 879 − 293

25. 519 − 276

26. 724 − 161

27. 757 − 682

28. 119 − 65

29. 916 − 75

30. $6.38 − $2.45

31. $4.63 − $1.80

32. $9.56 − $7.94

33. $4.36 − $.54

34. $6.17 − $.26

35. $5.11 − $.40

**PROBLEM SOLVING**

36. Mr. Neugent paid $4.29 to have a suit dry-cleaned and $1.35 to have a shirt cleaned. What is the difference in the two prices?

37. In one week 246 shirts and 163 suits were laundered. How many more shirts than suits were laundered?

## Calculator Activity

**Complete each number sentence.**

*Algebra* ✓

Use 565, 371, 253, and 166. Which two numbers have a difference of:

38. 118 = _?_ − _?_

39. 194 = _?_ − _?_

40. 205 = _?_ − _?_

41. 312 = _?_ − _?_

## 3-8  Regrouping Twice in Subtraction

Of 523 people traveling to Europe 376 went
by boat. How many people went by plane?

▶ First estimate: 500 − 400 = 100

▶ Then to find how many went
by plane, subtract: 523 − 376 = _?_

| More ones needed. Regroup tens. Subtract the ones. | More tens needed. Regroup hundreds. Subtract the tens. | Subtract the hundreds. |
|---|---|---|

| h | t | o |
|---|---|---|
|   | 1 | 13 |
| 5 | 2̸ | 3̸ |
| − 3 | 7 | 6 |
|   |   | 7 |

| h | t | o |
|---|---|---|
|   | 11 |   |
| 4 | 1̸ | 13 |
| 5̸ | 2̸ | 3̸ |
| − 3 | 7 | 6 |
|   | 4 | 7 |

| h | t | o |
|---|---|---|
|   | 11 |   |
| 4 | 1̸ | 13 |
| 5̸ | 2̸ | 3̸ |
| − 3 | 7 | 6 |
| 1 | 4 | 7 |

$$523$$
$$-376$$
$$\overline{147}$$

2 tens 3 ones =
1 ten 13 ones

5 hundreds 1 ten =
4 hundreds 11 tens

▶ To check subtraction, add the number
that was subtracted to the difference.

```
  1 1
  147
+ 376
  523  It checks!
```

Exactly 147 people went by plane.

**Study this example.**

| Line up the dollars and cents. | Subtract as usual. | Write $ and . in the difference. |
|---|---|---|

$8.53 − $2.65 = _?_

```
  $8.53
 − 2.65
```

```
     14
   7 ⁴13
  $8.5̸3̸
 − 2.6 5
   5 8 8
```

```
     14
   7 ⁴13
  $8.5̸3̸
 − 2.6 5
  $5.8 8
```

116

**Estimate. Then subtract and check.**

| 1. | 587<br>− 468 | 2. | 523<br>− 237 | 3. | 781<br>− 372 | 4. | 324<br>− 165 | 5. | 865<br>− 169 |
|---|---|---|---|---|---|---|---|---|---|

| 6. | 317<br>− 99 | 7. | 260<br>− 75 | 8. | 198<br>− 99 | 9. | 636<br>− 78 | 10. | 743<br>− 86 |
|---|---|---|---|---|---|---|---|---|---|

| 11. | $5.36<br>− 1.87 | 12. | $4.73<br>− 2.94 | 13. | $6.60<br>− 3.51 | 14. | $7.15<br>− 2.36 | 15. | $4.57<br>− 3.58 |
|---|---|---|---|---|---|---|---|---|---|

**Align, subtract, and check.**

**16.** 747 − 179      **17.** 643 − 335      **18.** 991 − 872

**19.** 562 − 98      **20.** 154 − 68      **21.** 240 − 76

**22.** $5.65 − $1.27      **23.** $9.36 − $2.49      **24.** $6.94 − $2.27

**25.** $1.16 − $.18      **26.** $2.23 − $1.78      **27.** $3.10 − $.84

**PROBLEM SOLVING**

**28.** Jenna is flying from the United States to Mexico. The airplane has traveled 315 miles. The entire trip is 825 miles. How many more miles does she have to fly?

**29.** Jenna bought two Mexican souvenirs. They cost a total of $8.89. She had $9.75 in her wallet. How much money did she have left?

 **Mental Math**

**Subtract $.99 from each.** Think: Subtract $1.00; add $.01.

**30.** $4.25          **31.** $7.48          **32.** $5.31          **33.** $1.64

# Regrouping with Zeros

Mill workers made 400 flags daily. Of these, 276 were
American flags. How many were not American flags?

▶ First estimate: 400 − 300 = 100

▶ Then to find how many were not American flags,
subtract: 400 − 276 = _?_

You can use base ten blocks to help subtract 276 from 400.

## Hands-On Understanding

**You Will Need:** base ten blocks

**Step 1** Model 400, using the
fewest blocks.

**Step 2** Subtract. Start with the ones.

Do you have enough
ones to take away
6 ones units?

Do you have enough tens
to regroup 1 ten as ones?

How can you make more
tens to regroup tens as ones?

Do you have enough hundreds
to regroup 1 hundred as tens?

**Step 3** Regroup 1 hundred as tens.

How many hundreds do you
have now?

How many tens do you have?

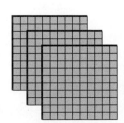

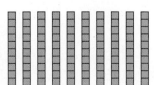

|  **Step 4** | Regroup 1 ten as ones. |
|---|---|

How many tens do you have now?

How many ones do you have?

Can you take away 6 ones now?

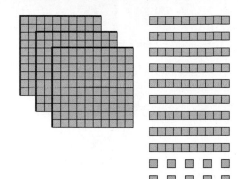

| **Step 5** | Take away 6 ones units. |
|---|---|

| **Step 6** | Take away 7 tens rods. |
|---|---|

| **Step 7** | Take away 2 hundreds flats. |
|---|---|

How many hundreds, tens, and ones do you have left?

> 124 is close to 100. The answer is reasonable.

There were 124 flags that were not American.

## Estimate. Then subtract. Use your base ten blocks to help.

1. 
$$\begin{array}{r} 600 \\ -246 \\ \hline \end{array}$$
2. 
$$\begin{array}{r} 500 \\ -192 \\ \hline \end{array}$$
3. 
$$\begin{array}{r} 300 \\ -191 \\ \hline \end{array}$$
4. 
$$\begin{array}{r} 400 \\ -385 \\ \hline \end{array}$$
5. 
$$\begin{array}{r} 700 \\ -413 \\ \hline \end{array}$$

6. 
$$\begin{array}{r} \$2.00 \\ -1.42 \\ \hline \end{array}$$
7. 
$$\begin{array}{r} \$9.00 \\ -1.36 \\ \hline \end{array}$$
8. 
$$\begin{array}{r} 300 \\ -\ 27 \\ \hline \end{array}$$
9. 
$$\begin{array}{r} 600 \\ -\ 59 \\ \hline \end{array}$$
10. 
$$\begin{array}{r} 200 \\ -\ 89 \\ \hline \end{array}$$

# Communicate

Discuss ✓

11. Explain why you sometimes need to regroup twice in the tens place when you subtract.

## Skills to Remember

**Align and add.**

12. 586 + 293

13. 841 + 75

14. $7.94 + $1.62

# 3-10 | Choose a Computation Method

You can use mental math or paper and pencil to subtract.

**Subtract:** $500 - 300 = \underline{?}$

$472 - 191 = \underline{?}$

To find each difference, first look at the numbers and determine which method is easiest for you. Then subtract.

▶ **Mental Math**    $500 - 300 = 200$

Think: Can I quickly solve $500 - 300$ in my head?

▶ **Paper and Pencil**    $472 - 191 = \underline{?}$

Think: Are the numbers too large to subtract mentally?
Yes. First estimate.
Then use paper and pencil.

$$472 \longrightarrow 500$$
$$-191 \longrightarrow -200$$
$$\text{about} \quad 300$$

$$\overset{3\ 17}{\cancel{4}\cancel{7}2}$$
$$-\ 1\ 9\ 1$$
$$2\ 8\ 1$$

281 is close to 300.
The answer is reasonable.

There is no one correct method to use.
Use whichever method is easiest for you.

---

**Subtract. Tell which method you used and why.**

*Communicate* ✓

**1.** $550 - 105 = \underline{?}$    **2.** $\$8.06 - \$1.39 = \underline{?}$    **3.** $946 - 521 = \underline{?}$

**4.** $500 - 498 = \underline{?}$    **5.** $\$6.00 - \$2.50 = \underline{?}$    **6.** $\$1.78 - \$1.29 = \underline{?}$

## 3-11 | Regrouping Thousands as Hundreds

Sometimes it is necessary to regroup thousands as hundreds.

1 thousand 2 hundreds = _?_

# Hands-On Understanding

**You Will Need:** base ten blocks

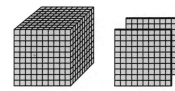

**Step 1** Model 1200 using the fewest blocks.

How did you model 1200?

**Step 2** Now model the same number regrouping one of the thousands as hundreds.

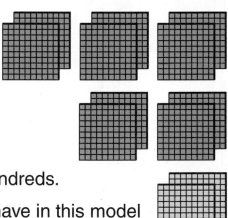

How many thousands do you have?

How many hundreds do you have?

So 1 thousand 2 hundreds can be regrouped as _?_ thousands _?_ hundreds.

How many more hundreds do you have in this model than in the model for 1200 with fewer blocks? Why?

**Model each number using the fewest blocks. Then model each number regrouping one of the thousands as hundreds. Describe the regrouped number.**

Discuss ✓

1. 8200    2. 2300    3. 5800    4. 1400    5. 4000

## Communicate

6. How is regrouping thousands as hundreds the same as regrouping hundreds as thousands? How is it different?

# Subtracting Larger Numbers

In Johnstown there are 9010 people.
If 6789 are adults, how many are children?

▶ First estimate: 9000 − 7000 = 2000

▶ Then to find how many are children,
subtract: 9010 − 6789 = _?_

| Regroup tens. Subtract ones. | → | Regroup thousands. Regroup hundreds. Subtract tens. | → | Subtract hundreds. | → | Subtract thousands. |
|---|---|---|---|---|---|---|

| th | h | t | o |
|----|---|---|---|
|    |   | 0 | 10 |
| 9  | 0 | 1̸ | 0̸ |
| − 6 | 7 | 8 | 9 |
|    |   |   | 1 |

| th | h | t | o |
|----|---|---|---|
|    | 9 | 10 |   |
| 8  | 1̸0̸ | 0̸ | 10 |
| 9̸  | 0̸ | 1̸ | 0̸ |
| − 6 | 7 | 8 | 9 |
|    |   | 2 | 1 |

| th | h | t | o |
|----|---|---|---|
|    | 9 | 10 |   |
| 8  | 1̸0̸ | 0̸ | 10 |
| 9̸  | 0̸ | 1̸ | 0̸ |
| − 6 | 7 | 8 | 9 |
|    | 2 | 2 | 1 |

| th | h | t | o |
|----|---|---|---|
|    | 9 | 10 |   |
| 8  | 1̸0̸ | 0̸ | 10 |
| 9̸  | 0̸ | 1̸ | 0̸ |
| − 6 | 7 | 8 | 9 |
| 2  | 2 | 2 | 1 |

1 ten =
0 tens 10 ones

9 thousands =
8 thousands 10 hundreds =
8 thousands   9 hundreds 10 tens

2221 is close
to the estimate
of 2000.

There are 2221 children in Johnstown.

## Estimate. Then subtract.

1. 9561
  − 4726

2. 3684
  − 2369

3. 5190
  − 4738

4. 7159
  − 6037

5. 5483
  − 2007

6. 8616
  − 1208

7. 1700
  − 1507

8. 9001
  − 4090

9. 8784
  − 1935

10. 6050
  − 4967

## Subtract. Then check by addition.

**11.**
```
  3535  ◄─────      3119
−  416         +  416
  3119    └────►  3535
```

**12.**
```
  2222
−   68
```

**13.**
```
  4318
−  108
```

**14.**
```
  6226
−  527
```

**15.**
```
  $74.19
−   1.27
```

**16.**
```
  $90.00
−   7.03
```

**17.**
```
  $17.63
−   2.96
```

**18.**
```
  $45.72
−  21.84
```

**19.**
```
  $50.43
−  18.52
```

---

### Making Change with Larger Numbers

Leroy buys a model jet for $17.50. He pays with two $10 dollar bills. What change can the cashier give him?

First add: $10 + $10 = $20.

Then subtract and count on to make change.

```
        9
   1  10 10
$2 0.0 0
−1 7.5 0
$  2.5 0
```

Think: How can I make $2.50?

The cashier can give Leroy 2 one-dollar bills and 2 quarters.

Remember: There are many ways to make change.

---

## PROBLEM SOLVING

**20.** Len buys a book for $13.49. He pays with a $5 and a $10 bill. How can the cashier give him change?

**21.** Amy buys a video for $16.99. She pays with two $10 bills. What change might Amy get?

**22.** Luna buys a compact disc for $11.98. She has a $5 bill, a $10 bill, and a $20 bill. Which two bills might Luna give the cashier? What change can she get?

**23.** Max buys a game for $15.98. He has two $10 bills and a $5 bill. Which two bills might Max give the cashier? What change can Max receive?

# 3-13 Problem Solving: Choose the Operation

*Algebra* ✓

| | |
|---|---|
| **Addition:** | Join equal or unequal sets, or quantities. |
| **Subtraction:** | Separate, or take away, equal or unequal sets, or quantities. |
| | Compare two sets, or quantities. |
| | Find part of a whole set. |
| | Find how many more are needed. |

**Problem:** There are 54 straw baskets and 28 painted baskets on sale. How many baskets are on sale?

**1 IMAGINE** Create a mental picture.

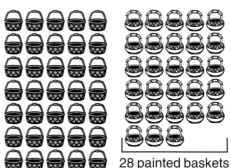

28 painted baskets

**2 NAME**

*Facts:* 54 straw baskets
28 painted baskets

*Question:* How many baskets are on sale?

**3 THINK** You are joining two unequal sets, add: 54 + 28 = ?

**4 COMPUTE** First estimate the sum.
50 + 30 = 80

Then add.
$$\begin{array}{r} \overset{1}{5}4 \\ +28 \\ \hline 82 \end{array}$$

54 straw baskets

There are 82 baskets on sale.

**5 CHECK** Compare the estimate and the actual sum.
The actual sum, 82, is close to the estimated sum, 80.
The answer is reasonable.

**Choose the operation to solve each problem.**

**1.** How many more straw baskets are there than painted baskets?

To estimate how many more: round, then find the difference:
$$50 - 30 = \underline{\ ?\ }$$

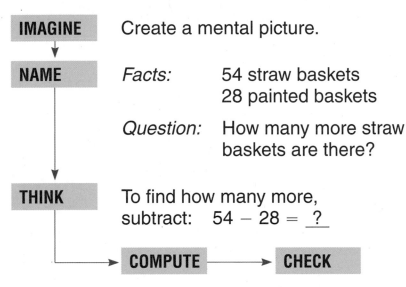

| IMAGINE | Create a mental picture. |
|---|---|

| NAME | *Facts:* | 54 straw baskets |
|---|---|---|
| | | 28 painted baskets |
| | *Question:* | How many more straw baskets are there? |

| THINK | To find how many more, subtract: $54 - 28 = \underline{\ ?\ }$ |
|---|---|

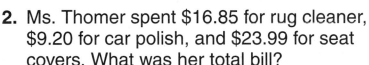

COMPUTE ⟶ CHECK

**2.** Ms. Thomer spent $16.85 for rug cleaner, $9.20 for car polish, and $23.99 for seat covers. What was her total bill?

**3.** Glen must stack 173 hubcaps on a rack. He has already stacked 96 hubcaps. How many more hubcaps must he stack?

**4.** On the first day of a sale, 1902 people came into the store. On the second day 1786 people came. How many more people came on the first day?

**5.** At the end of the sale, there were 92 cans of motor oil left. If 297 cans were sold, how many cans had been on sale?

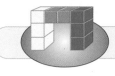

## Make Up Your Own

**6.** Choose an operation. Write your own problem. Then solve it.

# 3-14 Problem-Solving Applications

**Solve each problem and explain the method you used.**

1. Fifty-six students of Central School take music lessons. Twelve of them play more than one instrument. How many students play one instrument?

2. Leon practices 155 minutes a week. So far this week he has practiced 105 minutes. How much longer does he need to practice?

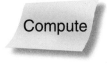

3. The school chorus has 37 third graders and 45 fourth graders. How many third and fourth graders are in the chorus?

Imagine

4. There are 53 students in the band. Twenty-six of them are boys. There are 5 drummers. How many girls are in the band?

Name

5. A shipment of 200 tapes was on sale. The music store sold 116 tapes. How many tapes were not sold?

Think

6. One year a music store sold 1005 instruments. It sold 568 instruments the next year. How many more instruments were sold in the first year?

Compute

Check

7. Rosa wants to buy a tape for $9.29. She has $6.35. About how much more money does she need?

8. Eric gave the cashier $35.00. He got $1.72 in change. How much money did he spend at the music store?

**Choose a strategy from the list or use another strategy you know to solve each problem.**

Use these strategies:

Choose the Operation
Use Simpler Numbers
Draw a Picture
Make a Table
Guess and Test
Extra Information

9. A music box usually sells for $6.00. The sale price is $1.05 less. How much does a music box cost on sale?

10. Seven students sold T-shirts to raise money for the band. The first student sold 3 shirts, the second sold 2 more than the first, the third sold 2 more than the second, and so on. How many shirts did the seventh student sell?

11. Marsha practiced the flute for double the minutes Silvio practiced. Together they practiced for 90 minutes. How long did each student practice?

12. Every fourth CD sold at the music store is country and western. One day 26 CDs were sold. How many were country and western?

13. The music store offers an entertainment center for 8500 coupons. The Adams family has saved 6985 coupons. How many more coupons do they need?

14. Frank bought a music book for $8.79. His cousin paid $9.75 for a tape. If Frank paid with ten dollars, how much change did he get?

**Use the graph for problems 15–16.**

15. How many more people liked country and western than classical music?

16. How many people were surveyed in all?

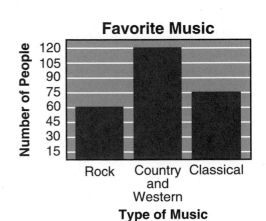

Favorite Music

# Chapter Review and Practice

## Regroup.
*(See pp. 112–113, 121.)*

**1.** 5 tens 6 ones =
  __4__ tens __16__ ones

**2.** 7 tens 1 one =
  __?__ tens __?__ ones

**3.** 6 hundreds 0 tens =
  __?__ hundreds __?__ tens

**4.** 3 hundreds 6 tens =
  __?__ hundreds __?__ tens

**5.** 9 thousands 0 hundreds =
  __?__ thousands __?__ hundreds

**6.** 8 thousands 4 hundreds =
  __?__ thousands __?__ hundreds

**7.** 5 dimes 3 pennies =
  __?__ dimes __?__ pennies

**8.** 3 dollars 5 dimes =
  __?__ dollars __?__ dimes

## Subtract and check.
*(See pp. 106–107, 110–119, 121–123.)*

**9.**  $\begin{array}{r} 55 \\ -15 \\ \hline \end{array}$  **10.**  $\begin{array}{r} 89 \\ -67 \\ \hline \end{array}$  **11.**  $\begin{array}{r} 325 \\ -219 \\ \hline \end{array}$  **12.**  $\begin{array}{r} \$7.66 \\ -\ 4.27 \\ \hline \end{array}$  **13.**  $\begin{array}{r} \$8.47 \\ -\ 5.38 \\ \hline \end{array}$

**14.**  $\begin{array}{r} 823 \\ -\ 36 \\ \hline \end{array}$  **15.**  $\begin{array}{r} 600 \\ -478 \\ \hline \end{array}$  **16.**  $\begin{array}{r} \$7.24 \\ -\ 4.93 \\ \hline \end{array}$  **17.**  $\begin{array}{r} \$5.36 \\ -\ .75 \\ \hline \end{array}$  **18.**  $\begin{array}{r} \$9.07 \\ -\ 7.17 \\ \hline \end{array}$

**19.**  $\begin{array}{r} 6875 \\ -1729 \\ \hline \end{array}$  **20.**  $\begin{array}{r} 8000 \\ -\ 75 \\ \hline \end{array}$  **21.**  $\begin{array}{r} 5702 \\ -2578 \\ \hline \end{array}$  **22.**  $\begin{array}{r} \$68.37 \\ -24.75 \\ \hline \end{array}$  **23.**  $\begin{array}{r} \$72.03 \\ -58.75 \\ \hline \end{array}$

## Estimate the difference.
*(See pp. 108–109.)*

**24.**  $\begin{array}{r} 38 \\ -17 \\ \hline \end{array}$  **25.**  $\begin{array}{r} 209 \\ -136 \\ \hline \end{array}$  **26.**  $\begin{array}{r} \$4.95 \\ -\ 2.31 \\ \hline \end{array}$  **27.**  $\begin{array}{r} \$97.04 \\ -53.68 \\ \hline \end{array}$  **28.**  $\begin{array}{r} 1061 \\ -\ 457 \\ \hline \end{array}$

## PROBLEM SOLVING
*(See pp. 104–105, 124–126.)*

**29.** Alexandra collected 100 marbles. Jeanelle collected 67 marbles. How many more marbles does Alexandra have than Jeanelle?

**30.** A school has 475 students. There are 248 girls. How many boys are there?

*(See Still More Practice, p. 445.)*

## ROMAN NUMERALS

The ancient Romans used letters for writing numbers. The numbers on the clock show one way Roman numerals can be used.

Each letter has a special value.

I = 1    V = 5    X = 10    L = 50

▶ Find the value of IX.

The value of I is less than X.

So to find the value of IX, subtract.

IX = 10 − 1 = 9

| | | | |
|---|---|---|---|
| I = 1 | V = 5 | X = 10 | L = 50 |
| II = 2 | VI = 6 | XI = 11 | LI = 51 |
| III = 3 | VII = 7 | XII = 12 | LII = 52 |
| IV = 4 | VIII = 8 | XIII = 13 | LIII = 53 |
| | IX = 9 | XIV = 14 | LIV = 54 |
| | | XV = 15 | LV = 55 |
| | | XVI = 16 | LVI = 56 |
| | | XVII = 17 | LVII = 57 |
| | | XVIII = 18 | LVIII = 58 |
| | | XIX = 19 | |
| | | XX = 20 | |

▶ Find the value of LI.

The value of L is greater than I.

So to find the value of LI, add.

LI = 50 + 1 = 51

For exercises 1–4 which operation(s), addition or subtraction, can be used to find the value of each?

**1.** XII          **2.** IV          **3.** LV          **4.** XIX

**5.** Write the Roman numerals from 21 to 40.

**6.** Write your age in Roman numerals.

**7.** What is the difference of XV − IX?

**8.** What is the sum of XXI + L?

129

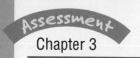

## Check Your Mastery

---

### Performance Assessment

**Use the numbers in the box to write subtraction sentences that will give a difference of:**

| 439 | 538 |
|-----|-----|
| 8020 | 638 |

**1.** 100      **2.** about 200      **3.** 7382

### Regroup.

**4.** 5 hundreds 5 tens =
      _?_ hundreds _?_ tens

**5.** 3 thousands 6 hundreds =
      _?_ thousands _?_ hundreds

**6.** 8 dimes 2 pennies =
      _?_ dimes _?_ pennies

**7.** 8 dollars 7 dimes =
      _?_ dollars _?_ dimes

### Subtract and check.

**8.**  83
     −25

**9.**  $.68
     − .49

**10.**  436
      − 27

**11.**  800
      −638

**12.**  $5.27
      − 1.36

**13.**  $8.05
      − 3.17

**14.**  3052
      − 121

**15.**  7246
      −3857

**16.**  4006
      −1728

**17.**  $42.36
      − 29.17

### Estimate the difference.

**18.**  87
      −37

**19.**  438
      − 162

**20.**  $6.99
      − 3.46

**21.**  $.53
      − .22

**22.**  5298
      −1910

### Find the difference.

**23.** 740 − 82

**24.** $8.75 − $1.79

### PROBLEM SOLVING      *Use a strategy you have learned.*

**25.** In March our school collected 890 pounds of paper. The eighth grade collected 346 pounds. How many pounds did the other grades collect altogether?

# Cumulative Review I

**Choose the best answer.**

**1.** Choose the standard form of the number.

7 hundreds 9 tens 0 ones

   **a.** 709
   **b.** 790
   **c.** 970
   **d.** 7090

**2.** Choose the standard form of the number.

6 ten thousands

   **a.** 6,000
   **b.** 16,000
   **c.** 60,000
   **d.** 600,000

**3.** Which statement is *not* true?

   **a.** $37 < 73$
   **b.** $7440 > 7420$
   **c.** $\$8.72 > \$7.99$
   **d.** $534 < 529$

**4.** Choose the order from least to greatest.

   **a.** 9325; 9436; 9427
   **b.** 9436; 9427; 9325
   **c.** 9436; 9325; 9427
   **d.** 9325; 9427; 9436

**5.** Round to the nearest thousand.

6583
   **a.** 5000   **b.** 6000
   **c.** 6600   **d.** 7000

**6.** Round to the nearest dollar.

$2.26
   **a.** $1.00   **b.** $2.00
   **c.** $2.30   **d.** $3.00

**7.** What is the missing number?

23, 27, 31, 35, __?__

   **a.** 21
   **b.** 36
   **c.** 37
   **d.** 39

**8.** Which number is at least 200 more than 5319 and has 5 in the thousands place?

   **a.** 5119
   **b.** 5518
   **c.** 5520
   **d.** 7519

**9.**    635
   + 195

   **a.** 760
   **b.** 820
   **c.** 830
   **d.** 730

**10.** Estimate.

   4739
   + 2647

   **a.** 2000
   **b.** 4000
   **c.** 5000
   **d.** 8000

**11.** $7.56 + $2.25 + $.30

   **a.** $9.11   **b.** $10.11
   **c.** $12.81   **d.** $10.01

**12.** 309 + 75 + 432

   **a.** 1491   **b.** 897
   **c.** 816   **d.** 806

**13.** $6.70 − $.98

   **a.** $5.72   **b.** $6.28
   **c.** $7.68   **d.** not given

**14.** 900 − 749

   **a.** 161   **b.** 249
   **c.** 251   **d.** not given

**15.**    $9.69
   − 3.78

   **a.** $3.00   **b.** $6.00
   **c.** $8.00   **d.** $9.00

**16.**    5007
   − 2839

   **a.** 2168   **b.** 2172
   **c.** 7846   **d.** 3832

# Ongoing Assessment I

## For Your Portfolio

**Solve each problem. Explain the steps and the strategy or strategies you used for each. Then choose one from problems 1–4 for your Portfolio.**

1. Jefferson County has 1497 more pets than Lansing County. Lansing County has 7342 pets. How many pets does Jefferson County have?

2. Ms. Costello bought a book marker for each student in her school. There are 223 boys and 316 girls. How many markers did she buy?

3. Alicia buys a bracelet for $4.65. She gives the cashier $5.00. What is her change?

4. There are 328 blue cars, 126 green cars, and 98 black cars in the parking lot. How many more blue cars than green cars are there?

**Tell about it.**

5. **a.** Assume that the cashier in problem 3 has no nickels. Use the strategy Draw a Picture to show six ways she could give the change.

   **b.** What other strategy can you use to show the six ways? Explain.

Communicate ✓

---

# For Rubric Scoring

**Listen for information on how your work will be scored.**

6. Count by 4s. Start at 23 and stop at 47. Write the six numbers you count.

7. Look at the ones digits of your answer in exercise 6. What do you notice about these digits? Write these digits in order: 7, 1, _?_, _?_, _?_, _?_.

8. Continue to count by 4s. List the first 15 numbers in 3 rows of 5 numbers each. Would 95, 100, or 105 be on the list? Explain.

# Multiplication Concepts and Facts

# 4

**In this chapter you will:**

Multiply numbers and
  cents by 0 through 5
Understand order in
  multiplication
Learn about missing
  factors
Solve multi-step
  problems

**Critical Thinking/
Finding Together**

Skip count to find how
many sheep the boy
counted before falling
asleep. Write an addition
sentence to show this.

## Sweet Dreams

I wonder as into bed I creep
What it feels like to fall asleep.
I've told myself stories, I've counted sheep,
But I'm always asleep when I fall asleep.
Tonight my eyes I will open keep,
And I'll stay awake till I fall asleep,
Then I'll know what it feels like to fall asleep,
Asleep,
Asleep,
Asleeep . . .

*Ogden Nash*

133

# Understanding Multiplication

There are 3 boxes of honey jars. Each box holds 2 jars.
How many jars of honey are there in all?

▶ To find how many jars in all,
model the problem.

## Hands-On Understanding

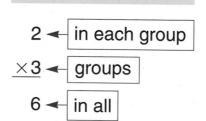

**You Will Need:** counters

| Step 1 | Show 3 groups of 2 counters each. |

| Step 2 | Write an addition sentence to show how many counters in all. |

What do the addends represent
in the addition sentence?

How many addends are in the addition sentence?
What does this number represent?

▶ You can also write a multiplication sentence
to show 3 groups of 2 counters.

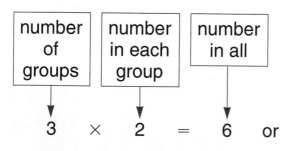

| number of groups | number in each group | number in all |
|---|---|---|

$$3 \times 2 = 6$$

3 twos = 6

Three times two equals six.

There are 6 jars in all.

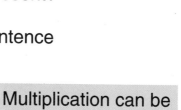

Multiplication can be
shown in two ways.

2 ← in each group
×3 ← groups
6 ← in all

**Use counters or draw dots to find how many in all.**

1. $4 + 4 + 4 =$ <u>12</u>

   3 fours = <u>12</u>

   $3 \times 4 =$ <u>12</u>

2. $5 + 5 =$ <u>?</u>

   2 fives = <u>?</u>

   $2 \times 5 =$ <u>?</u>

3. $3 + 3 + 3 + 3 =$ <u>?</u>

   4 threes = <u>?</u>

   $4 \times 3 =$ <u>?</u>

**Write an addition sentence and a multiplication sentence to show how many in all.**

4.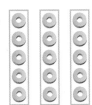

5.

6.

7.

**Draw dots or circles to show each multiplication. Solve each.**

8. 2 threes  6

9. 1 five

10. 5 threes

11. $4 \times 5 =$ <u>?</u>

12. $2 \times 2 =$ <u>?</u>

13. $1 \times 4 =$ <u>?</u>

14. $5 \times 5 =$ <u>?</u>

15.  4
   $\underline{\times 2}$

16.  3
   $\underline{\times 1}$

17.  2
   $\underline{\times 5}$

18.  4
   $\underline{\times 4}$

## Communicate

19. Look at exercises 1–7. Compare the addition sentences with the multiplication sentences. What do you notice?

20. What does the first number in each multiplication sentence represent?

21. What does the second number in each multiplication sentence represent?

22. How is multiplication like addition?

135

# 4-2 Zero and One as Factors

Algebra

There are 4 mailboxes.
There is 1 letter in each mailbox.
How many letters are there in all?

To find how many letters in all,
multiply:

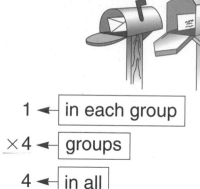

$$4 \times 1 = 4 \quad \text{or}$$

number of groups → 4

number in each group → 1

number in all → 4

1 ← in each group

×4 ← groups

4 ← in all

When you multiply any number and 1, the answer is that number.

There are 3 mailboxes. Each mailbox has
0 letters. How many letters are there in all?

To find how many letters in all,
multiply:

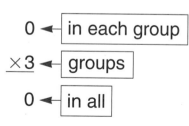

$$3 \times 0 = 0 \quad \text{or}$$

groups → 3

in each group → 0

in all → 0

0 ← in each group

×3 ← groups

0 ← in all

When you multiply any number and 0, the answer is 0.

## Multiply.

1. $3 \times 1$
2. $4 \times 0$
3. $2 \times 1$
4. $5 \times 0$
5. $2 \times 0$

6. $\begin{array}{r} 1 \\ \times 5 \\ \hline \end{array}$
7. $\begin{array}{r} 1 \\ \times 3 \\ \hline \end{array}$
8. $\begin{array}{r} 0 \\ \times 5 \\ \hline \end{array}$
9. $\begin{array}{r} 1 \\ \times 4 \\ \hline \end{array}$
10. $\begin{array}{r} 0 \\ \times 3 \\ \hline \end{array}$

11. $\begin{array}{r} 2 \\ \times 1 \\ \hline \end{array}$
12. $\begin{array}{r} 4 \\ \times 1 \\ \hline \end{array}$
13. $\begin{array}{r} 0 \\ \times 2 \\ \hline \end{array}$
14. $\begin{array}{r} 0 \\ \times 1 \\ \hline \end{array}$
15. $\begin{array}{r} 1 \\ \times 1 \\ \hline \end{array}$

## Multiplication Language

Here are some special words to learn.

**Factors** are the numbers you multiply.

$3 \times 2 = 6$    factors

$$\begin{array}{r} 2 \\ \times 3 \\ \hline 6 \end{array}$$ — factors

factors

**Product** is the answer when you multiply.

$4 \times 1 = 4$

$$\begin{array}{r} 1 \\ \times 4 \\ \hline 4 \end{array}$$ ← product

product

---

**Solve. Write each multiplication in two ways.**

**16.** My factors are 3 and 1.
What is my product?

**17.** My factors are 1 and 5.
What is my product?

**18.** One of my factors is 4.
My other factor is 0.
What is my product?

**19.** My product is 0. What must
one of my factors be?

**20.** My product is 1. What must
both of my factors be?

 **Connections: Art**

**Draw a picture and write a multiplication sentence to represent each.**

**21.** 5 groups of 1

**22.** 2 groups of 0

**23.** 1 group of 4

**24.** 0 groups of 3

# 4-3 Multiplying Twos

You can use counters to find the product of $2 \times 2$ or $\begin{array}{r} 2 \\ \underline{\times\, 2} \end{array}$.

## Hands-On Understanding

**You Will Need:** counters

**Step 1**  Model 2 groups of 2 counters.

How many:
   groups in all?
   counters in each group?
   altogether?

**Step 2**  Write an addition sentence and a multiplication sentence for your model.

**Step 3**  Repeat Steps 1 and 2 for these groups:

| Number of Groups | 2 | 3 | 4 | 5 | 6 | 7 | 8 | 9 |
|---|---|---|---|---|---|---|---|---|
| Counters in Each | 2 | 2 | 2 | 2 | 2 | 2 | 2 | 2 |
| Counters Altogether | 4 | ? | ? | ? | ? | ? | ? | ? |

How would you model 0 groups of 2 counters? 1 group?

## Communicate

1. What does each addend stand for in the addition sentences you wrote in Step 2?

2. What does each factor stand for in the multiplication sentences you wrote in Step 2?

3. Look at the factors in the multiplication sentences. How are they alike?

**4.** Look at the products. Describe any pattern you see.

Discuss ✓

**5.** Compare the addition sentences with the multiplication sentences. What do you notice?

**Write a multiplication sentence for each addition expression.**

**6.** $2 + 2 + 2$    **7.** $2 + 2 + 2 + 2 + 2$    **8.** $2 + 2$    **9.** $2 + 2 + 2 + 2$

**Find the product. You may use counters or draw dots.**

**10.** $3 \times 2$    **11.** $8 \times 2$    **12.** $9 \times 2$    **13.** $7 \times 2$

**14.** $1 \times 2$    **15.** $4 \times 2$    **16.** $0 \times 2$    **17.** $2 \times 2$

**18.** $\begin{array}{r} 2 \\ \times 4 \\ \hline \end{array}$    **19.** $\begin{array}{r} 2 \\ \times 6 \\ \hline \end{array}$    **20.** $\begin{array}{r} 2 \\ \times 5 \\ \hline \end{array}$    **21.** $\begin{array}{r} 2 \\ \times 2 \\ \hline \end{array}$    **22.** $\begin{array}{r} 2 \\ \times 1 \\ \hline \end{array}$    **23.** $\begin{array}{r} 2 \\ \times 3 \\ \hline \end{array}$

**Copy and complete the table.**

**24.**

| $0 \times 2$ | $1 \times 2$ | $2 \times 2$ | $3 \times 2$ | $4 \times 2$ | $5 \times 2$ | $6 \times 2$ | $7 \times 2$ | $8 \times 2$ | $9 \times 2$ |
|---|---|---|---|---|---|---|---|---|---|
| ? | ? | ? | ? | ? | ? | ? | ? | ? | ? |

**PROBLEM SOLVING**

**25.** Susan puts 9 packs of videotapes in her cart. There are two tapes in each pack. How many tapes are in her cart?

**26.** Carlos buys seven packs of videotapes. Each pack has two tapes in it. How many tapes does he buy in all?

**Critical Thinking**

**27.** When you multiply 2 by itself, what is the product?

**28.** When you multiply two by nine, what is the product?

**Multiplying Threes**

Each pack has 3 cartons of juice.
How many cartons of juice are there in all?

To find how many cartons in all,
multiply: $6 \times 3 = $ ?

product
↓
$6 \times 3 = 18$    or

$\begin{array}{r} 3 \\ \times 6 \\ \hline 18 \end{array}$ ← product

Six groups
of 3 are 18.

Six times three equals eighteen.

There are 18 cartons in all.

**Complete. Find the product.**

**1.**

4 threes = 12
$4 \times 3 = $ ?

**2.**

5 threes = 15
$5 \times 3 = $ ?

**3.**

8 threes = 24
$8 \times 3 = $ ?

**4.**

? threes = 6
$2 \times 3 = $ ?

**5.**

? threes = 21
$7 \times 3 = $ ?

**6.**

? threes = 27
$9 \times 3 = $ ?

**Multiply.**

**7.**

$$\begin{array}{r} 3 \\ \times 3 \\ \hline \end{array}$$

**8.**

$$\begin{array}{r} 3 \\ \times 6 \\ \hline \end{array}$$

**9.**

$$\begin{array}{r} 3 \\ \times 0 \\ \hline \end{array}$$

**Write a multiplication sentence for each addition expression.**

**10.** $3 + 3 + 3 + 3 + 3 + 3$     **11.** $3 + 3 + 3 + 3 + 3$     **12.** $3 + 3 + 3$

**Find the product. You may use counters or draw dots.**

**13.** $8 \times 3$     **14.** $4 \times 3$     **15.** $1 \times 3$     **16.** $9 \times 3$

**17.** $2 \times 3$     **18.** $5 \times 3$     **19.** $7 \times 3$     **20.** $6 \times 3$

**21.** $\begin{array}{r} 3 \\ \times 4 \\ \hline \end{array}$   **22.** $\begin{array}{r} 3 \\ \times 9 \\ \hline \end{array}$   **23.** $\begin{array}{r} 3 \\ \times 8 \\ \hline \end{array}$   **24.** $\begin{array}{r} 3 \\ \times 3 \\ \hline \end{array}$   **25.** $\begin{array}{r} 3 \\ \times 1 \\ \hline \end{array}$   **26.** $\begin{array}{r} 3 \\ \times 0 \\ \hline \end{array}$

**Copy and complete the table.**
**Describe any pattern you see.**

*Communicate* ✓

**27.**

| $0 \times 3$ | $1 \times 3$ | $2 \times 3$ | $3 \times 3$ | $4 \times 3$ | $5 \times 3$ | $6 \times 3$ | $7 \times 3$ | $8 \times 3$ | $9 \times 3$ |
|---|---|---|---|---|---|---|---|---|---|
| ? | ? | ? | ? | ? | ? | ? | ? | ? | ? |

**PROBLEM SOLVING**

**28.** Katy buys 7 packs of apple juice. There are 3 cartons in each pack. How many cartons does Katy buy in all?

**29.** Tom bought 4 packs of grape juice. Each pack had 3 cartons in it. How many cartons of grape juice did Tom buy?

141

Each box contains 4 golf balls.
How many golf balls are there in all?

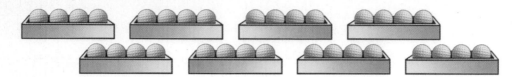

To find how many golf balls in all,
multiply: 8 × 4 = ?

product
↓
8 × 4 = 32     or
$$\begin{array}{r} 4 \\ \times 8 \\ \hline 32 \end{array}$$ ← product

Eight times four equals thirty-two.

There are 32 golf balls in all.

## Find the product.

**1.**

2 fours = 8
2 × 4 = ?

**2.**

1 four = 4
1 × 4 = ?

**3.**

7 fours = 28
7 × 4 = ?

**4.**

? fours = 0
0 × 4 = ?

**5.**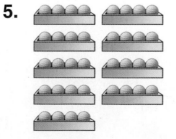

? fours = 36
9 × 4 = ?

**6.**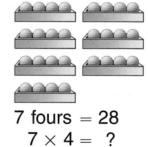

? fours = 16
4 × 4 = ?

**Write a multiplication sentence for each addition expression.**

**7.** 4 + 4      **8.** 4 + 4 + 4 + 4 + 4 + 4      **9.** 4 + 4 + 4

**Write a multiplication sentence for each model.**

**10.**      **11.**      **12.**

**Find the product. You may use counters or draw dots.**

**13.** $2 \times 4$    **14.** $5 \times 4$    **15.** $6 \times 4$    **16.** $7 \times 4$    **17.** $4 \times 0$

**18.**    4    **19.**    4    **20.**    4    **21.**    4    **22.**    4
      $\times 8$        $\times 9$        $\times 3$        $\times 1$        $\times 4$

**Copy and complete the table.**
**Describe any pattern you see.**

*Communicate* ✓

**23.**

| $0 \times 4$ | $1 \times 4$ | $2 \times 4$ | $3 \times 4$ | $4 \times 4$ | $5 \times 4$ | $6 \times 4$ | $7 \times 4$ | $8 \times 4$ | $9 \times 4$ |
|---|---|---|---|---|---|---|---|---|---|
| ? | ? | ? | ? | ? | ? | ? | ? | ? | ? |

## PROBLEM SOLVING

**24.** Mario gets 6 boxes of golf balls. There are 4 golf balls in each box. How many golf balls does Mario get in all?

**25.** Joan puts her tennis shirts in 3 drawers. Each drawer has 4 tennis shirts. How many tennis shirts does Joan have?

 **Challenge**

**26.** I am between $9 \times 4$ and $8 \times 4$. I am an even number. What am I?

**27.** I am greater than $5 \times 4$. I am less than $6 \times 4$. I am an odd number. I am not 21. What am I?

## 4-6 | Multiplying Fives

Each bowl holds 5 lemons.
How many lemons are there in all?

To find how many lemons in all,
multiply: $6 \times 5 = $ ?

$6 \times 5 = 30$ ←—product    or    $\begin{array}{r} 5 \\ \times 6 \\ \hline 30 \end{array}$ ←—product

Six times five equals thirty.

There are 30 lemons in all.

## Find the product.

**1.**

3 fives $= 15$
$3 \times 5 = $ ?

**2.**

1 five $= 5$
$1 \times 5 = $ ?

**3.**

5 fives $= 25$
$5 \times 5 = $ ?

---

**4.**

0 fives $= 0$
$0 \times 5 = $ ?

**5.**

? fives $= 10$
$2 \times 5 = $ ?

**6.**

? fives $= 20$
$4 \times 5 = $ ?

**Write a multiplication sentence for each addition expression.**

**7.** 5 + 5 + 5 + 5          **8.** 5 + 5 + 5          **9.** 5 + 5 + 5 + 5 + 5

**Write a multiplication sentence for each model.**

**10.** [model of counters]     **11.** [model of counters]     **12.** [model of counters]

**Find the product.**

**13.** 3 × 5     **14.** 9 × 5     **15.** 7 × 5     **16.** 8 × 5     **17.** 5 × 4

**18.**  5          **19.**  5          **20.**  5          **21.**  5          **22.**  5
    ×2               ×5               ×1               ×6               ×0

**Copy and complete the table.**
**Describe any pattern you see.**

Communicate ✓

**23.**

| 0 × 5 | 1 × 5 | 2 × 5 | 3 × 5 | 4 × 5 | 5 × 5 | 6 × 5 | 7 × 5 | 8 × 5 | 9 × 5 |
|-------|-------|-------|-------|-------|-------|-------|-------|-------|-------|
| ? | ? | ? | ? | ? | ? | ? | ? | ? | ? |

## PROBLEM SOLVING

**24.** Roberto buys 3 bags of limes. Each bag has 5 limes in it. How many limes are there in all?

**25.** Marie squeezes 5 lemons in each of 4 pitchers. How many lemons does she use?

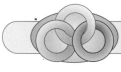

## Connections: Science

**26.** A wild rabbit can have babies 5 times in one year. Each time, the wild rabbit may have 3–7 baby rabbits. Suppose 2 wild rabbits had babies 3 times in one year, each time having 5 baby rabbits. How many baby rabbits would the two wild rabbits have in one year?

**Multiplying Cents**

Carmen groups her pennies by twos.
How much money does she have?

To find the amount of money:

- Multiply first.
- Then write the ¢ sign.

$$
\begin{array}{r}
2¢ \\
\times 4 \\
\hline
8¢
\end{array}
$$

2¢ ←——cents in each group
×4 ←——number of groups
8¢ ←——cents in all

Carmen has 8¢ in all.

## Multiply first. Then write the ¢ sign.

**1.**

$$
\begin{array}{r}
3¢ \\
\times 3 \\
\hline
\end{array}
$$

**2.**

$$
\begin{array}{r}
2¢ \\
\times 6 \\
\hline
\end{array}
$$

**3.**

$$
\begin{array}{r}
3¢ \\
\times 5 \\
\hline
\end{array}
$$

**4.** $\begin{array}{r} 5¢ \\ \times 6 \\ \hline \end{array}$   **5.** $\begin{array}{r} 2¢ \\ \times 7 \\ \hline \end{array}$   **6.** $\begin{array}{r} 4¢ \\ \times 8 \\ \hline \end{array}$   **7.** $\begin{array}{r} 5¢ \\ \times 7 \\ \hline \end{array}$   **8.** $\begin{array}{r} 4¢ \\ \times 6 \\ \hline \end{array}$

**9.** $\begin{array}{r} 3¢ \\ \times 4 \\ \hline \end{array}$   **10.** $\begin{array}{r} 2¢ \\ \times 3 \\ \hline \end{array}$   **11.** $\begin{array}{r} 4¢ \\ \times 4 \\ \hline \end{array}$   **12.** $\begin{array}{r} 1¢ \\ \times 9 \\ \hline \end{array}$   **13.** $\begin{array}{r} 0¢ \\ \times 3 \\ \hline \end{array}$

**Multiply mentally.**

**14.** 3 × 2       **15.** 4 × 3       **16.** 6 × 4       **17.** 3 × 5

**18.** 8 × 5¢       **19.** 9 × 2¢       **20.** 9 × 3¢       **21.** 7 × 4¢

**22.** 8 × 3¢       **23.** 5 × 5¢       **24.** 1 × 4¢       **25.** 8 × 2¢

**Find the product.**

| **26.** | **27.** | **28.** | **29.** | **30.** | **31.** |
|---|---|---|---|---|---|
| 0 | 3 | 4 | 5 | 4 | 3 |
| ×2 | ×2 | ×5 | ×2 | ×3 | ×7 |

| **32.** | **33.** | **34.** | **35.** | **36.** | **37.** |
|---|---|---|---|---|---|
| 2 | 3 | 2 | 5 | 4 | 5 |
| ×5 | ×0 | ×6 | ×9 | ×9 | ×1 |

| **38.** | **39.** | **40.** | **41.** | **42.** | **43.** |
|---|---|---|---|---|---|
| 5¢ | 3¢ | 5¢ | 4¢ | 5¢ | 4¢ |
| ×4 | ×8 | ×3 | ×4 | ×8 | ×2 |

## PROBLEM SOLVING

**44.** A canoe can carry 2 people. How many people can ride in 6 canoes?

**45.** A toy rowboat costs 5¢. What is the cost of 7 toy rowboats?

 **Skills to Remember**       Regroup.

5 thousands = 4 thousands + 1 thousand
= 4 thousands + 10 hundreds
= 4 thousands 10 hundreds

> 1 thousand = 10 hundreds

**46.** 15 hundreds =
_?_ thousand _?_ hundreds

**47.** 18 hundreds =
_?_ thousand _?_ hundreds

**48.** 7 thousands =
_?_ thousands _?_ hundreds

**49.** 9 thousands =
_?_ thousands _?_ hundreds

# 4-8 Sums, Differences, and Products

| Add to find sums. | Subtract to find differences. | Multiply to find products. |
|---|---|---|
| $\begin{array}{r} 1\ 1\ 1 \\ 2486 \\ +3867 \\ \hline 6353 \end{array}$ ← sum | $\begin{array}{r} 9\ 10 \\ 7\ 10\ \cancel{0}\ 10 \\ 8\ \cancel{0}\ \cancel{1}\ \cancel{0} \\ -5\ 9\ 8\ 2 \\ \hline 2\ 0\ 2\ 8 \end{array}$ ← difference | $\begin{array}{r} 3 \\ \times 9 \\ \hline 27 \end{array}$ ← product |

## Add, subtract, or multiply. Watch the signs.

**1.**  $\begin{array}{r} 234 \\ +545 \end{array}$
**2.**  $\begin{array}{r} 641 \\ -230 \end{array}$
**3.**  $\begin{array}{r} 423 \\ -115 \end{array}$
**4.**  $\begin{array}{r} 536 \\ +318 \end{array}$
**5.**  $\begin{array}{r} 948 \\ -376 \end{array}$

**6.**  $\begin{array}{r} 335 \\ +189 \end{array}$
**7.**  $\begin{array}{r} 2663 \\ +4728 \end{array}$
**8.**  $\begin{array}{r} 6574 \\ -2978 \end{array}$
**9.**  $\begin{array}{r} 4098 \\ +1927 \end{array}$
**10.**  $\begin{array}{r} 8406 \\ -3217 \end{array}$

**11.**  $\begin{array}{r} 5 \\ \times 7 \end{array}$
**12.**  $\begin{array}{r} 3 \\ \times 8 \end{array}$
**13.**  $\begin{array}{r} 4 \\ \times 0 \end{array}$
**14.**  $\begin{array}{r} 3 \\ \times 9 \end{array}$
**15.**  $\begin{array}{r} 0 \\ \times 1 \end{array}$

**16.**  $\begin{array}{r} 742 \\ +370 \end{array}$
**17.**  $\begin{array}{r} 4307 \\ -3244 \end{array}$
**18.**  $\begin{array}{r} 3¢ \\ \times 7 \end{array}$
**19.**  $\begin{array}{r} 2 \\ \times 6 \end{array}$
**20.**  $\begin{array}{r} 1178 \\ +6847 \end{array}$

**21.**  $\begin{array}{r} \$.89 \\ -.53 \end{array}$
**22.**  $\begin{array}{r} \$.47 \\ +.39 \end{array}$
**23.**  $\begin{array}{r} 4 \\ \times 7 \end{array}$
**24.**  $\begin{array}{r} 5 \\ \times 6 \end{array}$
**25.**  $\begin{array}{r} \$5.55 \\ -2.22 \end{array}$

**26.**  $\begin{array}{r} 5¢ \\ \times 8 \end{array}$
**27.**  $\begin{array}{r} 3¢ \\ \times 4 \end{array}$
**28.**  $\begin{array}{r} \$4.08 \\ +3.75 \end{array}$
**29.**  $\begin{array}{r} \$3.29 \\ -1.77 \end{array}$
**30.**  $\begin{array}{r} 5¢ \\ \times 1 \end{array}$

**31.**  $\begin{array}{r} 7215 \\ +1824 \end{array}$
**32.**  $\begin{array}{r} 5863 \\ -4778 \end{array}$
**33.**  $\begin{array}{r} 5 \\ \times 9 \end{array}$
**34.**  $\begin{array}{r} 2 \\ \times 8 \end{array}$
**35.**  $\begin{array}{r} 6374 \\ +1696 \end{array}$

## PROBLEM SOLVING

**36.** A violin has 4 strings. How many strings are there on 5 violins?

**37.** Mark's family traveled 296 miles in one day and 135 miles the next day. How many miles did the family travel altogether?

**38.** Emma put 75 pictures in her photo album. It holds 144 pictures. How many more pictures can Emma put in her album?

**39.** Jake has 9 packs of baseball cards. There are 5 cards in each pack. How many baseball cards does Jake have in all?

**40.** Eduardo bought gifts for three of his friends. The gifts cost $2.49, $3.29, and $2.98. What was the total cost of the gifts?

**41.** Dorotea bought a pencil eraser. She paid 5 nickels for the eraser. How much did she spend?

**42.** Chad bought 6 model sailboats at $5 each. Maris bought 8 model speedboats at $3 each. Who spent more? How much more?

## Choose a Computation Method

**Solve. Use paper and pencil or mental math.**
Tell your teacher which method you used.

**43.** $2 \times 4$

**44.**
$$
\begin{array}{r} 721 \\ + 333 \\ \hline \end{array}
$$

**45.**
$$
\begin{array}{r} 0 \\ \times 8 \\ \hline \end{array}
$$

**46.**
$$
\begin{array}{r} \$9.00 \\ - 2.31 \\ \hline \end{array}
$$

**47.**
$$
\begin{array}{r} 5263 \\ + 2986 \\ \hline \end{array}
$$

**48.**
$$
\begin{array}{r} 5¢ \\ \times 4 \\ \hline \end{array}
$$

**Order in Multiplication**    *Algebra* ✓

How many cars are there?

  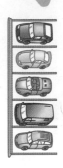

    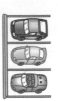

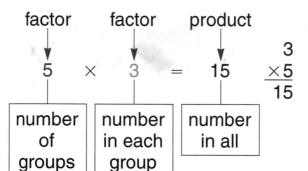

5 rows of 3 cars

factor    factor    product

$$5 \times 3 = 15$$

$$\begin{array}{r} 3 \\ \times 5 \\ \hline 15 \end{array}$$

| number of groups | number in each group | number in all |
|---|---|---|

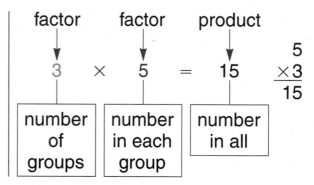

3 rows of 5 cars

factor    factor    product

$$3 \times 5 = 15$$

$$\begin{array}{r} 5 \\ \times 3 \\ \hline 15 \end{array}$$

| number of groups | number in each group | number in all |
|---|---|---|

There are 15 cars in all.

> When you change the **order** of the factors, the product stays the same.

**Find the product.**

1. $3 \times 2 = 6$
   $2 \times 3 = \underline{\ ?\ }$

2. $4 \times 1 = 4$
   $1 \times 4 = \underline{\ ?\ }$

3. $2 \times 5 = 10$
   $5 \times 2 = \underline{\ ?\ }$

4. $1 \times 2 = 2$
   $2 \times 1 = \underline{\ ?\ }$

5. $4 \times 3 = 12$
   $3 \times 4 = \underline{\ ?\ }$

6. $2 \times 0 = 0$
   $0 \times 2 = \underline{\ ?\ }$

7. $5 \times 1 = \underline{\ ?\ }$
   $1 \times 5 = \underline{\ ?\ }$

8. $4 \times 2 = \underline{\ ?\ }$
   $2 \times 4 = \underline{\ ?\ }$

9. $0 \times 5 = 0$
   $5 \times 0 = \underline{\ ?\ }$

10. $4 \times 5 = \underline{\ ?\ }$
    $5 \times 4 = \underline{\ ?\ }$

11. $8 \times 2 = \underline{\ ?\ }$
    $2 \times 8 = \underline{\ ?\ }$

12. $6 \times 5 = \underline{\ ?\ }$
    $5 \times 6 = \underline{\ ?\ }$

**Multiply.**

| 13. | 3 ×1 = 3 | 1 ×3 = 3 | 14. | 4 ×2 | 2 ×4 | 15. | 1 ×0 | 0 ×1 | 16. | 5 ×3 | 3 ×5 |

**13.** $\begin{array}{r}3\\\times1\\\hline 3\end{array}$ $\begin{array}{r}1\\\times3\\\hline 3\end{array}$   **14.** $\begin{array}{r}4\\\times2\\\hline\end{array}$ $\begin{array}{r}2\\\times4\\\hline\end{array}$   **15.** $\begin{array}{r}1\\\times0\\\hline\end{array}$ $\begin{array}{r}0\\\times1\\\hline\end{array}$   **16.** $\begin{array}{r}5\\\times3\\\hline\end{array}$ $\begin{array}{r}3\\\times5\\\hline\end{array}$

**17.** $\begin{array}{r}2\\\times3\\\hline\end{array}$ $\begin{array}{r}3\\\times2\\\hline\end{array}$   **18.** $\begin{array}{r}5\\\times1\\\hline\end{array}$ $\begin{array}{r}1\\\times5\\\hline\end{array}$   **19.** $\begin{array}{r}3\\\times4\\\hline\end{array}$ $\begin{array}{r}4\\\times3\\\hline\end{array}$   **20.** $\begin{array}{r}2\\\times5\\\hline\end{array}$ $\begin{array}{r}5\\\times2\\\hline\end{array}$

**21.** $\begin{array}{r}0\\\times4\\\hline\end{array}$ $\begin{array}{r}4\\\times0\\\hline\end{array}$   **22.** $\begin{array}{r}5\\\times4\\\hline\end{array}$ $\begin{array}{r}4\\\times5\\\hline\end{array}$   **23.** $\begin{array}{r}2\\\times7\\\hline\end{array}$ $\begin{array}{r}7\\\times2\\\hline\end{array}$   **24.** $\begin{array}{r}3\\\times9\\\hline\end{array}$ $\begin{array}{r}9\\\times3\\\hline\end{array}$

**Draw dots or circles to show each.**

**25.** $5 \times 2 = 2 \times 5$   **26.** $4 \times 1 = 1 \times 4$   **27.** $3 \times 2 = 2 \times 3$

**28.** $4 \times 3 = 3 \times 4$   **29.** $1 \times 3 = 3 \times 1$   **30.** $4 \times 5 = 5 \times 4$

**Find the missing factor.**

**31.** $4 \times 2 = \underline{?} \times 4$   **32.** $1 \times 2 = \underline{?} \times 1$   **33.** $3 \times 5 = \underline{?} \times 3$

**34.** $2 \times 5 = \underline{?} \times 2$   **35.** $\underline{?} \times 4 = 4 \times 5$   **36.** $\underline{?} \times 0 = 0 \times 1$

## PROBLEM SOLVING

*Communicate* ✓

**37.** José puts 5 cars in each of 4 large toy chests.
Laurie puts 4 cars in each of 5 small toy chests.
Which person has more cars? Explain your answer.

## Critical Thinking

**Compare. Write <, =, or >.**

**38.** $5 \times 3 \underline{\phantom{?}} 5 \times 4$   **39.** $3 \times 3 \underline{\phantom{?}} 4 \times 4$   **40.** $5 \times 2 \underline{\phantom{?}} 2 \times 5$

**41.** $2 \times 0 \underline{\phantom{?}} 1 \times 0$   **42.** $6 \times 4 \underline{\phantom{?}} 7 \times 2$   **43.** $0 \times 0 \underline{\phantom{?}} 1 \times 1$

**Missing Factors**

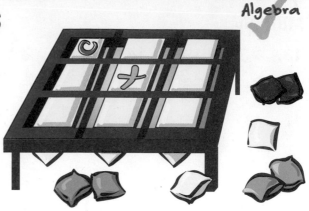

*Algebra* ✓

There are 8 bean bags in a bean bag toss game. Each player takes 2 bean bags. At most, how many can play the game?

To find the number of players, think:
What number times 2 equals 8?

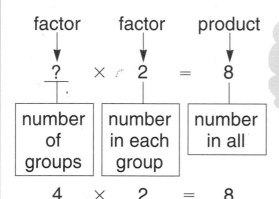

factor    factor    product

? × 2 = 8

| number of groups | number in each group | number in all |

4 × 2 = 8

**Think: Find the missing factor.**

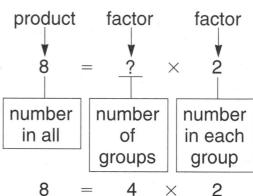

product    factor    factor

8 = ? × 2

| number in all | number of groups | number in each group |

8 = 4 × 2

The missing factor is 4.

Four players with 2 bean bags each can play the game.

## Find the missing factor.

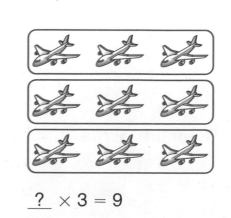

1.
___ × 3 = 12

2.
___ × 5 = 15

3.
___ × 3 = 9

**Find the missing factor.**

4. $\underline{?} \times 2 = 12$   5. $\underline{?} \times 3 = 15$   6. $\underline{?} \times 5 = 20$   7. $\underline{?} \times 4 = 12$

8. $2 \times \underline{?} = 10$   9. $2 \times \underline{?} = 8$   10. $1 \times \underline{?} = 5$   11. $4 \times \underline{?} = 4$

12. $\underline{?} \times 3 = 18$   13. $5 \times \underline{?} = 25$   14. $8 \times \underline{?} = 16$   15. $\underline{?} \times 4 = 24$

16. $20 = \underline{?} \times 4$   17. $16 = \underline{?} \times 4$   18. $14 = 7 \times \underline{?}$   19. $27 = 9 \times \underline{?}$

**Multiply.**

20. $\begin{array}{r} 4 \\ \times 9 \\ \hline \end{array}$   21. $\begin{array}{r} 0 \\ \times 8 \\ \hline \end{array}$   22. $\begin{array}{r} 5 \\ \times 9 \\ \hline \end{array}$   23. $\begin{array}{r} 4 \\ \times 7 \\ \hline \end{array}$   24. $\begin{array}{r} 5 \\ \times 6 \\ \hline \end{array}$   25. $\begin{array}{r} 2 \\ \times 5 \\ \hline \end{array}$

26. $\begin{array}{r} 3 \\ \times 7 \\ \hline \end{array}$   27. $\begin{array}{r} 5 \\ \times 0 \\ \hline \end{array}$   28. $\begin{array}{r} 1 \\ \times 6 \\ \hline \end{array}$   29. $\begin{array}{r} 4 \\ \times 8 \\ \hline \end{array}$   30. $\begin{array}{r} 2 \\ \times 9 \\ \hline \end{array}$   31. $\begin{array}{r} 5 \\ \times 8 \\ \hline \end{array}$

**Copy and complete each chart.**

32.

| factor | factor | product |
|--------|--------|---------|
| 7 | ? | 21 |
| ? | 5 | 30 |
| 1 | ? | 2 |
| 7 | ? | 35 |
| ? | 4 | 32 |
| 7 | ? | 14 |
| ? | 4 | 16 |

33.

| factor | factor | product |
|--------|--------|---------|
| 3 | ? | 3 |
| ? | 4 | 20 |
| 7 | ? | 28 |
| ? | 5 | 45 |
| ? | 2 | 0 |
| 9 | ? | 18 |
| ? | 3 | 24 |

## Mental Math

**Use doubles to find each missing factor.**

34. $2 \times \underline{?} = 10$   35. $\underline{?} \times 2 = 16$   36. $\underline{?} \times 2 = 12$

37. $18 = \underline{?} \times 2$   38. $14 = \underline{?} \times 2$   39. $0 = 2 \times \underline{?}$

# TECHNOLOGY

## Finding Missing Factors with a Calculator

Elise made 15 cinnamon rolls. She gave 3 rolls to each friend. How many of Elise's friends received cinnamon rolls?

To find how many, solve: __?__ × 3 = 15

You can use a calculator to find the missing factor.

Pressing the **=** key repeats the subtraction.

To solve __?__ × 3 = 15 on a calculator,

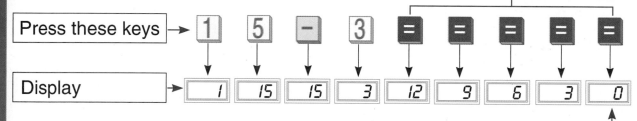

| Press these keys → | 1 | 5 | − | 3 | = | = | = | = | = |

| Display → | 1 | 15 | 15 | 3 | 12 | 9 | 6 | 3 | 0 |

Subtract 3s until you reach 0.

Count the number of times 3s were subtracted to reach 0.

3s were subtracted 5 times.

So the missing factor is 5.       5 × 3 = 15

Elise gave 5 friends 3 cinnamon rolls each.

---

**Use a calculator. Write the missing factor.**

1.

     __?__ × 2 = 8

2. 1 2 − 4 = = =

     __?__ × 4 = 12

**Use your calculator. Subtract to find the missing factor.**

3. _?_ × 4 = 36    4. 7 × _?_ = 35    5. 4 × _?_ = 16    6. _?_ × 2 = 12

7. 8 × _?_ = 8    8. 9 × _?_ = 27    9. _?_ × 4 = 20    10. 7 × _?_ = 21

11. 9 × _?_ = 36    12. 6 × _?_ = 24    13. _?_ × 5 = 45    14. _?_ × 5 = 40

---

### Factor Pairs

Use a calculator to test factor pairs.

Think: 2 is a factor of 12                    and    3 is a factor of 12

6 × 2 = 12                                            4 × 3 = 12

6, 2 and 4, 3 are
factor pairs of 12.

Both factor pairs have a product of 12.

---

**Use a calculator to find a factor pair for each product.**
Use each factor in the box only once.

| | | | |
|---|---|---|---|
| 15. 14 | 16. 20 | 17. 10 | 9 5 7 4 2 5 |
| 18. 32 | 19. 9 | 20. 40 | 2 4 8 5 1 8 |

**Use a calculator to find two factor pairs for each.**

21. _?_ × _?_ = 6    22. _?_ × _?_ = 8    23. _?_ × _?_ = 4

24. _?_ × _?_ = 0    25. _?_ × _?_ = 16    26. _?_ × _?_ = 18

**Write a multiplication sentence from each set of keys.**

27.

28.

29.

30.

# 4-12 Problem Solving: Multi-Step Problem

**Problem:** Tricia dropped 85 pieces of her puzzle on the floor. Mark found 23 pieces, José found 29 pieces, and Tricia found 26 pieces. How many pieces did the children *not* find?

**1 IMAGINE** Create a mental picture.

**2 NAME**

*Facts:*    85 pieces dropped
Mark found 23 pieces.
José found 29 pieces.
Tricia found 26 pieces.

*Question:*  How many pieces were *not* found?

**3 THINK** To find the number of pieces *not* found, subtract:

$$85 \quad - \quad \underline{\quad ? \quad} \quad = \quad \underline{\quad ? \quad}$$

in         number        number
all         found        *not* found

First find the number of pieces found. Add:

$$23 \quad + \quad 29 \quad + \quad 26 \quad = \quad \underline{\quad ? \quad}$$

Mark's    José's    Tricia's    number
pieces    pieces    pieces    found

Then subtract the sum from 85.

**4 COMPUTE** $23 + 29 + 26 = 78$
They found 78 pieces.

$85 - 78 = 7$
There were 7 pieces *not* found.

**5 CHECK** Use addition to check your answer.

$$78 \quad + \quad 7 \quad = \quad 85$$

pieces    pieces    pieces
found    *not* found    in all

## Use the Multi-Step strategy to solve each problem.

**1.** Cassy bought a box of crayons for 79¢ and four stickers for 5¢ each. How much money did she spend altogether?

| | | |
|---|---|---|
| **IMAGINE** | Create a mental picture. | |
| **NAME** | *Facts:* | crayons — 79¢ a box<br>4 stickers — 5¢ each |
| | *Question:* | How much did Cassy spend? |
| **THINK** | First find the cost of 4 stickers.<br>Multiply:   4 × 5¢ = _?_ | |
| | Then find how much she spent altogether.<br>Add:   79¢ + cost of stickers = _?_ | |

**COMPUTE** ⟶ **CHECK**

79¢

5¢

5¢

5¢

5¢

**2.** Marlo's Pet Shop has 5 angelfish in each of 4 tanks. It has 4 guppies in each of 3 tanks. Ms. Fin bought all the guppies and angelfish. How many fish did she buy?

**3.** Jon needs 30 stamps to fill his stamp album. Each of 3 friends gives him 7 stamps. How many more stamps does Jon need?

**4.** Ken scored 9 points more than Josh. Renee scored two times the number of points scored by Josh. Josh scored 6 points. How many points did they score altogether?

**5.** Lauren collected 4 cans of food from each of 7 neighbors. Her goal is 35 cans of food for the poor. How many more cans does she need?

**Solve each problem and explain the method you used.**

1. Ana buys 3 favors for each of her 7 guests. How many favors does Ana have for her party?

2. Mrs. Yanni sets up 6 tables. She puts 4 chairs at each table. How many chairs are there in all?

3. Ana has a package of 32 balloons. Fifteen of them have stripes. The rest are yellow. How many yellow balloons does Ana have?

4. Darryl, Neil, and Russ each won 2 prizes. How many prizes did the boys win altogether?

5. Kimo found 5¢. Carl found 5 times as much money. How much money did Carl find?

6. Ana received $25 from her grandparents. She also received $8 from each of three friends. How much money did Ana receive?

7. Six children each won one prize in the first game. Three children won two prizes each in the second game. How many prizes in all did the children win in the two games?

8. Each of Marta's 5 baskets has 4 flowers. Each of Cara's 6 baskets has 3 flowers. Which girl has more flowers? How many more?

Imagine

Name

Think

Compute

Check

**Choose a strategy from the list or use another strategy you know to solve each problem.**

9. Ms. Reis has 35 guests. She seats 4 guests at each of the first 8 tables. How many guests sit at the ninth table?

10. José buys 4 baseball cards. Each card costs 5¢. He pays a quarter. How much change should he get?

11. Ned spent $2.80 more than Fred. Ken spent $1.25 less than Fred. Fred spent $5.00. How much did Ned spend? How much did Ken spend?

Use these strategies:

Hidden Information
Find a Pattern
Draw a Picture
Guess and Test
Use Simpler Numbers
Multi-Step Problem

12. A magician pulled 12 rabbits from a hat. She pulled 1 more gray rabbit than white rabbits, and 1 more black rabbit than gray rabbits. How many rabbits of each color did she pull out of the hat?

13. A party store had 30 party hats. It sold 7 hats on Monday, 6 hats on Tuesday, 5 hats on Wednesday, 4 hats on Thursday, and so on. How many hats will be left at the end of the day on Saturday?

**Use the bar graph for problems 14–16.**

14. This graph shows the number of birthdays in 3 months. How many more birthdays are there in May than in April?

15. There are three times as many birthdays in June as in April. How many birthdays are there in June?

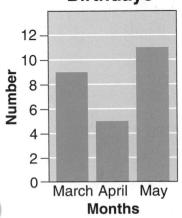

**Birthdays**

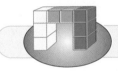

**Make Up Your Own**

16. Use the data on the birthday graph to make up a problem. Then solve it.

# Chapter 4

# Chapter Review and Practice

## Find the product.

*(See pp. 134–147.)*

1.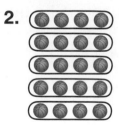
6 threes = _?_
6 × 3 = _?_

2.
5 fours = _?_
5 × 4 = _?_

## Write a multiplication sentence for each.

**3.** 5 + 5 + 5          **4.** 2 + 2          **5.** 4 + 4 + 4 + 4

## Multiply.

**6.** 2 × 4¢          **7.** 6 × 4          **8.** 3 × 2          **9.** 7 × 5¢

**10.** 4 × 0          **11.** 5 × 3¢          **12.** 9 × 4          **13.** 9 × 1

| 14. 3 ×4 | 15. 3 ×9 | 16. 5¢ ×9 | 17. 4 ×7 | 18. 2¢ ×9 | 19. 0 ×8 |

## Find the missing factor.

*(See pp. 150–153.)*

**20.** 8 × 4 = _?_ × 8          **21.** 7 × 3 = _?_ × 7          **22.** 2 × 3 = _?_ × 2

**23.** _?_ × 3¢ = 24¢          **24.** 4 × _?_ = 16          **25.** _?_ × 3¢ = 0¢

**26.** 3 × _?_ = 9          **27.** _?_ × 5¢ = 20¢          **28.** 16 = 8 × _?_

## PROBLEM SOLVING

*(See pp. 148–149, 156–159.)*

**29.** There are 5 goldfish in each bowl. How many goldfish are in 6 bowls?

**30.** Rhea bought a guppy. She gave the clerk 98¢ and 5 nickels. How much did the guppy cost?

(See *Still More Practice*, p. 446.)

PREDICT PATTERNS OF SUMS

**Find the pattern. Copy and complete each sentence.**

1.  $1 = \underline{1} \times \underline{1}$

    Think: $1 + 3 = 4$
    $4 = 2 \times 2$

2.  $1 + 3 = \underline{2} \times \underline{2}$

3.  $1 + 3 + 5 = \underline{3} \times \underline{?}$

    Think: $1 + 3 + 5 = 9$

4.  $1 + 3 + 5 + 7 = \underline{?} \times \underline{?}$

5.  $1 + 3 + 5 + 7 + 9 = \underline{?} \times \underline{?}$

**Find the pattern. Copy and complete.**

6.  2,    4,    6,    8, $\underline{10}$ , $\underline{12}$ , $\underline{14}$

    Think: $5 \times 2 = 10$
    $6 \times 2 = 12$
    $7 \times 2 = 14$

    $1 \times 2 \quad 2 \times 2 \quad 3 \times 2 \quad 4 \times 2$

7.  0, 3, 6, 9, $\underline{?}$ , $\underline{?}$ , $\underline{?}$

8.  10, 15, 20, 25, $\underline{?}$ , $\underline{?}$ , $\underline{?}$

9.  0, 4, 8, 12, $\underline{?}$ , $\underline{?}$ , $\underline{?}$

10. 3, 5, 7, 9, $\underline{?}$ , $\underline{?}$ , $\underline{?}$ ◄─────

    Think: $1 \times 2 + 1 = 3$
    $2 \times 2 + 1 = 5$
    $3 \times 2 + 1 = 7$
    $4 \times 2 + 1 = 9$

11. 4, 7, 10, 13, $\underline{?}$ , $\underline{?}$ , $\underline{?}$

12. 6, 11, 16, 21, $\underline{?}$ , $\underline{?}$ , $\underline{?}$

**Look at the pattern in exercises 1–5.**
**Predict the factors.** Check using a calculator.

13. $1 + 3 + 5 + 7 + 9 + 11 = \underline{?} \times \underline{?}$

# Check Your Mastery

## Performance Assessment

**Use pennies. Model and write a multiplication sentence for each.**

1. order in multiplication    2. 0 as a factor    3. 1 as a factor

**Multiply.**

4. $4 \times 3$    5. $7 \times 2$    6. $5 \times 3¢$    7. $6 \times 4$

8. $5 \times 5$    9. $9 \times 2¢$    10. $7 \times 4$    11. $8 \times 5$

12. $\begin{array}{r} 5 \\ \times 2 \\ \hline \end{array}$  13. $\begin{array}{r} 4¢ \\ \times 9 \\ \hline \end{array}$  14. $\begin{array}{r} 5 \\ \times 6 \\ \hline \end{array}$  15. $\begin{array}{r} 4 \\ \times 2 \\ \hline \end{array}$  16. $\begin{array}{r} 3 \\ \times 7 \\ \hline \end{array}$  17. $\begin{array}{r} 5¢ \\ \times 4 \\ \hline \end{array}$

18. $\begin{array}{r} 2 \\ \times 5 \\ \hline \end{array}$  19. $\begin{array}{r} 4¢ \\ \times 4 \\ \hline \end{array}$  20. $\begin{array}{r} 3¢ \\ \times 9 \\ \hline \end{array}$  21. $\begin{array}{r} 4¢ \\ \times 8 \\ \hline \end{array}$  22. $\begin{array}{r} 4 \\ \times 5 \\ \hline \end{array}$  23. $\begin{array}{r} 2 \\ \times 8 \\ \hline \end{array}$

**Find the missing factor.**

24. $\underline{\;?\;} \times 5 = 35$    25. $7 \times \underline{\;?\;} = 28$    26. $\underline{\;?\;} \times 2 = 12$

27. $3 \times 5 = \underline{\;?\;} \times 3$    28. $2 \times 4 = \underline{\;?\;} \times 2$    29. $\underline{\;?\;} \times 3 = 3 \times 4$

**PROBLEM SOLVING**    *Use a strategy you have learned.*

30. There are 5 players on a basketball team. Each player scores 5 points. What is the final score?

31. There are 4 cassettes in each case. How many cassettes are in 6 cases?

32. There are 5 books on each shelf. How many books are on 9 shelves?

33. My factors are 6 and 3. What is my product?

# Division Concepts and Facts

# 5

## In this chapter you will:

Explore both meanings of division
Relate multiplication and division
Learn about 0 and 1 in division
Divide numbers and cents by 2 through 5
Solve problems by writing a number sentence

### Critical Thinking/Finding Together

Explain how you can arrange the orange
poppies on the page in equal rows and
columns with none left over. Use counters
to show each model.

## NATURE KNOWS ITS MATH

*Divide*
the year
into seasons,
four,
*subtract*
the snow then
*add*
some more
green,
a bud,
a breeze,
a whispering
behind
the trees,
and here
beneath the
rain-scrubbed
sky
orange poppies
*multiply.*

*Joan Bransfield Graham*

## 5-1 Understanding Division   *Algebra* ✓

Jimmy has 15 fish. He puts 3 fish in each bowl. How many bowls does he use?

To find how many bowls, divide. One way to divide is to **separate** into equal groups.

## Hands-On Understanding

**You Will Need:** counters, record sheet, 5 blank sheets of paper

**Step 1**   Model 15 counters.

What do the 15 counters represent?

**Step 2**   Take away as many groups of 3 as you can.

How many counters are in each group?
How many groups did you take away?
How many bowls does Jimmy use?

You can also write a division sentence to show 15 counters separated into equal groups of 3.

**Write**: $15 \div 3 = 5$.

**Read as**: "Fifteen divided by three equals five."

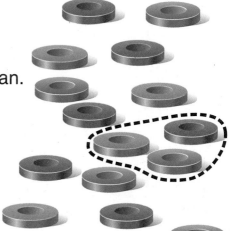

Suppose Jimmy has 15 fish and 5 bowls. He puts an equal number of fish in each bowl. At most, how many fish can he put in each bowl?

To find how many fish in each bowl, divide. Another way to divide is to **share** equally.

**Step 3**   Model 15 counters.

What do the 15 counters represent?

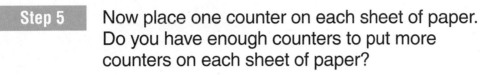

| Step 4 | Use 5 blank sheets of paper. |
|---|---|

What do the 5 sheets of paper stand for?

| Step 5 | Now place one counter on each sheet of paper. Do you have enough counters to put more counters on each sheet of paper? |
|---|---|

| Step 6 | Continue to place one counter on each sheet of paper until you can no longer give them out equally. |
|---|---|

How many counters are on each sheet of paper?

How many fish can Jimmy put in each bowl?

You can also write a division sentence to show 15 counters shared equally among 5 groups.

**Write**: $15 \div 5 = 3$.

**Read as**: "Fifteen divided by five equals three."

**Use counters or draw dots to separate into equal groups. Find how many groups.**

**1.** 14 in all
2 in each group

**2.** 18 in all
6 in each group

**3.** 24 in all
4 in each group

**4.** Write a division sentence for each model in exercises 1–3.

**Use counters or draw dots to share. Find how many in each group.**

**5.** 15 in all
5 groups

**6.** 8 in all
2 groups

**7.** 12 in all
6 groups

**8.** Write a division sentence for each model in exercises 5–7.

# Communicate

Discuss

**9.** What is the difference between separating and sharing?

# Relating Multiplication and Division *Algebra*

Ms. Hardy has 21 tennis balls in all.
There are 3 tennis balls in each can.
How many cans are there?

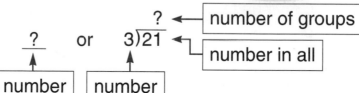

To find how many cans,
divide: $21 \div 3 = \underline{?}$

$$21 \div 3 = \underline{?} \quad \text{or} \quad 3\overline{)21}$$

$\underline{?}$ ← number of groups

number in all

| 21 | 3 | $\underline{?}$ | $3\overline{)21}$ |
|---|---|---|---|
| number in all | number in each group | number of groups | number in each group |

▶ Every division fact has a related multiplication fact.

$$21 \div 3 = \underline{?} \quad\longrightarrow\quad \underline{?} \times 3 = 21$$

| 21 | 3 | $\underline{?}$ | $\underline{?}$ | 3 | 21 |
|---|---|---|---|---|---|
| in all | in each | groups | groups | in each | in all |

$21 \div 3 = \underline{?}$       Think: $\underline{?} \times 3 = 21 \longrightarrow 7 \times 3 = 21$

So   $21 \div 3 = 7$.     There are 7 cans.

**Study this example.**

$20 \div 5 = \underline{?}$   or   $5\overline{)20}$       Think: $\underline{?} \times 5 = 20$
$20 \div 5 = 4$                                             $4 \times 5 = 20$

---

**Write a multiplication fact for each.**

1.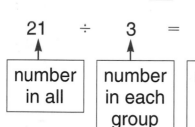

   $6 \div 2 = 3$

   $3 \times 2 = \underline{?}$

2. 

   $8 \div 4 = 2$

   $\underline{?} \times \underline{?} = \underline{?}$

3. 

   $9 \div 3 = 3$

   $\underline{?} \times \underline{?} = \underline{?}$

**Write a multiplication sentence and a division sentence for each.**

**4.** ⬚⬚⬚
⬚⬚⬚
⬚⬚⬚
⬚⬚⬚

**5.** ⬚⬚⬚⬚⬚
⬚⬚⬚⬚⬚
⬚⬚⬚⬚⬚
⬚⬚⬚⬚⬚
⬚⬚⬚⬚⬚

**6.** ⬚⬚⬚⬚⬚
⬚⬚⬚⬚⬚
⬚⬚⬚⬚⬚
⬚⬚⬚⬚⬚

**Complete each multiplication and division sentence.**

**7.** $9 \times 3 = \underline{?}$
$27 \div 3 = \underline{?}$

**8.** $7 \times 5 = \underline{?}$
$35 \div 5 = \underline{?}$

**9.** $6 \times 3 = \underline{?}$
$18 \div 3 = \underline{?}$

**10.** $4 \times 2 = \underline{?}$
$8 \div 2 = \underline{?}$

**11.** $6 \times 4 = \underline{?}$
$24 \div 4 = \underline{?}$

**12.** $8 \times 2 = \underline{?}$
$16 \div 2 = \underline{?}$

**13.** $9 \times 4 = \underline{?}$
$36 \div 4 = \underline{?}$

**14.** $3 \times 5 = \underline{?}$
$15 \div 5 = \underline{?}$

**15.** $\begin{array}{r} 2 \\ \times 7 \\ \hline ? \end{array}$  $2\overline{)14}$

**16.** $\begin{array}{r} 4 \\ \times 8 \\ \hline ? \end{array}$  $4\overline{)32}$

**17.** $\begin{array}{r} 5 \\ \times 9 \\ \hline ? \end{array}$  $5\overline{)45}$

**18.** $\begin{array}{r} 3 \\ \times 4 \\ \hline ? \end{array}$  $3\overline{)12}$

**19.** How can knowing the related multiplication fact help you complete a division sentence? Give an example.

*Math Journal* ✓

**PROBLEM SOLVING**

**20.** There are 20 tennis players in all. Five players are on each team. How many teams are there?

**21.** Laura has 40 tennis balls. She puts 5 balls into each box. How many boxes does she use?

 **Critical Thinking**

**22.** Twenty stickers are shared equally among 6 friends. How many stickers does each receive? Are they able to share all the stickers? Draw a picture to explain your answer.

# Zero and One in Division

Here are some special words to learn.

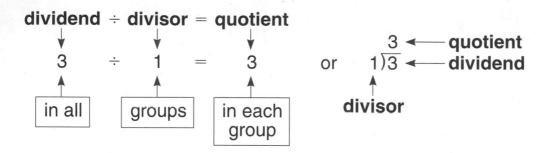

**dividend ÷ divisor = quotient**

3 ÷ 1 = 3     or     $1\overline{)3}$

quotient
dividend
divisor

in all     groups     in each group

The **dividend** is the number you divide.
The **divisor** is the number you divide by.
The **quotient** is the answer.

Read $3 \div 1 = 3$ as: "Three divided by one equals three."

▶ You can model $3 \div 1 = 3$ using counters.

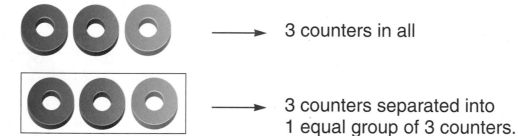

⟶ 3 counters in all

⟶ 3 counters separated into
1 equal group of 3 counters.

▶ You can model $3 \div 3 = 1$ using counters.

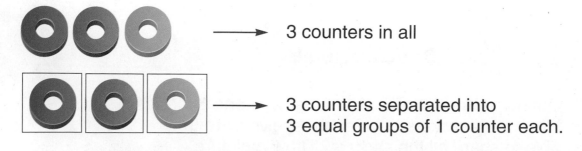

⟶ 3 counters in all

⟶ 3 counters separated into
3 equal groups of 1 counter each.

**Use counters or draw dots to find each quotient.**

**1.** $4 \div 1 = \underline{\ ?\ }$

**2.** $7 \div 1 = \underline{\ ?\ }$

**3.** $1\overline{)9}$ with $?$ above

**4.** $2 \div 1 = \underline{\ ?\ }$

**5.** $1\overline{)6}$ with $?$ above

**6.** $8 \div 1 = \underline{\ ?\ }$

**7.** Look at exercises 1–6. What do you notice about the dividends, divisors, and quotients?

*Communicate* ✓

**8.** Write a rule for dividing any number by 1.

**Use counters or draw dots to find each quotient.**

**9.** $4 \div 4 = \underline{\ ?\ }$

**10.** $6\overline{)6}$ with $?$ above

**11.** $8 \div 8 = \underline{\ ?\ }$

**12.** $1 \div 1 = \underline{\ ?\ }$

**13.** $5 \div 5 = \underline{\ ?\ }$

**14.** $7\overline{)7}$ with $?$ above

**15.** Look at exercises 9–14. What do you notice?

*Communicate* ✓

**16.** Write a rule for dividing any number by itself.

---

### Zero in Division

$0 \div 3 = 0$ or $3\overline{)0}$ with $0$ above

When you divide 0 by any number, the answer is 0.

---

**17.** $0 \div 4 = \underline{\ ?\ }$

**18.** $0 \div 2 = \underline{\ ?\ }$

**19.** $0 \div 1 = \underline{\ ?\ }$

**20.** $5\overline{)0}$ with $?$ above

---

**Skills to Remember**   **Multiply.**

**21.** $3 \times 2$

**22.** $7 \times 5$

**23.** $8 \times 3$

**24.** $4 \times 0$

**25.** $4 \times 4$

**26.** $9 \times 2$

**27.** $5 \times 1$

**28.** $9 \times 5$

## 5-4 Dividing by 2

Algebra ✓

You can use counters or skip counting
to help you divide by 2.

### Hands-On Understanding

**You Will Need:** counters, record sheet

**Step 1**    Label the columns on your record
sheet with these headings:

| Counters in all | Counters in each group | Number of groups | Division sentence<br>$\underline{\,?\,} \div \underline{\,?\,} = \underline{\,?\,}$ | Multiplication sentence<br>$\underline{\,?\,} \times \underline{\,?\,} = \underline{\,?\,}$ |
|---|---|---|---|---|
|  |  |  |  |  |

**Step 2**    Model 18 counters.
What is the total number of counters?

**Step 3**    Now make as many groups of 2 as you can.

How many counters are in each group.

How many groups did you make?

Write a division sentence to represent
your model.

Now write a multiplication sentence
to represent your model.

**Step 4**    Repeat Steps 2 and 3 for these groups:

| Counters in all |
|---|
| 16 |
| 14 |
| 12 |
| 10 |
| 8 |
| 6 |
| 4 |
| 2 |

**You can also skip count to help you divide by 2.**

Step 5 Count back by 2s. Start at 8 and stop at 0.

How many 2s are in 8?

Write a division sentence to show how many 2s are in 8.

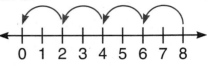

Step 6 Count back by 2s. Start at 12 and stop at 0.

How many 2s are in 12?

Write a division sentence to show how many 2s are in 12.

## Communicate

Discuss ✓

1. Describe what happens when you divide a number of objects by 2.

2. How can skip counting help you find the quotient in a division sentence?

**Write a division sentence for each model.**

3. ⊙⊙⊙⊙⊙⊙⊙⊙⊙⊙
   ⊙⊙⊙⊙⊙⊙⊙⊙⊙⊙

4. ⊙⊙⊙⊙⊙⊙⊙⊙⊙
   ⊙⊙⊙⊙⊙⊙⊙⊙⊙

5. ⊙⊙⊙⊙⊙⊙⊙
   ⊙⊙⊙⊙⊙⊙⊙

**Find the quotient. Skip count or use counters to help you.**

6. $4 \div 2 =$ ___?   7. $8 \div 2 =$ ___?   8. $10 \div 2 =$ ___?   9. $2 \div 2 =$ ___?

10. $2\overline{)18}$   11. $2\overline{)14}$   12. $2\overline{)6}$   13. $2\overline{)0}$   14. $2\overline{)16}$

**Copy and complete the table.**
**Describe the pattern you see.**

Communicate ✓

15.

| $0 \div 2$ | $2 \div 2$ | $4 \div 2$ | $6 \div 2$ | $8 \div 2$ | $10 \div 2$ | $12 \div 2$ | $14 \div 2$ | $16 \div 2$ | $18 \div 2$ |
|---|---|---|---|---|---|---|---|---|---|
| ? | ? | ? | ? | ? | ? | ? | ? | ? | ? |

171

**Dividing by 3**

An art class has 12 paintings. The students hang the paintings in 3 equal groups. How many paintings are in each group?

▶ To find how many paintings are in each group, divide: $12 \div 3 =$ _?_

| **dividend** | $\div$ | **divisor** | = | **quotient** |
|---|---|---|---|---|
| 12 | $\div$ | 3 | = | _?_ |
| ↑ | | ↑ | | ↑ |
| in all | | groups | | in each |

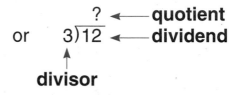

or     $3\overline{)12}$     ?← **quotient**     dividend

**divisor**

▶ You can skip count to find how many are in each group.

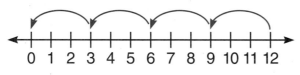

0  1  2  3  4  5  6  7  8  9  10  11  12

Think: What number times 3 equals 12?
$$\underline{?} \times 3 = 12$$
$$4 \times 3 = 12$$

So $12 \div 3 = 4$     or     $3\overline{)12}^{\,4}$ .

There are 4 paintings in each group.

---

**Write a division sentence for each model.**

1. $9 \div 3 =$ _?_

2. $6 \div 3 =$ _?_

3. $0 \div 3 =$ _?_

4. 

5. 

6.

**Write a division sentence for each model.**

7. ○○○○○○○○
   ○○○○○○○○
   ○○○○○○○○

8. ○○○○○○○
   ○○○○○○○
   ○○○○○○○

9. ○○○○○○○○○
   ○○○○○○○○○
   ○○○○○○○○○

**Find the quotient. Skip count or use counters to help you.**

10. $6 \div 3$     11. $24 \div 3$     12. $3 \div 3$     13. $12 \div 3$     14. $18 \div 3$

15. $27 \div 3$    16. $21 \div 3$    17. $15 \div 3$    18. $0 \div 3$     19. $9 \div 3$

20. $3\overline{)6}$    21. $3\overline{)12}$    22. $3\overline{)0}$    23. $3\overline{)3}$    24. $3\overline{)27}$

25. $3\overline{)18}$   26. $3\overline{)24}$   27. $3\overline{)21}$   28. $3\overline{)15}$   29. $3\overline{)9}$

**Copy and complete the table.**
**Describe the pattern you see.**

*Communicate* ✓

30.

| $0 \div 3$ | $3 \div 3$ | $6 \div 3$ | $9 \div 3$ | $12 \div 3$ | $15 \div 3$ | $18 \div 3$ | $21 \div 3$ | $24 \div 3$ | $27 \div 3$ |
|---|---|---|---|---|---|---|---|---|---|
| ? | ? | ? | ? | ? | ? | ? | ? | ? | ? |

## PROBLEM SOLVING

31. The art class needs 18 tubes of paint. The tubes come in packs of 3. How many packs does the class need?

32. Pablo draws a comic strip. The strip has 12 pictures. The pictures are in 3 rows. The same number of pictures are in each row. How many pictures are in each row?

## Share Your Thinking

33. Look at problem 32. Which words were most important in solving the problem? Why?

**Dividing by 4**

Dick puts 28 shells on a tray. He puts 4 shells in each row. How many rows are there?

▶ To find how many rows,
divide: $28 \div 4 = \underline{\ ?\ }$    or    $4\overline{)28}^{\ ?}$

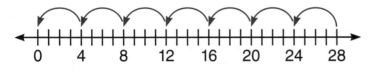

Think: $\underline{\ ?\ } \times 4 = 28 \longrightarrow 7 \times 4 = 28$

So $28 \div 4 = 7$. There are 7 rows of shells.

The 28 shells are divided into 4 equal rows. How many shells are there in each row?

▶ To find how many in each row,
divide: $28 \div 4 = \underline{\ ?\ }$    or    $4\overline{)28}$

You can also skip count to find how many are in each row.

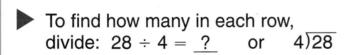

Think: $4 \times \underline{\ ?\ } = 28$
$4 \times 7 = 28$

So $28 \div 4 = 7$. There are 7 shells in each row.

---

**Write a division sentence for each model.**

1.

$4 \div 4 = \underline{\ ?\ }$

2.

$8 \div 4 = \underline{\ ?\ }$

3.

$0 \div 4 = \underline{\ ?\ }$

4.

5.

6.

**Find the quotient. Skip count or use counters to help.**

**7.** $8 \div 4$      **8.** $12 \div 4$      **9.** $4 \div 4$      **10.** $20 \div 4$      **11.** $24 \div 4$

**12.** $0 \div 4$      **13.** $28 \div 4$      **14.** $32 \div 4$      **15.** $36 \div 4$      **16.** $16 \div 4$

**17.** $4\overline{)8}$      **18.** $4\overline{)20}$      **19.** $4\overline{)4}$      **20.** $4\overline{)28}$      **21.** $4\overline{)32}$

**22.** $4\overline{)12}$      **23.** $4\overline{)0}$      **24.** $4\overline{)24}$      **25.** $4\overline{)16}$      **26.** $4\overline{)36}$

**Copy and complete the table.**
**Describe the pattern you see.**

*Communicate*

**27.**

| $0 \div 4$ | $4 \div 4$ | $8 \div 4$ | $12 \div 4$ | $16 \div 4$ | $20 \div 4$ | $24 \div 4$ | $28 \div 4$ | $32 \div 4$ | $36 \div 4$ |
|---|---|---|---|---|---|---|---|---|---|
| ? | ? | ? | ? | ? | ? | ? | ? | ? | ? |

## PROBLEM SOLVING

**28.** Thirty-six shells were collected by 4 students. Each student collected the same number of shells. How many shells did each student collect?

**29.** Mary, Dan, Fay, and Bob want to share 16 cookies. At most, how many cookies should each child receive?

**30.** There are 12 children playing horseshoes. There are 4 teams with the same number of children on each team. How many children are on each team?

**31.** If you buy 4 shirts, the store gives you another shirt for free. You buy 8 shirts. How many shirts will you take home?

## Share Your Thinking

**Without finding the quotients, tell which quotient will be the greatest.**

**32.** $6 \div 1 = \underline{\ ?\ }$      $6 \div 2 = \underline{\ ?\ }$      $6 \div 3 = \underline{\ ?\ }$

**33.** $12 \div 2 = \underline{\ ?\ }$      $12 \div 3 = \underline{\ ?\ }$      $12 \div 4 = \underline{\ ?\ }$

**Dividing by 5**

Terry has 25 books. She puts the same number of books on each of 5 shelves. At most, how many books does she put on each shelf?

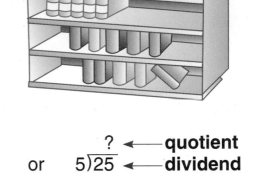

▶ To find how many books on each shelf, divide: $25 \div 5 = \underline{?}$

**dividend ÷ divisor = quotient**

$$25 \quad \div \quad 5 \quad = \quad \underline{?}$$

| in all | groups | in each |

or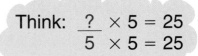

$$5)\overline{25}$$

? ⟵ **quotient**
⟵ **dividend**

**divisor**

▶ You can skip count to find how many books are on each shelf.

Think: $\underline{?} \times 5 = 25$
$5 \times 5 = 25$

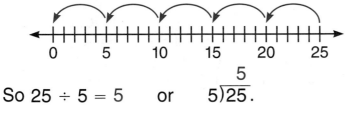

So $25 \div 5 = 5$ or $5)\overline{25}$ with quotient $5$.

She puts 5 books on each shelf.

---

**Write a division sentence for each model.**

**1.** ⬭⬤⬤⬤⬤⬤⬭

$5 \div 5 = \underline{?}$

**2.** ⬭⬤⬤⬤⬤⬤⬭
⬭⬤⬤⬤⬤⬤⬭

$10 \div 5 = \underline{?}$

**3.**

$0 \div 5 = \underline{?}$

**4.** ⬭⬤⬤⬤⬤⬤⬭
⬭⬤⬤⬤⬤⬤⬭
⬭⬤⬤⬤⬤⬤⬭

**5.** ⬭⬤⬤⬤⬤⬤⬭⬤⬤⬤⬤⬤⬭
⬭⬤⬤⬤⬤⬤⬭⬤⬤⬤⬤⬤⬭
⬭⬤⬤⬤⬤⬤⬭⬤⬤⬤⬤⬤⬭

**6.**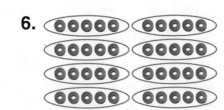

**Divide. Skip count or use counters to help.**

**7.** 40 ÷ 5      **8.** 30 ÷ 5      **9.** 5 ÷ 5      **10.** 35 ÷ 5      **11.** 45 ÷ 5

**12.** 25 ÷ 5      **13.** 20 ÷ 5      **14.** 15 ÷ 5      **15.** 10 ÷ 5      **16.** 0 ÷ 5

**17.** 5)30̄      **18.** 5)20̄      **19.** 5)35̄      **20.** 5)25̄      **21.** 5)10̄

**22.** 5)45̄      **23.** 5)40̄      **24.** 5)5̄      **25.** 5)0̄      **26.** 5)15̄

**Copy and complete the table.
Describe the pattern you see.**

Communicate

**27.**

| 0 ÷ 5 | 5 ÷ 5 | 10 ÷ 5 | 15 ÷ 5 | 20 ÷ 5 | 25 ÷ 5 | 30 ÷ 5 | 35 ÷ 5 | 40 ÷ 5 | 45 ÷ 5 |
|---|---|---|---|---|---|---|---|---|---|
| ? | ? | ? | ? | ? | ? | ? | ? | ? | ? |

## PROBLEM SOLVING

**28.** A book has 45 pages. Each chapter is 5 pages long. How many chapters are there in the book?

**29.** Marnie buys 30 wood shelves. The store ties them up in bundles of 5. How many bundles are there?

 **Number Sense**

**30.** How many 5s are in 25?

**31.** Use repeated subtraction to prove that there are 5 fives in 25.

**32.** How is subtraction like division?

**33.** How is subtraction different than division?

## 5-8 Dividing Cents

Daniela has 15 pennies. She gives the pennies to her 3 sisters. Each of them receives the same amount. At most, how much money will each sister get?

To find the amount of money:

- First divide.

- Then write the ¢ sign.

$$15¢ \quad ÷ \quad 3 \quad = \quad 5¢ \quad \text{or} \quad 3\overline{)15¢}^{\,5¢}$$

| cents in all | number of groups | cents in each group |

Think: $3 × \underline{\ ?\ } = 15¢$
$3 × \underline{5¢} = 15¢$

Each sister will get 5¢.

**Divide.** Remember to write the ¢ sign.

1.

$12¢ ÷ 4 = \underline{\ ?\ }$

2.

$8¢ ÷ 2 = \underline{\ ?\ }$

**3.** 10¢ ÷ 5          **4.** 4¢ ÷ 2          **5.** 9¢ ÷ 3          **6.** 20¢ ÷ 4

**7.** 9¢ ÷ 1          **8.** 30¢ ÷ 5          **9.** 24¢ ÷ 3          **10.** 0¢ ÷ 4

**11.** $2\overline{)6¢}$          **12.** $4\overline{)28¢}$          **13.** $3\overline{)21¢}$          **14.** $5\overline{)45¢}$

**Divide mentally.**

**15.** $10 \div 2$     **16.** $4¢ \div 4$     **17.** $6 \div 3$     **18.** $15¢ \div 5$

**19.** $8 \div 1$     **20.** $12 \div 2$     **21.** $9¢ \div 3$     **22.** $8¢ \div 4$

**23.** $2\overline{)14}$     **24.** $5\overline{)5}$     **25.** $4\overline{)16¢}$     **26.** $5\overline{)20¢}$

**Multiply or divide.**

**27.** $1 \times 2$     **28.** $4 \times 3$     **29.** $6 \times 4$     **30.** $8 \times 1$

**31.** $5 \times 5$     **32.** $0 \times 5$     **33.** $7 \times 3¢$     **34.** $2 \times 4¢$

**35.** $\begin{array}{r} 3 \\ \times 9 \\ \hline \end{array}$     **36.** $\begin{array}{r} 1 \\ \times 7 \\ \hline \end{array}$     **37.** $\begin{array}{r} 5¢ \\ \times 8 \\ \hline \end{array}$     **38.** $\begin{array}{r} 2¢ \\ \times 9 \\ \hline \end{array}$

**39.** $18 \div 3$     **40.** $28 \div 4$     **41.** $10 \div 5$     **42.** $16 \div 2$

**43.** $9 \div 3$     **44.** $35 \div 5$     **45.** $6¢ \div 1$     **46.** $36¢ \div 4$

**47.** $2\overline{)12}$     **48.** $4\overline{)24}$     **49.** $5\overline{)40¢}$     **50.** $3\overline{)27¢}$

## PROBLEM SOLVING

**51.** There are 20 bikes. The same number of bikes are in each of 4 racks. At most, how many bikes are in each rack?

**52.** Francis rents his bike to Ann for 5 hours. Ann pays the same amount for each hour. The total bill is 35¢. How much money does Francis get for each hour?

 **Connections: Language Arts**     Communicate ✓

**53.** Write a problem that can be solved using the division sentence $36¢ \div 4 =$ _?_ . Then explain how you can use coins or skip counting on a number line to solve it.

# 5-9 Problem Solving: Write a Number Sentence

Algebra

| Number Sentence | Definition |
|---|---|
| □ + □ = □ | Join sets, or quantities. |
| □ − □ = □ | Separate, or take away, from a set.<br>Compare two sets, or quantities.<br>Find part of a set.<br>Find how many more are needed. |
| □ × □ = □ | Join only equal sets, or quantities. |
| □ ÷ □ = □ | Separate a set into equal groups.<br>Share a set equally. |

**Problem:** There are 20 smiley faces in rows on the computer screen. Each row has 4 faces. How many rows of faces are there on the screen?

**1 IMAGINE**    Look at the faces on the screen.

**2 NAME**

*Facts:*    20 smiley faces in all
4 faces in each row

*Question:*    How many rows are there?

**3 THINK**    To find how many rows there are, divide.

$20 \div 4 = \underline{\ ?\ }$ ⟵—— Division number sentence

**4 COMPUTE**   
$$\begin{array}{r} 5 \\ 4\overline{)20} \end{array}$$
There are 5 rows of smiley faces.

**5 CHECK**    Multiply to check.    or    Count the rows of faces.

$5 \times 4 = 20$          5 rows of 4 = 20

**Write a number sentence to solve each problem.**

1. There are 6 baseball teams in the league. Each team has 2 pitchers. How many pitchers are there in the league?

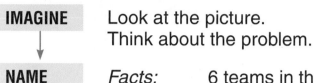

| | |
|---|---|
| **IMAGINE** | Look at the picture. Think about the problem. |
| **NAME** | *Facts:* 6 teams in the league<br>2 pitchers on each team<br><br>*Question:* How many pitchers are there in the league? |
| **THINK** | Look at the picture.<br>To find the number of pitchers in all, multiply: $6 \times 2 =$ _?_ |

**COMPUTE** ⟶ **CHECK**

2. Sixty-eight people were watching a soccer game. Twenty-nine of them were children. How many were *not* children?

3. The coach bought a baseball mitt, a jersey, and a batting helmet. What was the total cost?

$17.68

4. The coach bought 7 boxes of baseballs. Each box held 4 baseballs. How many baseballs did the coach buy?

$19.98

5. The visiting team scored 98 points. The home team scored 9 points more than the visiting team. How many points did the home team score?

$7.39

6. The snack-bar manager assigned 27 workers equally over 3 workstations. How many workers did she put at each?

# 5-10 Problem-Solving Applications

**Solve each problem and explain the method you used.**

1. Thirty students are playing a math game. The students are grouped equally on 5 teams. At most, how many students are on each team?

2. A team of 4 students had a high score of 36 points. Each student scored the same number of points. How many points did each student score?

Imagine

3. Ms. Doyle has 35 students in her class. The same number of students sit in each of 5 rows. How many students sit in each row?

Name

4. There are 4 tables in the art room. Five students are at each table. How many students are in the art room?

Think

5. The art teacher has 18 markers. She gives an equal number of them to each of 3 students. At most, how many markers does each student get?

Compute

6. Mr. Ross gave 1 book to each of 28 students in his class. The students are arranged in 4 equal rows. How many students are in each row?

Check

7. A box of books weighs 18 pounds. Each of the books in the box weighs 2 pounds. How many books are in the box?

8. Lenore spent 32¢ for 4 identical sheets of colored paper. How much did each sheet cost?

**Choose a strategy from the list or use another strategy you know to solve each problem.**

Use these strategies:

9. Mr. Ross graded 3 papers in 27 minutes. He spent the same amount of time on each paper. How much time did he take on 1 paper?

10. Two numbers have a sum of 9 and a product of 14. What are the numbers?

11. The art teacher paired off 16 students. How many pairs did she make?

12. Each morning Ms. Doyle gives award stickers to the same number of students. On Monday she had 32 stickers. On Tuesday she had 28 stickers. On Wednesday she had 24 stickers. On what day will she have 16 stickers?

13. Tara buys 5 sheets of art paper for 3¢ each and 4 sheets of parchment for 5¢ each. How much does she spend in all?

14. The science class has 12 fish, 2 fish tanks, and 1 ant farm. The students put the same number of fish in each tank. How many fish are in each tank?

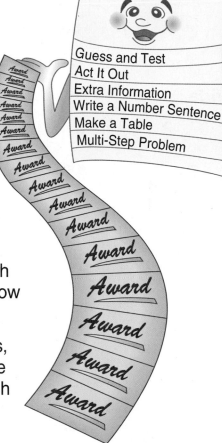

Guess and Test
Act It Out
Extra Information
Write a Number Sentence
Make a Table
Multi-Step Problem

**Use the graph for problems 15 and 16.**

15. The art class used 4 jars of blue paint and 4 jars of green paint. How many jars of each color are now left?

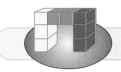

**Make Up Your Own**

16. Use the graph. Write a problem. Then solve it.

**Art Supplies**

| Color | Number of Jars |
|-------|----------------|
| blue | 🏺🏺🏺🏺🏺 |
| red | 🏺🏺🏺 |
| green | 🏺🏺🏺🏺 |
| white | 🏺 |
| black | 🏺🏺 |
| Key: Each 🏺 = 3 jars. | |

## Write a division sentence for each model.  *(See pp. 164–165, 168–179.)*

1.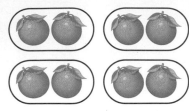

2.

## Divide.

3. 8 ÷ 2      4. 6 ÷ 3      5. 16 ÷ 4      6. 10¢ ÷ 5      7. 14 ÷ 2

8. 12 ÷ 4      9. 15¢ ÷ 3      10. 18¢ ÷ 3      11. 24 ÷ 4      12. 25 ÷ 5

13. 2)‾4‾      14. 4)‾8‾      15. 5)‾15‾      16. 3)‾21¢‾      17. 4)‾36¢‾

18. 5)‾20‾      19. 4)‾32¢‾      20. 2)‾2‾      21. 4)‾20‾      22. 1)‾0‾

## Complete each multiplication and division sentence.  *(See pp. 166–167.)*

23. 5 × 2 = __?__      24. 4 × 3 = __?__      25. 6 × 4 = __?__      26. 6 × 5 = __?__

      10 ÷ 2 = __?__            12 ÷ 3 = __?__            24 ÷ 4 = __?__            30 ÷ 5 = __?__

27.   5                28.   4                29.   4                30.   3
    ×5  5)‾25‾              ×8  4)‾32‾              ×3  4)‾12‾              ×7  3)‾21‾

## PROBLEM SOLVING  *(See pp. 180–183.)*
### Write a number sentence to solve.

31. The bookstore is open 7 days a week for 5 hours each day. How many hours is the bookstore open in one week?

32. Pam bought 5 of the same pens for a total of 45¢. How much did each pen cost?

*(See Still More Practice, pp. 446–447.)*

## CHAIN OPERATIONS

▶ **Compute from left to right.**          **Compute from left to right.**

$2 + 4 - 3 = \underline{\ ?\ }$   Think: $2 + 4 = 6$          $16 \div 4 \times 2 = \underline{\ ?\ }$   Think: $16 \div 4 = 4$

$\phantom{2+4}6 \quad - 3 = 3$          $\phantom{16\div4}4 \quad \times 2 = 8$

**Compute.**

**1.** $6 + 8 - 3$          **2.** $18 - 3 + 2$          **3.** $9 \times 2 - 3$

**4.** $7 \times 3 + 1$          **5.** $24 \div 4 \times 2$          **6.** $35 \div 5 \times 3$

**7.** $12 \div 3 + 9$          **8.** $16 \div 2 - 8$          **9.** $4 \times 5 - 9$

**10.** $8 \times 3 + 6$          **11.** $45 \div 5 - 7$          **12.** $27 \div 3 + 6$

**13.** $6 + 8 + 2 - 4$          **14.** $10 - 2 - 3 + 5$          **15.** $3 \times 3 \times 2 - 7$

**16.** $2 \times 3 \times 1 + 5$          **17.** $12 \div 3 \div 2 - 1$          **18.** $8 \div 2 \div 2 + 1$

▶ **Compare.**     $\underbrace{8 \times 3}_{24} \ \underline{\ ?\ } \ \underbrace{6 \times 5}_{30}$
$\phantom{xxxxxxxxx}24 \quad < \quad 30$

**Copy and compare. Write $<$, $=$, or $>$.**

**19.** $2 \times 5 + 8 \ \underline{\ ?\ } \ 3 \times 4 - 3$          **20.** $14 \div 2 + 3 \ \underline{\ ?\ } \ 16 \div 4 + 6$

**21.** two plus twelve $\underline{\ ?\ }$ fifteen minus three

**22.** four plus zero $\underline{\ ?\ }$ ten minus five

**23.** three times two plus seven $\underline{\ ?\ }$ twelve divided by three plus nine

**24.** twenty-four divided by four minus two $\underline{\ ?\ }$ three times five minus ten

# Check Your Mastery

## Performance Assessment

**Use drawings to model each meaning of division.**

**1.** sharing: $15 \div 5 = \underline{\ ?\ }$       **2.** separating: $12 \div 4 = \underline{\ ?\ }$

**Divide.**

**3.** $9 \div 3$    **4.** $0 \div 1$    **5.** $16 \div 4$    **6.** $10¢ \div 2$    **7.** $8¢ \div 4$

**8.** $27 \div 3$    **9.** $18 \div 2$    **10.** $32¢ \div 4$    **11.** $30¢ \div 5$    **12.** $3\overline{)24}$

**13.** $5\overline{)40¢}$    **14.** $4\overline{)24¢}$    **15.** $2\overline{)14¢}$    **16.** $3\overline{)18}$    **17.** $5\overline{)0}$

**18.** $5\overline{)25¢}$    **19.** $2\overline{)4}$    **20.** $5\overline{)45}$    **21.** $2\overline{)16¢}$    **22.** $4\overline{)4¢}$

**Complete each multiplication and division sentence.**

**23.** $7 \times 4 = \underline{\ ?\ }$      **24.** $9 \times 4¢ = \underline{\ ?\ }$      **25.** $6 \times 2¢ = \underline{\ ?\ }$

      $28 \div 4 = \underline{\ ?\ }$        $36¢ \div 4 = \underline{\ ?\ }$        $12¢ \div 2 = \underline{\ ?\ }$

**26.**   $\begin{array}{r} 5 \\ \times 3 \\ \hline \end{array}$   $3\overline{)15}$      **27.**   $\begin{array}{r} 5¢ \\ \times 7 \\ \hline \end{array}$   $7\overline{)35¢}$      **28.**   $\begin{array}{r} 5¢ \\ \times 4 \\ \hline \end{array}$   $4\overline{)20¢}$

**PROBLEM SOLVING**    *Use a strategy you have learned.*

**29.** Jill had 32 books. She put 4 books in each box. How many boxes did she use?

**30.** There are 14 strawberries. They are shared equally by 2 girls. What is the greatest number each girl can get?

**31.** Bill bought 24 pencils. He gave each of his friends 3. How many friends received pencils?

**32.** Adam spent $27 for tickets. He bought 3 tickets. How much did each ticket cost?

**33.** Aunt Eva paid $40 for books. She bought 5 books. Each book cost the same amount. How much did each book cost?

# Cumulative Review II

**Choose the best answer.**

**1.** Choose the standard form of the number.

$9,000 + 500 + 30 + 7$

   **a.** 9,537
   **b.** 9,573
   **c.** 14,037
   **d.** 90,537

**2.** Round to the nearest thousand.

7091

   **a.** 6,000
   **b.** 7,000
   **c.** 7,090
   **d.** 7,100

**3.** What is the amount?

2 ten-dollar bills,
1 five-dollar bill,
6 nickels

   **a.** $15.30
   **b.** $25.30
   **c.** $25.60
   **d.** $30.30

**4.** Estimate.

$5.09 + $4.51

   **a.** $10.00
   **b.** $9.00
   **c.** $11.00
   **d.** $9.60

**5.**
$$\begin{array}{r} 4347 \\ + 1686 \end{array}$$

   **a.** 5033
   **b.** 5343
   **c.** 6033
   **d.** 5933

**6.** The quotient is 9.
The divisor is 4.
What is the dividend?

   **a.** 13
   **b.** 30
   **c.** 36
   **d.** 32

**7.** Estimate.
$$\begin{array}{r} 886 \\ - 321 \end{array}$$

   **a.** 600   **b.** 800
   **c.** 900   **d.** 200

**8.**
$$\begin{array}{r} \$28.20 \\ - 19.85 \end{array}$$

   **a.** $8.35   **b.** $9.35
   **c.** $11.65   **d.** $48.05

**9.**
$$\begin{array}{r} 3 \\ \times 9 \end{array}$$

   **a.** 6   **b.** 12
   **c.** 27   **d.** 18

**10.** $8 \times 5¢$

   **a.** 12¢   **b.** 32¢
   **c.** 40¢   **d.** 13¢

**11.** $5 \times 1$

   **a.** 0   **b.** 6
   **c.** 1   **d.** 5

**12.** $4\overline{)28}$

   **a.** 7   **b.** 14
   **c.** 32   **d.** 8

**13.** $2\overline{)2}$

   **a.** 0   **b.** 2
   **c.** 4   **d.** 1

**14.** $0 \div 5$

   **a.** 0   **b.** 1
   **c.** 5   **d.** 2

**15.** $3\overline{)27¢}$

   **a.** 6¢   **b.** 8¢
   **c.** 9¢   **d.** 7¢

**16.** $\underline{\phantom{?}}\ \div 1 = 1$

   **a.** 0   **b.** 1
   **c.** 2   **d.** not given

## Ongoing Assessment II

### For Your Portfolio

Solve each problem. Explain the steps and the strategy or strategies you used for each. Then choose one from problems 1–4 for your Portfolio.

1. Gil's teacher has 35 base ten blocks for her class. She has given 7 blocks to each of 4 groups of students. How many more blocks does she need to give out?

2. Mr. Winters found 40 marbles in a box. He gave the same number of marbles to each of the 5 boys. At most, how many marbles did he give to each boy?

3. Blanca gives the cashier $25. About how much does she spend if the cashier gives her $3.79 change?

4. Saul drinks 3 glasses of milk every day. How many glasses of milk does he drink a week?

### Tell about it.

5. In problem 3, what tells you that you need to estimate to find the answer? Explain how you estimated the answer.

Communicate

6. What is the hidden information in problem 4?

---

### For Rubric Scoring

**Listen for information on how your work will be scored.**

7. Study the multiplication fact. Explain how you would find the missing factor. Draw a picture to show this multiplication fact.

$$\square \times 2 = 6 \qquad \begin{array}{r} 2 \\ \times\,\square \\ \hline 6 \end{array}$$

8. Study the division fact. Explain how you would find the missing divisor. Draw a picture to show this division fact.

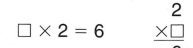

$$28 \div \square = 7 \qquad \square\overline{)28}^{\,7}$$

# Statistics and Probability

## FIVE CENT BALLOONS

Pietro has twenty red and blue balloons on a string.
They flutter and dance pulling Pietro's arm.
A nickel apiece is what they sell for.

Wishing children tag Pietro's heels.

He sells out and goes the streets alone.

*Carl Sandburg*

### In this chapter you will:
Collect, organize, and display data on graphs
Study arrangements and combinations
Conduct probability experiments
Learn about computer programs
Solve problems by using a graph

### Critical Thinking/Finding Together
If you were to choose a balloon from the boy's left hand without looking, would you have an equally likely chance of choosing a red or a blue balloon? Explain. What if you chose a balloon from his right hand?

# 6-1 Collecting Data

Ms. Avillo asked her third-grade class to name their favorite type of book. She recorded her results in a **tally chart**. Which type was the most favorite? Which type was the least favorite?

Count the tallies to find the totals.

| Type of Book | Tally | Total |
|---|---|---|
| Mystery | ⅢⅡ | 5 |
| Adventure | ⅢⅡ II | 7 |
| Science Fiction | IIII | 4 |
| Biography | III | 3 |
| Nature | ⅢⅡ I | 6 |
| Drama | IIII | 4 |

Remember: I stands for 1 student.
ⅢⅡ stands for 5 students.

The most favorite type is the book with the highest total.

The least favorite type is the book with the lowest total.

Adventure books are favored the most.

Biography books are favored the least.

## PROBLEM SOLVING

This tally chart shows the number of items to be recycled that Dorothy collected in one weekend.

**Use the tally chart to the right to answer each question.**

| Items To Be Recycled | Tally | Total |
|---|---|---|
| Plastic bottles | ⅢⅡ ⅢⅡ I | ? |
| Aluminum cans | ⅢⅡ ⅢⅡ ⅢⅡ | ? |
| Glass bottles | ⅢⅡ I | ? |
| Cardboard boxes | ⅢⅡ III | ? |
| Newspapers | ⅢⅡ ⅢⅡ IIII | ? |

1. Copy and complete the tally chart.

2. What items were collected to be recycled?

3. Which item did Dorothy collect the most of?

4. Which item did Dorothy collect the fewest of?

5. How many items did Dorothy collect altogether?

6. Dorothy's sister helped her collect 6 more aluminum cans. How many cans does Dorothy have now?

This list shows favorite after-school activities.

| | | |
|---|---|---|
| Mark — football | Chris — volleyball | Anne Marie — soccer |
| Janice — soccer | Joanne — track | Anthony — football |
| Kara — track | David — basketball | Ben — basketball |
| Vinny — soccer | Roberto — football | Julio — football |
| Jonathan — football | Stephen — football | Jennifer — volleyball |
| Lucy — volleyball | Alexandra — soccer | Donato — football |
| Patel — soccer | Jeanelle — volleyball | Grace — basketball |
| Kim — volleyball | Michael — football | James — volleyball |

**Use the list above to answer each question.**

7. Use the list to make a tally chart.

8. What activities are listed in your tally chart?

9. Which activity is the most favorite?

10. Which activity is the second favorite?

11. Which activity is the least favorite?

12. How many people were surveyed altogether?

## Share Your Thinking

Math Journal

13. Write 2 or 3 sentences to describe the data in your tally chart in exercise 7.

14. Explain in your Math Journal some advantages of representing your data in a tally chart instead of a list.

15. Would you use the list or the tally chart to answer the questions below? Explain why you made your choice.

   a. How many more students like football than basketball?

   b. Did more boys or girls choose volleyball as their favorite after-school activity?

# 6-2 | Making Pictographs

A **graph** shows information. Pictures or symbols are used to represent numbers in a **pictograph**.

Tawana made a pictograph from this tally chart, which shows the colors of her friends' bicycles.

| Color | Tally | Total |
|-------|-------|-------|
| Blue | ЖН ЖН II | 12 |
| Purple | ЖН III | 8 |
| Silver | ЖН ЖН | 10 |
| Pink | III | 3 |

▶ Tawana used these steps to make a pictograph.

- List each color.

- Choose a picture or symbol to represent the number of friends for each color.

- Choose a key. Let ☺ = 2 friends.

- Draw pictures to represent the total number of friends for each color.

- Label the pictograph. Write the title and the key.

| Colors of Bicycles | |
|---|---|
| Blue | ☺ ☺ ☺ ☺ ☺ ☺ |
| Purple | ☺ ☺ ☺ ☺ |
| Silver | ☺ ☺ ☺ ☺ ☺ |
| Pink | ☺ ◖ |
| Key: Each ☺ = 2 friends. | |

If ☺ = 2 friends, then ◖ = 1 friend.

How many of Tawana's friends have pink bicycles?

▶ To find how many, count the number of pictures for the color pink. Then use the key.

There are 1 and $\frac{1}{2}$ pictures for the color pink.

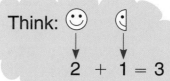

Think: ☺ ◖
2 + 1 = 3

Three of Tawana's friends have pink bicycles.

1. Use the tally chart at the right to make a pictograph.

**PROBLEM SOLVING**
**Use your pictograph to answer each question.**

2. How many stamps are represented by each symbol?

3. How many stamps does Stacy have from each country?

4. From which country does Stacy have the most stamps?

5. From which country does Stacy have the least stamps?

6. How many stamps does Stacy have in all?

7. How many more stamps does Stacy have from Brazil than from Canada?

8. List the countries from the one with the least number of stamps to the one with the most.

**Stacy's Stamp Collection**

| Country | Tally | Total |
|---------|-------|-------|
| Italy | ЖТ ЖТ | 10 |
| Ghana | ЖТ ЖТ ЖТ ЖТ ЖТ | 25 |
| France | ЖТ ЖТ ЖТ ЖТ | 20 |
| Canada | ЖТ | 5 |
| Brazil | ЖТ ЖТ ЖТ | 15 |

**Use each tally chart to make a pictograph. Then write 3 or 4 sentences describing the information you can read from your pictograph.**

Communicate

9.

| Favorite Lunch | Tally |
|----------------|-------|
| Hamburger | IIII |
| Macaroni | ЖТ I |
| Taco | II |
| Grilled cheese | ЖТ III |

10.

| Favorite TV Show | Tally |
|------------------|-------|
| Comedy | ЖТ ЖТ |
| Sports | ЖТ ЖТ ЖТ |
| Nature | ЖТ |
| Quiz | ЖТ ЖТ ЖТ ЖТ |

 **Project**

11. Take a survey to see how many of your classmates' birthdays are in the winter, spring, summer, and fall. Record your data in a tally chart. Then make a pictograph to show your results.

193

# 6-3 Making Bar Graphs

A **bar graph** can be used to report or compare data.

Neil surveyed students in his school to find out the kinds of fruit they eat at lunch. This tally chart shows the results of his survey.

| Fruit | Tally | Total |
|-------|-------|-------|
| Pears | IIII | 4 |
| Bananas | ⊥⊦⊦⊤ ⊥⊦⊦⊤ II | 12 |
| Apples | ⊥⊦⊦⊤ ⊥⊦⊦⊤ ⊥⊦⊦⊤ | 15 |
| Oranges | ⊥⊦⊦⊤ ⊥⊦⊦⊤ | 10 |

▶ Neil used these steps to show his results on a bar graph.

- List each kind of fruit.

- Use the data from the tally chart to make an appropriate scale.

- Draw bars to represent the number of students for each fruit.

- Label the bar graph.

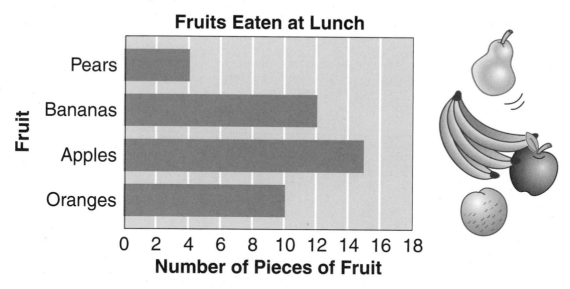

**Fruits Eaten at Lunch**

How many students eat apples at lunch?

▶ To find how many, read the number on the scale at the end of the bar for apples.

The bar for apples is *halfway* between 14 and 16.

So 15 students eat apples at lunch.

1. Use the tally chart at the right to make a bar graph.

**PROBLEM SOLVING**
**Use your bar graph to answer these questions.**

| Items Sold | Tally | Total |
|---|---|---|
| Folders | JHT II | 7 |
| Pencils | JHT JHT II | 12 |
| Pens | JHT JHT JHT | 15 |
| Markers | JHT III | 8 |

2. How many pencils were sold?

3. How many pens were sold?

4. Were more pencils or markers sold?

5. How many more markers were sold than folders?

6. How many more pens were sold than markers?

7. How many items were sold in all?

**Use each chart to make a bar graph. Then write 3 or 4 sentences describing the information you can read from your bar graph.**

Communicate ✓

8.

| Items in Stock | Tally |
|---|---|
| Sleds | JHT JHT JHT |
| Tents | JHT JHT |
| Bikes | JHT JHT JHT JHT |
| Treadmills | JHT |

9.

| Week | Distance Run |
|---|---|
| 1 | 4 km |
| 2 | 8 km |
| 3 | 13 km |
| 4 | 16 km |

## Skills to Remember

10. List at least 2 other coin combinations that would make the same amount of change shown.

## 6-4 | Arrangements and Combinations

Mark and Varel have a set of
4 geometric shapes: a circle, a triangle,
a rectangle, and a square.

How many ways can they arrange
the 4 shapes *in a line* so that the
rectangle and triangle are *not* side
by side?

To find how many ways, make an organized list
of the possible arrangements.

> An organized list can help you see
> if all possibilities have been tried
> and that no solution is repeated.

Here is part of the list they made.

| Line | Shapes | | | |
|------|--------|--------|--------|--------|
| 1 | circle | triangle | square | rectangle |
| 2 | circle | rectangle | square | triangle |
| 3 | triangle | circle | square | rectangle |
| 4 | triangle | circle | rectangle | square |
| 5 | triangle | square | circle | rectangle |
| 6 | triangle | square | rectangle | circle |

## PROBLEM SOLVING
### Make an organized list of arrangements.

1. Complete the list above. How
   many ways can Mark and Varel
   arrange the shapes?

2. How many ways can you arrange
   in a line a circle, a rectangle, and
   2 different-sized squares if the
   squares are *not* side by side?

**Make an organized list of arrangements for each.**

**3.** How many ways can you arrange in a line a red square, a green triangle, a red circle, and a green rectangle if the shapes of the same color are *not* side by side?

**4.** How many ways can you arrange in a line a circle, a square, a rectangle, and 2 triangles if you always begin with the triangles?

## Tree Diagrams

Nicky has a red, a yellow, and a green sweatshirt.
He also has a pair of blue and a pair of black jeans.
How many different outfits can Nicky make?

To find how many outfits, make an organized list or
a **tree diagram**.
A tree diagram shows different combinations.

| **Tree Diagram** | **Organized List** |
|---|---|
| | |

**Tree Diagram**

blue jeans
- red sweatshirt
- yellow sweatshirt
- green sweatshirt

black jeans
- red sweatshirt
- yellow sweatshirt
- green sweatshirt

**Organized List**

blue jeans, red sweatshirt
blue jeans, yellow sweatshirt
blue jeans, green sweatshirt

black jeans, red sweatshirt
black jeans, yellow sweatshirt
black jeans, green sweatshirt

Nicky can make 6 different outfits.

**Make an organized list and a tree diagram for each.**

**5.** Catherine has a brown, a plaid, and a black skirt. She also has a white, a red, and a tan sweater. How many different outfits can Catherine make?

**6.** Dean makes a sandwich with 1 slice of meat and 1 slice of cheese. He has ham, bologna, and turkey. He also has white and yellow cheese. How many different sandwiches can he make?

# Probability Experiments

**Probability** is the chance that something will happen.

## Hands-On Understanding

**You Will Need:** 2 spinners, record sheet, crayons

**Step 1**  Make a spinner like the one shown here.

What are the possible colors the spinner can land on?

Is it **possible** or **impossible** that the spinner will land on purple?

Is it **certain** that the spinner will land on red, blue, green, or yellow?

Is there an **equally likely** chance of landing on each color? Explain your answer.

Since there is 1 red section out of a total of 4 sections, the probability of the spinner landing on red is **1 out of 4**.

**Step 2**  Spin the spinner 10 times and record your results in a table or bar graph.

Write one or two sentences that describe the data.

**Step 3**  Predict which color the spinner will land on if you spin the spinner again.

**Step 4**  Test your prediction. Record your results in a table or bar graph.

Compare your results with your prediction.

| Step 5 | Now make a spinner like the one shown here. |
|---|---|

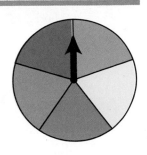

Is there an equally likely chance of landing on each color? Explain your answer.

> The probability of landing on green is **2 out of 5** (2 green out of 5 in all).

Predict which color the spinner will land on the most.

Predict how many times the spinner will land on this color if you spin the spinner 10 times.

| Step 6 | Spin the spinner 10 times and record your results. |
|---|---|

Compare your results with your predictions.

## Communicate

Discuss ✓

1. What does it mean if situations are equally likely to occur? not equally likely? certain? impossible?

**Write *equally likely* or *not equally likely* for the probability of each color occurring.**

2.

3.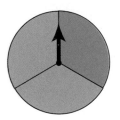

4.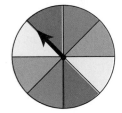

5. Find the probability of landing on green for each spinner above.

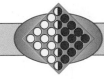 **Finding Together**

6. Toss a coin 10 times. Record and describe your data. Predict how many times the coin will land on heads or tails if you toss the coin 20 times. 30 times. Test your predictions. Compare your results with your predictions. What do you notice?

# TECHNOLOGY

## Computer Programs

Lori is making a pictograph to show the pencils the school store sold in September. How many pictures should Lori use for red pencils?

To find how many pictures, you can write a **computer program**.

| Color Pencil | Tally | Total |
|---|---|---|
| Blue | ЖТ IIII | 9 |
| Gold | ЖТ I | 6 |
| Red | ЖТ ЖТ ЖТ | 15 |
| Green | ЖТ ЖТ II | 12 |

A computer program is a set of statements that tells the computer what to do.

This computer program allows Lori to enter or **input** information to determine how many pictures are needed for the color red.

| Colors of Pencils | |
|---|---|
| Blue | ✏ ✏ ✏ |
| Gold | ✏ ✏ |
| Red | |
| Green | ✏ ✏ ✏ ✏ |
| Key: Each ✏ = 3 pencils. | |

Numbers tell the order of the steps the computer will follow.

Remember: Press the [ENTER↵] key after each line of instruction.

```
10    PRINT    "How many pencils does each picture stand for?"
20    INPUT    P
30    PRINT    "How many red pencils were sold?"
40    INPUT    S
50    PRINT    S/P; "pictures are needed for the color red."
60    END
      RUN
```

means ÷

A command that tells the computer to process the information.

**Output:** How many pencils does each picture stand for?
?3 ◄

How many red pencils were sold?
?15 ◄

Press [ENTER↵] to continue the program.

5 pictures are needed for the color red.

**Match each word with the correct definition.**

1. computer program

    **a.** the results of a computer program

2. INPUT

    **b.** a command that tells the computer to process information

3. RUN

    **c.** a set of statements that tells the computer what to do

4. OUTPUT

    **d.** information entered into the computer

**Use the program below to answer each question.**

```
50  PRINT "How many green balloons does the clown have?"
60  INPUT G
30  PRINT "How many blue balloons does the clown have?"
40  INPUT B
10  PRINT "How many red balloons does the clown have?"
80  END
20  INPUT R
70  PRINT "The total number of balloons is "; G + B + R
```

5. How many lines of instruction does this program have?

6. What command would you enter after line 80 to process the information?

7. Which line will the computer process first? last?

8. What is the output for line 10? line 30? line 50?

9. What number would you enter for line 20? line 40? line 60?

10. What is the output for line 70?

11. If the clown received 2 yellow balloons, write a PRINT and an INPUT statement to follow the program.

12. How would line 70 change if yellow balloons were added to the program?

## 6-7 Problem Solving: Use a Graph

**Problem:** Joe made a pictograph to display the number of coins from each country in his coin collection. How many coins does Joe have?

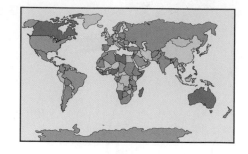

| 1 IMAGINE | Look at the graph. Think about the problem. |
|---|---|

| Country | Number of Coins |
|---|---|
| Brazil | ⊙ ⊙ ⊙ |
| Canada | ⊙ ⊙ ⊙ ⊙ |
| Germany | ⊙ ⊙ |
| Ireland | ⊙ ⊙ ⊙ ⊙ |
| Jamaica | ⊙ ⊙ ⊙ ⊙ ◖ |
| Key: Each ⊙ = 4 coins. Each ◖ = 2 coins. | |

**2 NAME**

*Facts:*  12 coins from Brazil
16 coins each from Canada and Ireland
8 coins from Germany
18 coins from Jamaica

*Question:* How many coins does Joe have?

**3 THINK**  First count on to find the number of coins from each country.
Then add to find the total number of coins.

**4 COMPUTE**

**Coins from Each Country**

Brazil    — 3 ⊙ = 12
Canada   — 4 ⊙ = 16
Germany — 2 ⊙ =  8
Ireland    — 4 ⊙ = 16
Jamaica  — 4 ⊙ and 1 ◖ = 18

Joe has 70 coins from different countries.

**Total Number**

$$
\begin{array}{r}
3 \\
12 \\
16 \\
8 \\
16 \\
+18 \\
\hline
70 \text{ coins}
\end{array}
$$

10

**5 CHECK**  Count on by 4 and by 2 to total the coins on the graph.

**Use the graph to solve each problem.**

1. This bar graph shows the results
of a survey of the states visited
by students during their vacations.
How many students were
surveyed? What states had
more than 15 visiting students?

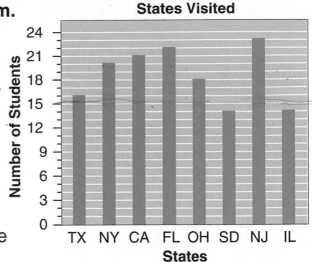

**States Visited**

| IMAGINE | Look at the graph. |

| NAME | *Facts:* Shown on the bar graph |

*Questions:* How many students were surveyed?
What states had more than 15 visiting students?

| THINK | Add the number of students that visited the eight states.
List the states that had more than 15 student visitors. |

COMPUTE → CHECK

2. Use this pictograph. How many more
citizens prefer summer than winter?
What seasons are enjoyed by less than
50 citizens? Survey your classmates for
the season they prefer.

3. Use the key symbol to increase the
number of citizens who prefer
autumn by 15.

| Season | Citizens |
|--------|----------|
| Spring | x x x x x x |
| Summer | x x x x x x / |
| Autumn | x x x / |
| Winter | x x x x / |
| Key: Each x = 10 citizens. | |
| Each / = 5 citizens. | |

**Connections: History**

*Discuss* ✓

4. Research information on the various colors used in
state flags. Organize your data in a tally chart, then
make a pictograph or a bar graph to show your results.
Compare your data with that of your classmates.

# Problem-Solving Applications

**Solve each problem and explain the method you used. Use the bar graph for problems 1–4.**

Imagine

Name

Think

Compute

Check

1. What types of air travel did the following numbers of students like?

   a. 40 students
   b. 45 students
   c. 15 students
   d. *about* 30 students

2. What types of air travel did more than 30 students like?

3. What type of air travel is liked by the least number surveyed?

4. How many more students liked the space shuttle than the helicopter?

**Favorite Air Travel**

Type

Jet
Helicopter
Hot-Air Balloon
Blimp
Space Shuttle

10 20 30 40 50
**Number Surveyed**

**Use the pictograph for problems 5–10.**

5. In what two months were the same number of cards sold?

6. How many cards were sold in November and December?

7. In what month was the greatest number of cards sold?

8. In what months were between 200 and 500 cards sold?

9. The number of cards sold in May was double the number sold in April. How many cards were sold in May?

| Months | Cards Sold |
|--------|-----------|
| Apr. | ▯ ▯ ▭ |
| Mar. | ▯ ▯ ▯ |
| Feb. | ▯ ▯ ▯ ▯ ▯ ▭ |
| Jan. | ▯ ▯ |
| Dec. | ▯ ▯ ▯ ▯ ▯ ▯ ▯ |
| Nov. | ▯ ▯ ▯ |

Key: Each ▯ = 100 cards.
Each ▭ = 50 cards.

10. The store sold about 450 cards in June. What key symbols would be used on the pictograph for 450 cards?

**Choose a strategy from the list or use another strategy you know to solve each problem.**

Use these strategies:

Logical Reasoning
Multi-Step Problem
Write a Number Sentence
Make an Organized List
Use a Graph

**11.** Margo has 3 posters: a dancer, a skater, and a cheerleader. In how many different ways can she arrange the posters left to right on her wall?

**Use bags 1 and 2 for problems 12–14.**

**12.** From which bag is Terri more likely to pull a red marble?

**13.** From which bag is Terri less likely to pull a blue marble?

**14.** From which bag does Terri have an equally likely chance of pulling a red or a blue marble?

**Use the bar graph for problems 15–18.**

**15.** How many more students liked black rather than white sneakers?

**16.** What sneaker color(s) were liked by between 16 and 22 students?

**17.** How many students were surveyed for their favorite sneaker color?

**18.** What sneaker color was preferred by 2 more than twice the number that preferred white?

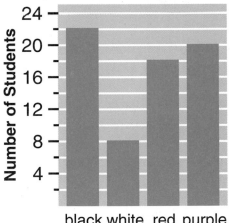

**Favorite Sneakers**

Number of Students: 4, 8, 12, 16, 20, 24

Color: black, white, red, purple

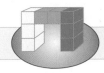

## Make Up Your Own

**19.** Use the data from the tally chart to make a graph. Then write a problem for a classmate to solve.

**Future Careers**

| | |
|---|---|
| Medicine | JHT |
| Firefighter | JHT JHT I |
| Sports | JHT JHT IIII |
| Teacher | JHT I |
| Artist | JHT IIII |

1. Beverly has a black and a white T-shirt. She also has a pair of navy and a pair of gray sweatpants. Make an organized list or tree diagram to find how many different outfits Beverly can make.

*(See pp. 196–199.)*

2. Use the spinner. Which has a greater chance of occurring, an odd number or an even number? Explain your answer. What is the probability of landing on an odd number?

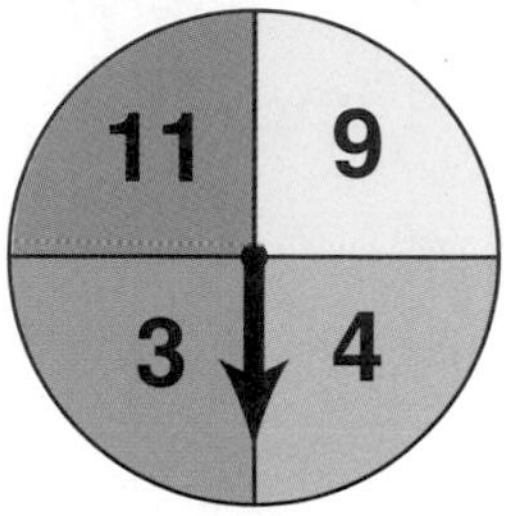

**PROBLEM SOLVING**

*(See pp. 190–195, 202–205.)*

3. Complete the tally chart at the right. Then make a pictograph.

| Baseball Team | Tally of Wins | Total Wins |
|---|---|---|
| Gold |卌 卌 卌 | ? |
| White | 卌 I | ? |
| Green | 卌 卌 卌 卌 I | ? |
| Blue | 卌 卌 II | ? |

**Use your pictograph to answer problems 4–6.**

4. How many games did the Blue team win?

5. Which team won exactly 15 games?

6. Write 2 sentences explaining the data displayed in your graph.

7. Complete the tally chart at the right. Then make a bar graph.

| TVs Repaired by | Tally of Repairs | Total Repairs |
|---|---|---|
| Len | 卌 卌 II | ? |
| Adia | 卌 IIII | ? |
| Hoy | 卌 I | ? |
| Tim | 卌 卌 IIII | ? |

**Use your bar graph to answer problems 8 and 9.**

8. Which bar on your graph is the shortest? Why?

9. How many TVs were repaired altogether?

*(See Still More Practice, p. 447.)*

## AVERAGES

An **average** is a quotient found by dividing the sum of a group of addends by the number of addends.

A group of students made a bar graph to show the amount of time each student spends watching television. What is the average amount of time spent watching television?

To find the average amount of time, first add the total number of hours. Then divide by the number of students.

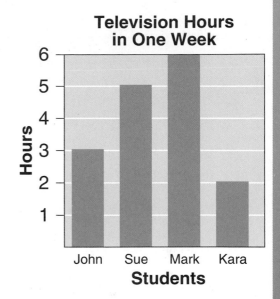

**Television Hours in One Week**

**Add:**
```
  3
  5
  6
+ 2
 16
```

**Divide:**
```
      4  ◄──── average number
  4)1 6  ◄──── of hours
  -1 6
     0
```
number of students

total number of hours

The group spends an average of 4 hours watching television.

## PROBLEM SOLVING

1. Alexa bought games costing $7, $12, and $8. What is the average cost per game?

2. Scott hiked 6 mi, 8 mi, 12 mi, and 10 mi over a 4-day period. What is the average number of miles per day he hiked?

3. On Monday Jerry filled 4 bags of leaves. On Tuesday he filled 3 bags and on Wednesday he filled 5 bags of leaves. What is the average number of bags Jerry fills with leaves each day?

4. During a gymnastic tournament Jenna received the following scores: 10, 9, 10, 8, 8. What is her average score?

## Performance Assessment

**Use the tally chart for exercises 1–3.**

1. Complete the tally chart.

2. Use the tally chart to make a pictograph.

3. Write 2 sentences describing the data from your pictograph.

| Favorite Musical Instruments | | Total |
|---|---|---|
| Flute | ЖI | ? |
| Guitar | ЖI ЖI ЖI III | ? |
| Violin | ЖI ЖI I | ? |
| Drums | ЖI ЖI ЖI II | ? |

4. In the spinner at the right, does each color have an equally likely chance of occurring? Explain why or why not.

5. Suppose you spin the spinner twice. List all the different color combinations you can get.

6. Find the probability of landing on red.

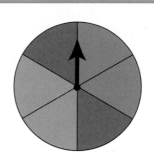

**PROBLEM SOLVING**   *Use a strategy you have learned.*

7. Use the tally chart at the right to make a bar graph.

8. Which bar is twice the length of the bar for fiction?

9. In the tally chart at the right, how many students preferred baseball? basketball or football?

10. Which sport was the least favorite?

11. Which sport was the favorite of exactly 15 students?

| Favorite Books | |
|---|---|
| Fiction | ЖI ЖI |
| Biography | ЖI |
| Sports | ЖI ЖI ЖI |
| Animals | ЖI ЖI ЖI ЖI |

| Favorite Sports | |
|---|---|
| Basketball | ЖI ЖI ЖI ЖI |
| Baseball | ЖI ЖI ЖI II |
| Tennis | ЖI ЖI ЖI |
| Football | ЖI II |

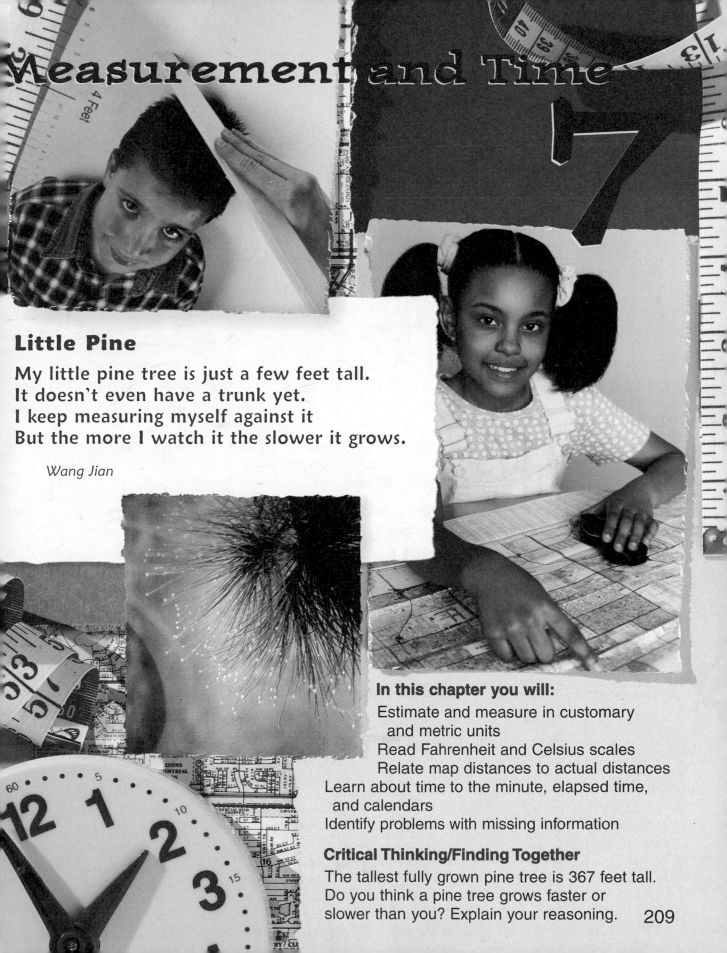

# Measurement and Time

## 17

## Little Pine

My little pine tree is just a few feet tall.
It doesn't even have a trunk yet.
I keep measuring myself against it
But the more I watch it the slower it grows.

*Wang Jian*

### In this chapter you will:

Estimate and measure in customary
  and metric units
Read Fahrenheit and Celsius scales
Relate map distances to actual distances
Learn about time to the minute, elapsed time,
  and calendars
Identify problems with missing information

### Critical Thinking/Finding Together

The tallest fully grown pine tree is 367 feet tall.
Do you think a pine tree grows faster or
slower than you? Explain your reasoning.

# 7-1 Quarter Inch, Half Inch, Inch

The **quarter inch**, **half inch**, and **inch** are customary units used to measure length.

Look at the markings on the ruler.

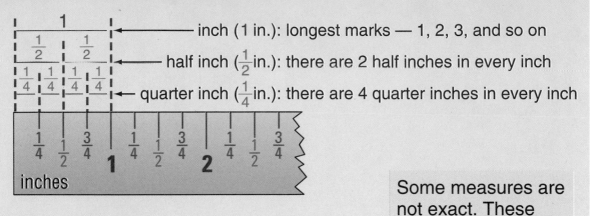

inch (1 in.): longest marks — 1, 2, 3, and so on

half inch ($\frac{1}{2}$ in.): there are 2 half inches in every inch

quarter inch ($\frac{1}{4}$ in.): there are 4 quarter inches in every inch

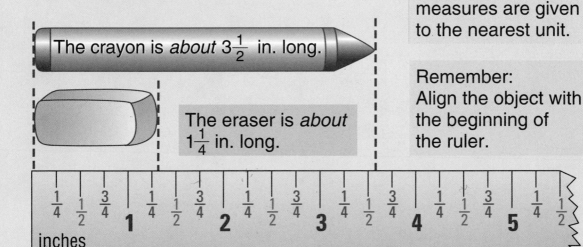

The crayon is *about* $3\frac{1}{2}$ in. long.

The eraser is *about* $1\frac{1}{4}$ in. long.

Some measures are not exact. These measures are given to the nearest unit.

Remember: Align the object with the beginning of the ruler.

## Hands-On Understanding

**You Will Need:** inch ruler, record sheet

**Step 1**   Label the columns of your record sheet with these headings: Object, Nearest Quarter Inch, Nearest Half Inch, Nearest Inch.

**Step 2**    Choose five objects from your classroom to measure to the nearest quarter inch, half inch, and inch.

Write the names of these objects on your record sheet.

**Step 3**    Align the end of one of the objects with the beginning of the ruler.

**Step 4**    Look at the other end of the object.

Record the length of the object to the nearest quarter inch.

Record the length of the object to the nearest half inch.

Record the length of the object to the nearest inch.

**Step 5**    Repeat Steps 3 and 4 for the remainder of the objects listed on your record sheet.

# Communicate

Discuss

1. When measuring an object to the nearest inch, how do you decide which is the correct measurement?

2. List the steps a classmate should use to measure an object's length to the nearest quarter inch.

3. How many $\frac{1}{2}$ in. units are between $1\frac{1}{2}$ in. and 3 in. on a ruler? How many $\frac{1}{4}$ in. units?

**Draw a line for each length.**

4. 3 in.

5. 7 in.

6. $3\frac{3}{4}$ in.

7. $6\frac{1}{2}$ in.

8. $8\frac{1}{4}$ in.

9. $10\frac{1}{4}$ in.

10. $4\frac{3}{4}$ in.

11. $12\frac{1}{2}$ in.

 **Share Your Thinking**

12. Explain how you used your ruler to draw the line in exercise 11.

# Foot, Yard

The **foot** and **yard** are also customary units used to measure length.

▶ The width of your hands spread is *about* 1 foot.

> 1 foot = 12 inches
> 1 ft = 12 in.

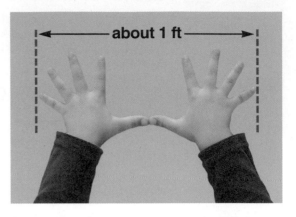
about 1 ft

▶ The distance from the tip of your nose to your fingertip is *about* 1 yard.

> 1 yard = 36 inches
> 36 in. = 3 ft
> So 1 yd = 3 ft

about 1 yd

The foot and the yard are used to measure large objects.

**Study these examples.**

about 1 ft

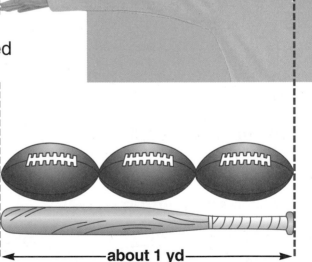
about 1 yd

**Which unit is used to measure each: in., ft, or yd?**

1. width of a book

2. height of a person

3. length of a crayon

4. length of a soccer field

**Write the letter of the best estimate.**

5. height of a glass       **a.** 8 in.     **b.** 80 in.     **c.** 8 yd

6. length of a football field    **a.** 10 ft     **b.** 100 ft     **c.** 100 yd

7. length of a truck       **a.** 5 ft     **b.** 20 ft     **c.** 20 yd

**Copy and complete the table.**

8.

| inches | 12 | 24 | ? | ? | 60 | ? |
|--------|----|----|----|----|----|----|
| feet | 1 | 2 | ? | 4 | ? | ? |
| yards | | | 1 | | | ? |

Think:
2 ft = 1 ft + 1 ft
2 ft = 12 in. + 12 in.
2 ft = 24 in.

9. Explain how you completed the table in exercise 8.

*Algebra* ✓

**Compare. Write <, =, or >.**

10. 3 in. _?_ 3 ft      11. 4 yd _?_ 4 ft      12. 1 ft _?_ 13 in.

13. 2 ft _?_ 1 yd      14. 6 ft _?_ 72 in.      15. 2 yd _?_ 5 ft

16. 2 yd _?_ 74 in.      17. 30 in. _?_ 1 yd      18. 108 in. _?_ 3 yd

## Project

**Copy and complete the chart. Use a yardstick or tape measure to find each length.**

| | Nearest foot | Nearest yard |
|---|---|---|
| 19. length of your desk | ? | ? |
| 20. length of the board | ? | ? |
| 21. length of the classroom | ? | ? |

## 7-3 Mile

The **mile** (**mi**) is a customary unit used to measure distance.

A mile is *about* how far you can walk in 25 minutes.

> 1 mi = 5280 ft
> 1 mi = 1760 yd

**Which unit is used to measure each: in., ft, yd, or mi?**

1. distance across town

2. length of a room

3. width of a quarter

4. distance to the Moon

**Write the letter of the best estimate.**

5. length of a kite's string     **a.** 30 in.    **b.** 30 yd   **c.** 30 mi

6. length of a river     **a.** 400 ft    **b.** 1 mi    **c.** 400 mi

7. distance from New York to Florida    **a.** 1200 ft   **b.** 50 mi   **c.** 1200 mi

**Compare. Write <, =, or >.**

8. 600 yd __?__ 1 mi      9. 2 mi __?__ 2000 yd      10. 5280 ft __?__ 2 mi

11. 1 mi __?__ 3000 ft      12. 1000 yd __?__ 1000 ft    13. 1760 ft __?__ 5280 yd

14. Explain the method you used to compare measurements in exercises 8–13.

*Communicate* ✓

15. A football field is 100 yards long. About how many times, up and back, would you have to walk to walk one mile? Explain your reasoning.

**Complete. Write *feet*, *yards*, or *miles*.**

**16.** The length of the school bus is about 40 ___ long.

**17.** Tommy walked about 4 ___ in 2 hours.

---

## Distance in Miles

To find *about* how many miles it is from Bluewater to Little Town, and from Little Town to Boatwood, estimate the distance.

$103 \text{ mi} + 228 \text{ mi} = \underline{\ ?\ }$

$100 \text{ mi} + 200 \text{ mi} = 300 \text{ mi}$

The distance is about 300 miles.

---

**PROBLEM SOLVING** Use the map above.

**18.** About how many miles is Bluewater from Fintail?

**19.** Which distance is shorter: from Riverfield to Little Town or from Bluewater to Deepsea?

**20.** What is the shortest distance from Fintail to Deepsea?

**21.** What is the shortest distance from Boatwood to Riverfield?

**22.** The Troy family drove from Little Town to Deepsea. They went through 2 cities. About how many miles did the Troy family drive?

**23.** The state is building two roads: one directly from Bluewater to Boatwood, the other directly from Little Town to Deepsea. Which will be longer?

**24.** Describe a route that is about 300 miles. Explain how you calculated this route.

**25.** Describe a route that is about 500 miles. Explain how you calculated this route.

# 7-4  Measurement Sense

Roger and Ted are comparing their heights. Roger says that he is about 4 ft tall. Ted says that he is about $4\frac{1}{2}$ ft tall.

You can estimate to find *about* how long something is. Check your estimate by finding the actual measure.

**Tell which object is described.**

1. The height of this object is about 1 yd.

2. The width of this object is less than 3 in.

3. The height of this object is greater than 4 ft.

**Copy and complete. Estimate. Then use a ruler, a yardstick, or a tape measure to find the actual measure.**

| | Object | Estimate | Actual |
|---|---|---|---|
| 4. | length of a pen | ? | ? |
| 5. | height of a chair | ? | ? |
| 6. | height of a classmate | ? | ? |
| 7. | length of a window | ? | ? |
| 8. | length of two desks | ? | ? |

**Name two objects having each length.**

**9.** about 5 in.    **10.** about 5 ft    **11.** about 2 yd    **12.** about 2 ft

## Map Distance

Bryan drew a map of his neighborhood. He wanted to show how far he lives from the library.

Measure the **map distance**.

The distance on the map from Bryan's house to the library is 2 inches.

Use the scale to find the **actual distance**.

2 in. = 1 in. + 1 in.

2 in. = 1 mi + 1 mi

2 in. = 2 mi

The actual distance from Bryan's house to the library is 2 miles.

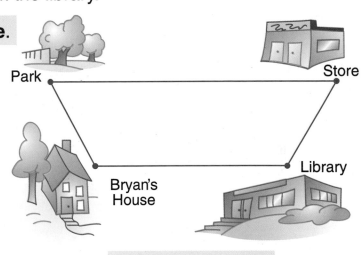

Scale: 1 in. = 1 mi

## PROBLEM SOLVING

**13.** What is the map distance between Bryan's house and the park?

**14.** What is the actual distance between Bryan's house and the park?

**15.** What is the actual distance between the park and the store?

## Critical Thinking

Communicate

**16.** How would each actual distance change if the scale is 1 in. = 2 mi?

## 7-5 Customary Units of Capacity

The cup, pint, quart, **half gallon**, and **gallon** are customary units used to measure liquids.

Remember: 1 pt = 2 c
1 qt = 2 pt

1 half gallon          1 gallon (gal)

 =

2 qt = 1 half gallon

4 qt = 1 gal

---

**Which unit is used to measure each: c, pt, qt, or gal?**

**1.** milk in a small glass

**2.** water in a bathtub

**3.** water in a large vase

**4.** juice in a small container

**Write the letter of the best estimate.**

**5.**

Ice Cream

**a.** 2 c
**b.** 1 qt
**c.** 1 half gallon

**6.**

**a.** 10 pt
**b.** 10 qt
**c.** 10 gal

218

**Complete.**

**7.** If 1 pt = 2 c, then

    **a.** 2 pt = _4_ c

    **b.** 3 pt = _?_ c

    **c.** 8 pt = _?_ c

Think:
2 pt = 1 pt + 1 pt
2 pt = 2 c + 2 c
2 pt = 4 c

**8.** If 1 qt = 2 pt, then

    **a.** 2 qt = _?_ pt

    **b.** 3 qt = _?_ pt

    **c.** 6 qt = _?_ pt

**9.** If 1 half gallon = 2 qt, then

    **a.** 2 half gallons = _?_ qt

    **b.** 3 half gallons = _?_ qt

    **c.** 4 half gallons = _?_ qt

**10.** If 1 gal = 4 qt, then

    **a.** 3 gal = _?_ qt

    **b.** 4 gal = _?_ qt

    **c.** 5 gal = _?_ qt

**Compare. Write <, =, or >.**

**11.** 3 c _?_ 1 pt

**12.** 2 pt _?_ 5 c

**13.** 2 pt _?_ 2 qt

**14.** 2 qt _?_ 4 pt

**15.** 3 qt _?_ 3 gal

**16.** 1 gal _?_ 3 qt

**17.** 1 half gallon _?_ 3 qt

**18.** 5 qt _?_ 2 half gallons

**19.** 4 c _?_ 1 qt

**20.** 2 qt _?_ 5 c

## PROBLEM SOLVING

**21.** Carolyn is making 4 milk shakes. She needs 2 c of milk for each. How many pints of milk does Carolyn need?

**22.** Joyce used 1 gal of apple juice for punch. She used 2 qt of orange juice. How much more apple juice than orange juice did she use?

**23.** Tony fills his 2-gal fish tank with water. He uses a container that holds 1 qt of water. How many times does Tony fill the container?

# Ounce, Pound

The **ounce** and **pound** are customary units used to measure weight.

▶ The ounce (**oz**) can be used to weigh small objects.

A slice of bread can be used as a benchmark for 1 ounce (1 oz). A slice of bread weighs *about* 1 ounce.

▶ The pound (**lb**) can be used to weigh large objects.

A loaf of bread can be used as a benchmark for 1 pound (1 lb). A loaf of bread weighs *about* 1 pound.

$$1 \text{ lb} = 16 \text{ oz}$$

## Which unit is used to measure each: oz or lb?

**1.** feather

**2.** chicken

**3.** pen

**4.** television

**5.** table-tennis ball

**6.** toaster

## Write the letter of the best estimate.

**7.** stick of butter     **a.** 4 oz     **b.** 4 lb     **c.** 8 oz

**8.** apple     **a.** 5 oz     **b.** 5 lb     **c.** 20 oz

**9.** dog     **a.** 50 oz     **b.** 5 lb     **c.** 50 lb

**10.** bag of potatoes     **a.** 10 oz     **b.** 1 lb     **c.** 10 lb

**Copy and complete the table.
Describe any pattern you see.**

**11.**

| ounces | 16 | 32 | 48 | 64 | ? | 96 | ? |
|--------|----|----|----|----|----|----|----|
| pounds | 1 | 2 | ? | ? | 5 | ? | 7 |

Think:
  1 lb = 16 oz
  2 lb = 16 oz + 16 oz
  2 lb = 32 oz

**Compare. Write <, =, or >.**

**12.** 3 lb _?_ 64 oz

**13.** 16 oz _?_ 2 lb

**14.** 15 oz _?_ 1 lb

**15.** 2 lb _?_ 33 oz

**16.** 96 oz _?_ 6 lb

**17.** 6 lb _?_ 80 oz

**18.** 5 lb _?_ 76 oz

**19.** 108 oz _?_ 7 lb

**20.** 110 oz _?_ 6 lb

**21.** 19 oz _?_ 2 lb

**22.** 4 lb _?_ 70 oz

**23.** 3 lb _?_ 45 oz

## PROBLEM SOLVING

**24.** A jar of nails weighs 2 lb. How much
do 2 jars of nails weigh?

**25.** Mike uses 8 oz of cheese for a cake.
How many pounds of cheese does
he need for 2 cakes?

**26.** Matt used 3 lb of bananas for 3 cakes.
How many ounces is this?

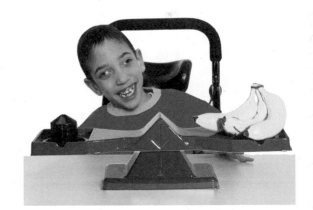

## Challenge

**27.** If an object weighs 30 oz, it is *about* _?_ lb.

**28.** If an object weighs 46 oz, it is *about* _?_ lb.

## 7-7 Metric Units of Length

The **centimeter** and **decimeter** are metric units used to measure length.

Remember: The width of your thumb is *about* 1 centimeter (1 cm).

A crayon is about 1 decimeter (1 dm) long.

1 decimeter = 10 centimeters
1 dm = 10 cm

You can use a crayon as a benchmark for 1 dm.

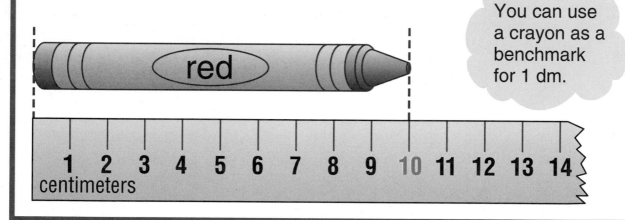

**Which unit is used to measure each: cm or dm?**

**1.** length of an eraser

**2.** length of a book

**3.** width of a cassette tape

**4.** length of a calculator

**Write the letter of the best estimate.**

**5.** length of a mosquito    **a.** 1 cm    **b.** 1 dm    **c.** 3 cm

**6.** length of a garden snake    **a.** 3 cm    **b.** 3 dm    **c.** 1 cm

**7.** length of a paper clip    **a.** 3 cm    **b.** 8 cm    **c.** 1 dm

**8.** length of a cassette tape    **a.** 4 cm    **b.** 4 dm    **c.** 1 dm

**Copy and complete the table.
Describe any pattern you see.**

9.

| centimeters | 10 | 20 | ? | 40 | 50 | ? | 70 |
|---|---|---|---|---|---|---|---|
| decimeters | 1 | 2 | 3 | ? | ? | 6 | ? |

Think:
2 dm = 1 dm + 1 dm
2 dm = 10 cm + 10 cm
2 dm = 20 cm

**Compare. Write <, =, or >.**

10. 4 dm __?__ 43 cm        11. 18 cm __?__ 1 dm        12. 30 cm __?__ 3 dm

13. 45 cm __?__ 5 dm        14. 4 dm __?__ 40 cm        15. 38 cm __?__ 4 dm

**Without using a ruler, draw a line segment for each
given length. Then measure to find the actual length.**

16. 1 cm                17. 5 cm                18. 1 dm                19. 3 dm

## Connections: Science

Did you know your heart is about the same size as your
fist? Your heart and fist grow at about the same rate.

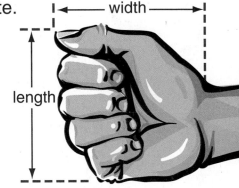
width
length

**Work with a partner.**

20. Cut a piece of string about 50 centimeters
long. Use your string and a centimeter ruler
to measure:
  **a.** the width of your fist.
  **b.** the length of your fist.
  **c.** the distance around your fist.
  **d.** the width, length, and distance around an adult's fist.

21. Estimate the size of your heart.

22. Compare your measurements and estimates
with classmates'. What do you notice?

23. How big do you think your fist and heart will be
in 10 years? Explain your reasoning.

Communicate

223

# Meter

A **meter** (**m**) is a metric unit used to measure long lengths.

A door can be used as a benchmark for 1 meter (1 m). A door is *about* 1 meter wide.

> 1 meter = 100 centimeters
>
> 1 m = 100 cm
>
> 1 meter = 10 decimeters
>
> 1 m = 10 dm

## Hands-On Understanding

**You Will Need:** meterstick, record sheet

**Step 1**  Label the columns of your record sheet with the following headings:

| Item | Nearest Meter |
|------|---------------|
| 1. | |
| 2. | |

**Step 2**  List the following items to measure under "Item" on your record sheet:

- length of the classroom
- width of the board
- length of a desk
- height of a classmate

| **Step 3** | Align the end of one of the items with the beginning of the meterstick. |
| **Step 4** | Look at the other end of the item.<br><br>What is the measure of the item to the nearest meter? |
| **Step 5** | Repeat Steps 3 and 4 for the remainder of the items on your record sheet. |

# Communicate

Math Journal ✓

1. List in your Math Journal the steps a classmate should use to measure an object's length to the nearest meter.

2. Explain why you would not want to use a meterstick to measure the thickness of your math book.

**Copy and complete the table.**
**Describe any pattern you see.**

Algebra ✓

3.

| centimeters | 100 | 200 | 300 | ? | ? |
|---|---|---|---|---|---|
| decimeters | 10 | ? | ? | 40 | ? |
| meters | 1 | 2 | ? | ? | ? |

Think:
1 m = 100 cm
2 m = 100 cm + 100 cm
2 m = 200 cm

**Compare. Write <, =, or >.**

4. 100 cm ? 1 m

5. 3 m ? 300 cm

6. 40 dm ? 3 m

7. 5 m ? 48 dm

8. 35 cm ? 2 dm

9. 6 dm ? 61 cm

10. 300 cm ? 2 m

11. 33 dm ? 3 m

12. 5 m ? 500 cm

## PROBLEM SOLVING

13. A car is 3 m long. A boat is 9 m long. Which is longer? How much longer?

14. In gym class Janet jumped 8 dm. Sharon jumped 1 m. Who jumped farther?

## 7-9 Kilometer

The **kilometer** (**km**) is a metric unit used to measure long distances.

A kilometer is *about* how far you can walk in 15 minutes.

150 km    Orlando
Tampa

| 1 kilometer = 1000 meters |
| 1 km = 1000 m |

Remember:   10 cm = 1 dm
            100 cm = 1 m
            10 dm = 1 m

---

**Which unit is used to measure each: cm, m, or km?**

**1.** width of a small picture

**2.** distance from home to school

**3.** length of a football field

**4.** length of a flashlight

**5.** distance between 2 cities

**6.** length of a truck

**Copy and complete the table.
Describe any pattern you see.**

*Communicate* ✓

**7.**

| meters | 1000 | 2000 | 3000 | ? | ? |
|---|---|---|---|---|---|
| kilometers | 1 | 2 | ? | 4 | ? |

Think:
1 km = 1000 m
2 km = 1000 m + 1000 m
2 km = 2000 m

**Compare. Write <, =, or >.**

**8.** 1 km  ?  1000 m

**9.** 100 m  ?  1 km

**10.** 100 cm  ?  2 m

**11.** 3 m  ?  300 cm

**12.** 3888 m  ?  4 km

**13.** 5 km  ?  4500 m

226

Niguel collects cans and newspapers for recycling. Here is a map of his city.

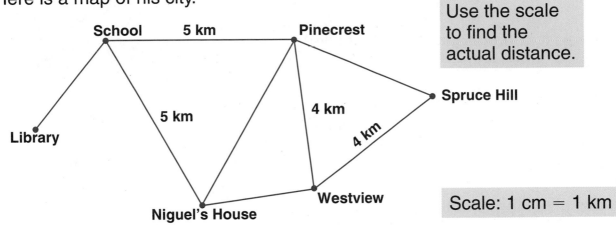

Use the scale to find the actual distance.

Scale: 1 cm = 1 km

## PROBLEM SOLVING
**For problems 14–21 use the scale and a centimeter ruler.**

14. Is the actual distance from Pinecrest to Spruce Hill about 2 km, about 4 km, or about 7 km?

15. Is the actual distance from Niguel's house to Westview longer or shorter than 5 km?

16. What is the actual distance from school to the library?

17. Niguel traveled from school to Pinecrest to Spruce Hill. About how many kilometers did Niguel travel in all?

18. What is the map distance from Niguel's house to Pinecrest?

19. What is the actual distance from Pinecrest to Spruce Hill?

20. On the way to school, Niguel stopped at Pinecrest. How many kilometers did he travel in all?

21. What is the shortest route from Spruce Hill to school? How long is the route?

## Challenge

22. Katie walked 2600 m on Saturday and 3200 m on Sunday. About how many kilometers did Katie walk altogether?

# Milliliter, Liter

The **milliliter** and **liter** are metric units used to measure the amount of liquid a container holds.

▶ The milliliter (**mL**) can be used to measure small amounts of liquid. There are about 20 drops of water in 1 mL.

▶ The liter (**L**) can be used to measure large amounts of liquid. The liter is the amount of liquid that fills about 4 glasses.

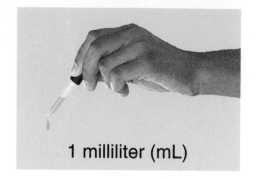

1 milliliter (mL)

1 liter = 1000 milliliters
1 L = 1000 mL

1 liter (L)

**Write *less than a liter* or *more than a liter* for the amount of liquid each real object holds.**

**1.**

glass of juice

**2.**

pool

**3.**

watering can

**4.**

jar of bubbles

**5.**

milk truck

**6.**

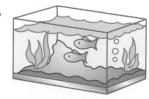

fish tank

**Which unit is used to measure each: mL or L?**

**7.** liquid in a test tube

**8.** water in a bucket

**9.** water in a bathroom cup

**10.** paint in a small can

**11.** bottle of detergent

**12.** fruit punch in a bowl

**Write the letter of the best estimate.**

**13.** water in a kitchen sink    **a.** 20 mL    **b.** 2 L    **c.** 20 L

**14.** water in a wading pool    **a.** 50 mL    **b.** 100 mL    **c.** 100 L

**15.** juice in a picnic jug    **a.** 3 mL    **b.** 3 L    **c.** 30 L

**Copy and complete the table.**
**Describe any pattern you see.**

*Algebra*

**16.**

| milliliters | 1000 | 2000 | 3000 | ? | ? |
|---|---|---|---|---|---|
| liters | 1 | 2 | ? | 4 | ? |

Think:
$1 L = 1000 mL$
$2 L = 1 L + 1 L$
$2 L = 1000 mL + 1000 mL$
$2 L = 2000 mL$

**Compare. Write $<$, $=$, or $>$.**

**17.** 1423 mL _?_ 1 L    **18.** 3 L _?_ 2500 mL    **19.** 3 L _?_ 3000 mL

**20.** 2 L _?_ 3500 mL    **21.** 4000 mL _?_ 4 L    **22.** 5 L _?_ 4500 mL

**23.** 5000 mL _?_ 6 L    **24.** 6 L _?_ 5100 mL    **25.** 4860 mL _?_ 5 L

**PROBLEM SOLVING**

**26.** Sam is having a party with 10 friends. Does he need 5 L or 5 mL of juice?

**27.** Laurie drinks about 50 mL of milk each day. About how much does she drink in 5 days?

# Gram, Kilogram

The **gram** and **kilogram** are metric units used to measure mass.

▶ The gram (**g**) can be used to weigh light objects. A small feather can be used as a benchmark for 1 gram (1 g). A small feather has a mass of *about* 1 gram (1 g).

1 gram (g)

▶ The kilogram (**kg**) can be used to weigh heavy objects. A textbook can be used as a benchmark for 1 kilogram. A textbook has a mass of *about* 1 kilogram (1 kg).

1 kilogram (kg)

> 1 kilogram = 1000 grams
> 1 kg = 1000 g

Write *more than a kilogram* or *less than a kilogram* for the mass of each real object.

**1.**

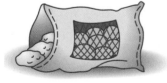

bag of potatoes

**2.**

grapes

**3.**

chicken

**4.**

cracker

**5.**

large fish

**6.**

muffin

**Which unit is used to measure each: g or kg?**

**7.** computer        **8.** comb        **9.** classmate

**10.** toothbrush      **11.** bowling ball      **12.** letter

**Write the letter of the best estimate.**

| | | | |
|---|---|---|---|
| **13.** fly | **a.** 2 g | **b** 200 g | **c.** 20 kg |
| **14.** horse | **a.** 30 g | **b.** 30 kg | **c.** 300 kg |
| **15.** dog | **a.** 25 g | **b.** 25 kg | **c.** 250 kg |

**Copy and complete the table.**
**Describe any pattern you see.**

*Algebra*

**16.**

| grams | 1000 | 2000 | 3000 | ? | ? |
|---|---|---|---|---|---|
| kilograms | 1 | 2 | ? | 4 | ? |

> Think:
> 1 kg = 1000 g
> 2 kg = 1 kg + 1 kg
> 2 kg = 1000 g + 1000 g
> 2 kg = 2000 g

**Compare. Write <, =, or >.**

**17.** 3000 g <u>?</u> 3 kg     **18.** 2500 g <u>?</u> 2 kg     **19.** 4 kg <u>?</u> 4100 g

**20.** 2900 g <u>?</u> 3 kg     **21.** 5000 g <u>?</u> 5 kg     **22.** 1500 g <u>?</u> 2000 kg

**PROBLEM SOLVING**

**23.** Jason's pet has a mass of 8 kg. June's pet has a mass of 8120 g. Whose pet weighs more? How much more?

### Challenge

**24.** Melissa has 4 pieces of chicken. They have a total mass of 675 g. Each of 3 of the pieces of chicken has a mass of 175 g. What is the mass of the fourth piece of chicken?

231

# Measuring Tools

Each **measuring tool** is used for a different purpose.

Jena is measuring different objects. Which tools can she use to measure each?

| Measuring Tools | |
| --- | --- |
| meterstick | yardstick |
| ruler (in.) | ruler (cm) |
| balance | scale |
| cup | liter |
| gallon | tape measure |

- To find the length of her bed, she can use a tape measure, a yardstick, or a meterstick.

- To find the mass of her dog, she can use a scale or a balance.

- To find how much water is in her fish tank, she can use a quart, a liter, or a gallon.

## Which tools could you use to measure each?

**1.** length of a classroom

**2.** weight of an apple

**3.** amount of juice in a jug

**4.** height of a doorway

**5.** amount of water in a birdbath

**6.** width of your thumbnail

**Match each object with the tool you would use to find each measure.**

**7.** length of a pool        **a.** centimeter ruler

**8.** pitcher of punch        **b.** inch ruler

**9.** width of a book        **c.** tape measure

**10.** thickness of a quarter        **d.** scale

**11.** weight of a fish        **e.** gallon

 **Project**

**Copy and complete the table for exercises 12–14.**

| Object | Estimate | Measuring Tool | Actual Measure |
|---|---|---|---|
| **1.** | | | |

**12.** Choose 3 objects in the classroom that have different weights. Estimate the weight or mass of each object. Arrange the objects in order from lightest to heaviest. Then find the actual measure of each.

**13.** Choose 3 objects in the classroom that have different lengths. Estimate the length of each object. Arrange the objects in order from shortest to longest. Then find the actual measure of each.

**14.** Choose 3 objects that hold different amounts of liquid. Estimate the amount of liquid each object will hold. Arrange the objects in order from the greatest amount to the least amount. Then find the actual measure of each.

**15.** Explain how it is possible for two objects of the same length to have different weights. Give an example.

*Communicate* ✓

233

# Fahrenheit Thermometer

A **thermometer** is used to measure temperature.

The **degree Fahrenheit (°F)** is used to measure temperature.

The temperature on a very cold day may be 2°F.

The temperature on a very hot day may be 96°F.

Each line on this thermometer stands for 2 degrees Fahrenheit.

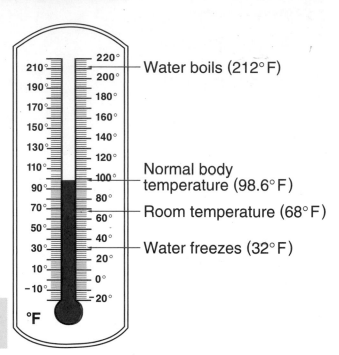

Water boils (212°F)

Normal body temperature (98.6°F)

Room temperature (68°F)

Water freezes (32°F)

°F

## Write the letter of the most reasonable temperature.

1. ice cube      **a.** 30°F      **b.** 60°F      **c.** 90°F

2. hot cocoa      **a.** 23°F      **b.** 48°F      **c.** 120°F

3. classroom      **a.** 12°F      **b.** 70°F      **c.** 109°F

## Write each temperature.

4.  °F

5.  °F

6.  °F

7. Record the temperature every day for one week. Display your results in a bar graph. What was the highest temperature? the lowest?

# Celsius Thermometer

The **degree Celsius** (°C) is the metric unit used to measure temperature.

The temperature on a very cold day may be ⁻10°C.

The temperature on a very hot day may be 30°C.

Each line on this thermometer stands for 1 degree Celsius.

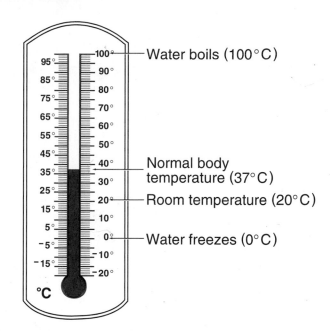

Water boils (100°C)

Normal body temperature (37°C)

Room temperature (20°C)

Water freezes (0°C)

°C

**Write the letter of the most reasonable temperature.**

1. frozen yogurt      **a.** 0°C      **b.** 20°C      **c.** 32°C

2. bowl of hot cereal      **a.** 45°C      **b.** 0°C      **c.** 75°C

3. warm bath      **a.** 90°C      **b.** 40°C      **c.** 65°C

**Write each temperature.**

**4.**

**5.**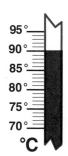

**6.**

7. Describe what the weather is like today. Estimate the temperature in degrees Celsius. Then ask your teacher what the actual temperature is. How close was your estimate?

Communicate

235

## 7-15 | Quarter Hour

Remember: 1 hour = 60 minutes
         1 h = 60 min

▶ There are 15 minutes in one quarter of an hour.

Read this time as: two fifteen,
                or quarter past two,
                or quarter after two.

Write in standard form as: 2:15.

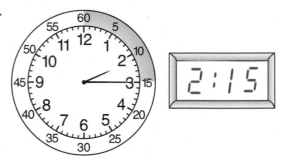

▶ There are 30 minutes in one half hour.

Read this time as: two thirty,
                or half past two,
                or thirty minutes after two.

Write in standard form as: 2:30.

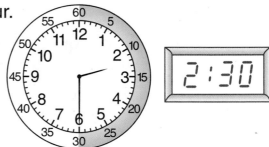

▶ There are 45 minutes in three quarters of an hour.

Read this time as: two forty-five,
                or quarter to three.

Write in standard form as: 2:45.

**Tell how many minutes are in:**

**1.** 1 hour 15 minutes     **2.** 1 hour 45 minutes     **3.** 2 hours

# Write each time in standard form.

**4.**

**5.**

**6.**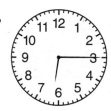

**7.**

**8.**

**9.**

**10.**

**11.**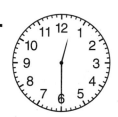

# Draw the time. Show the hour and minute hands.

**12.**

**13.**

**14.**

**15.**

A.M. **and** P.M.

There are 12 hours between 9:00 A.M. and 9:00 P.M.

A.M. means the time after 12:00 midnight and before 12:00 noon.

P.M. means the time after 12:00 noon and before 12:00 midnight.

# Write the time in words. Use A.M. or P.M.

**16.**

dinnertime

**17.**

lunchtime

**18.**

breakfast time

**19.**

time school begins

Each mark on this clock stands for 1 **minute**. It takes five minutes for the minute hand to move from one number to the next.

This clock shows 5 minutes after 10. Write in standard form as: 10:05.

Remember: 60 min = 1 h

This clock shows 10:36. Read this time as: ten thirty-six, or 36 minutes after 10, or 24 minutes before 11.

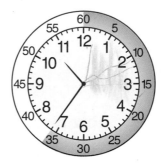

## Write each time.

**1.**

<u>15</u> minutes after <u>12</u>

<u>45</u> minutes before <u>1</u>

**2.**

<u>?</u> minutes after <u>?</u>

<u>?</u> minutes before <u>?</u>

**3.**

<u>?</u> minutes after <u>?</u>

<u>?</u> minutes before <u>?</u>

**4.**

<u>?</u> minutes after <u>?</u>

<u>?</u> minutes before <u>?</u>

**5.**

<u>?</u> minutes after <u>?</u>

<u>?</u> minutes before <u>?</u>

**6.**

<u>?</u> minutes after <u>?</u>

<u>?</u> minutes before <u>?</u>

**Write the time in standard form.**

**7.** 25 minutes before 9

<u> 8 </u> : <u> 35 </u>

**8.** 53 minutes past 2

<u> ? </u> : <u> ? </u>

**9.** 3 minutes after 3

<u> ? </u> : <u> ? </u>

**10.** 1 minute before 10

<u> ? </u> : <u> ? </u>

**11.** 40 minutes past 5

<u> ? </u> : <u> ? </u>

**12.** 15 minutes before 7

<u> ? </u> : <u> ? </u>

**Draw the time. Show the hour and minute hands.**

**13.** 4:12          **14.** 6:53          **15.** 3:41          **16.** 8:27

---

### Estimating Time

Curt's school bus came at 7:25 A.M. this morning. At about what time did Curt's school bus come?

Estimate time to the nearest half hour or hour.

Think: 7:25 is close to 7:30

Curt's school bus came at *about* 7:30 A.M.

---

**PROBLEM SOLVING**

**17.** Margaret practices the piano everyday at 3:55. About what time does she practice piano?

**18.** Louis watches the news at 6:35 P.M. Is this time closer to 6:00 P.M. or to 7:00 P.M.?

**19.** Recess begins at 10:20. About what time does recess begin?

**20.** Jonathan takes a bath at 7:25. Rose takes a bath at 7:55. Who takes a bath closer to 7:30?

The amount of time between two given times is **elapsed time.**

Gina leaves school at 3:05. She arrives home at 3:25. How long does it take Gina to get home from school?

▶ To find out how long it takes Gina to get home, skip count by 5.

Start at 3:05.

Skip count by 5 to 3:25.

3:05
3:10    5 minutes
3:15    10 minutes
3:20    15 minutes
3:25    20 minutes

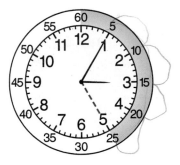

It takes Gina 20 minutes to get home.

Jim has recess at 10:20 A.M. He eats lunch at 12:30 P.M. How much time is between recess and lunchtime?

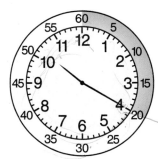

▶ To find out how much time is between recess and lunchtime, count by ones. Then skip count by 5.

Start at 10:20 A.M.

Count by ones to 12:20 P.M.

Skip count by 5 to 12:30 P.M.

10:20
11:20    1 hour
12:20    2 hours
12:25    5 minutes
12:30    10 minutes

There are 2 hours, 10 minutes between recess and lunchtime.

**What time will it be in 15 minutes if it is now:**

**1.** 3:30       **2.** 10:05       **3.** 6:35       **4.** 9:45

**What time will it be in 2 hours?**

**5.** 6:30       **6.** 5:15       **7.** 9:50       **8.** 7:45

**What time will it be in 1 hour, 20 minutes?**

**9.** 1:10       **10.** 3:30       **11.** 5:05       **12.** 8:35

**How much time is between:**

**13.** 6:05 A.M.       **14.** 7:10 P.M.       **15.** 2:30 A.M.       **16.** 3:20 P.M.
6:25 A.M.            7:40 P.M.            2:45 A.M.            3:50 P.M.

**17.** 1:05 A.M.       **18.** 4:20 P.M.       **19.** 11:30 A.M.       **20.** 10:45 A.M.
3:05 A.M.            9:20 P.M.            1:00 P.M.            1:30 P.M.

**Estimate the time it will be in 20 minutes.**

**21.** 2:15       **22.** 4:05       **23.** 7:35       **24.** 11:30
about 2:30

**How long does it take to go from:**

| Schedule | |
|---|---|
| *City* | *Arrives* |
| Orlando . . . . . . . . . . 9:05 A.M. | |
| Rockford . . . . . . . . . 10:15 A.M. | |
| Maywood . . . . . . . . . 10:45 A.M. | |
| Bay City . . . . . . . . . 12:15 P.M. | |

**25.** Rockford to Maywood?

**26.** Orlando to Rockford?

**27.** Orlando to Maywood?

 **Skills to Remember**

**Write the ordinal number or word name for each.**

**28.** 15th       **29.** twentieth       **30.** 72nd       **31.** thirty-third

241

**Calendar**

A **calendar** organizes the days of the week into months and years. This is the calendar for the month of November.

| NOVEMBER (Nov.) | | | | | | |
|---|---|---|---|---|---|---|
| S | M | T | W | TH | F | S |
|  |  |  | 1 | 2 | 3 | 4 |
| 5 | 6 | 7 | 8 | 9 | 10 | 11 |
| 12 | 13 | 14 | 15 | 16 | 17 | 18 |
| 19 | 20 | 21 | 22 | 23 | 24 | 25 |
| 26 | 27 | 28 | 29 | 30 |  |  |

Look at the days and dates of the week.

- There are 30 days in November.
- The day of the week for November 10th is Friday.
- The date of the first Monday is November 6.

■ Election day   ■ Thanksgiving day

1 day (d) = 24 hours (h)     1 year (y) = 365 days

1 week (wk) = 7 days                  = 52 weeks

                                       = 12 months (mo)

1 leap year = 366 days

**Complete.**

1. Each year has  ?  months.

2. Every week has  ?  days.

3. Each year has about  ?  weeks.

4. Every year usually has  ?  days.

**Use the November calendar above.**

5. Give the day of the week for:

  a. November 22nd     b. Election day     c. the last day of November

6. Give the date in November for the:

  a. 4th Monday          b. last Saturday     c. day before Thanksgiving

## Calendar

**JANUARY (Jan.)**

| S | M | T | W | TH | F | S |
|---|---|---|---|----|---|---|
|   |   |   |   |    |   | 1 |
| 2 | 3 | 4 | 5 | 6 | 7 | 8 |
| 9 | 10 | 11 | 12 | 13 | 14 | 15 |
| 16 | 17 | 18 | 19 | 20 | 21 | 22 |
| 23/30 | 24/31 | 25 | 26 | 27 | 28 | 29 |

**FEBRUARY (Feb.)**

| S | M | T | W | TH | F | S |
|---|---|---|---|----|---|---|
|   |   | 1 | 2 | 3 | 4 | 5 |
| 6 | 7 | 8 | 9 | 10 | 11 | 12 |
| 13 | 14 | 15 | 16 | 17 | 18 | 19 |
| 20 | 21 | 22 | 23 | 24 | 25 | 26 |
| 27 | 28 | 29 |   |   |   |   |

**MARCH (Mar.)**

| S | M | T | W | TH | F | S |
|---|---|---|---|----|---|---|
|   |   |   | 1 | 2 | 3 | 4 |
| 5 | 6 | 7 | 8 | 9 | 10 | 11 |
| 12 | 13 | 14 | 15 | 16 | 17 | 18 |
| 19 | 20 | 21 | 22 | 23 | 24 | 25 |
| 26 | 27 | 28 | 29 | 30 | 31 |   |

**APRIL (Apr.)**

| S | M | T | W | TH | F | S |
|---|---|---|---|----|---|---|
|   |   |   |   |    |   | 1 |
| 2 | 3 | 4 | 5 | 6 | 7 | 8 |
| 9 | 10 | 11 | 12 | 13 | 14 | 15 |
| 16 | 17 | 18 | 19 | 20 | 21 | 22 |
| 23/30 | 24 | 25 | 26 | 27 | 28 | 29 |

**MAY (May)**

| S | M | T | W | TH | F | S |
|---|---|---|---|----|---|---|
|   | 1 | 2 | 3 | 4 | 5 | 6 |
| 7 | 8 | 9 | 10 | 11 | 12 | 13 |
| 14 | 15 | 16 | 17 | 18 | 19 | 20 |
| 21 | 22 | 23 | 24 | 25 | 26 | 27 |
| 28 | 29 | 30 | 31 |   |   |   |

**JUNE (June)**

| S | M | T | W | TH | F | S |
|---|---|---|---|----|---|---|
|   |   |   |   | 1 | 2 | 3 |
| 4 | 5 | 6 | 7 | 8 | 9 | 10 |
| 11 | 12 | 13 | 14 | 15 | 16 | 17 |
| 18 | 19 | 20 | 21 | 22 | 23 | 24 |
| 25 | 26 | 27 | 28 | 29 | 30 |   |

**JULY (July)**

| S | M | T | W | TH | F | S |
|---|---|---|---|----|---|---|
|   |   |   |   |    |   | 1 |
| 2 | 3 | 4 | 5 | 6 | 7 | 8 |
| 9 | 10 | 11 | 12 | 13 | 14 | 15 |
| 16 | 17 | 18 | 19 | 20 | 21 | 22 |
| 23/30 | 24/31 | 25 | 26 | 27 | 28 | 29 |

**AUGUST (Aug.)**

| S | M | T | W | TH | F | S |
|---|---|---|---|----|---|---|
|   |   | 1 | 2 | 3 | 4 | 5 |
| 6 | 7 | 8 | 9 | 10 | 11 | 12 |
| 13 | 14 | 15 | 16 | 17 | 18 | 19 |
| 20 | 21 | 22 | 23 | 24 | 25 | 26 |
| 27 | 28 | 29 | 30 | 31 |   |   |

**SEPTEMBER (Sept.)**

| S | M | T | W | TH | F | S |
|---|---|---|---|----|---|---|
|   |   |   |   |    | 1 | 2 |
| 3 | 4 | 5 | 6 | 7 | 8 | 9 |
| 10 | 11 | 12 | 13 | 14 | 15 | 16 |
| 17 | 18 | 19 | 20 | 21 | 22 | 23 |
| 24 | 25 | 26 | 27 | 28 | 29 | 30 |

**OCTOBER (Oct.)**

| S | M | T | W | TH | F | S |
|---|---|---|---|----|---|---|
| 1 | 2 | 3 | 4 | 5 | 6 | 7 |
| 8 | 9 | 10 | 11 | 12 | 13 | 14 |
| 15 | 16 | 17 | 18 | 19 | 20 | 21 |
| 22 | 23 | 24 | 25 | 26 | 27 | 28 |
| 29 | 30 | 31 |   |   |   |   |

**NOVEMBER (Nov.)**

| S | M | T | W | TH | F | S |
|---|---|---|---|----|---|---|
|   |   |   | 1 | 2 | 3 | 4 |
| 5 | 6 | 7 | 8 | 9 | 10 | 11 |
| 12 | 13 | 14 | 15 | 16 | 17 | 18 |
| 19 | 20 | 21 | 22 | 23 | 24 | 25 |
| 26 | 27 | 28 | 29 | 30 |   |   |

**DECEMBER (Dec.)**

| S | M | T | W | TH | F | S |
|---|---|---|---|----|---|---|
|   |   |   |   |    | 1 | 2 |
| 3 | 4 | 5 | 6 | 7 | 8 | 9 |
| 10 | 11 | 12 | 13 | 14 | 15 | 16 |
| 17 | 18 | 19 | 20 | 21 | 22 | 23 |
| 24/31 | 25 | 26 | 27 | 28 | 29 | 30 |

## Use the calendar above.

7. Which months have 30 days?

8. Which months have 31 days?

9. Which is the third month?

10. Which is the last month?

11. Which day is October 28th?

12. Which day is June 25th?

13. Give the date for the:
   a. first Sunday in August
   b. fourth Monday in September

---

### Writing Dates

You can write dates using numbers.

November 30, 2000

Use 11 for the 11th month. → 11 / 30 / 00 ← Use 00 for the year 2000.

Use 30 for the 30th day.

**Study this example:** Write May 28, 2004 as 5/28/04.

---

## Write each date in two ways.

14. your birth date

15. today's date

16. last day of school

## 7-19 | Problem Solving: Missing Information

**Problem:** The car-wash vacuum costs $.75 for 7 minutes. Janell went to the car wash at 9:00 A.M. How much money did she spend for the vacuum?

CAR WASH

$.75 for 7 min

VACUUM

**1 IMAGINE**  Create a mental picture.

**2 NAME**

*Facts:*  Vacuum costs $.75 for 7 minutes. Janell went at 9:00 A.M.

*Question:*  How much money did she spend?

**3 THINK**  Can you find the amount of money spent? Do you have enough information to solve the problem?

No, you do not know *how long Janell used the vacuum.*

As stated, the problem can not be solved. There is missing information.

Information is missing.

To find a possible answer, make up the amount of time, such as 14 minutes. Fourteen minutes is double the original time of 7 minutes. So double $.75 to find how much 14 minutes cost.

```
                                    1
                                $ .75
                                + .75
Add: $.75 + $.75 = _?_           $1.50
      7 min   7 min   14 min
```

**4 COMPUTE**

Janell spent $1.50 for 14 minutes.

**5 CHECK**  Estimate to check your answer.

**Make up the missing information to solve each problem.**

1. Ken wants to visit his grandmother in Nicetown. If it takes him 45 minutes to get there, will he arrive at his grandmother's house by 9:00?

**IMAGINE**     Create a mental picture.

**NAME**     *Facts:*    Ken takes 45 minutes to get to Nicetown.

         *Question:*    Will Ken arrive at his grandmother's by 9:00?

**THINK**     What information is missing?

The time Ken left to go to his grandmother's.
Make up a time: 8:30.
Ask: What is the time 45 minutes later?

**COMPUTE** ⟶ **CHECK**

2. Ken is in Nicetown and needs to call a friend in Joyland. He has 3 quarters, 2 dimes, and a nickel. Is this enough money?

3. The Pleasantville Museum charges children $4.00 admission. How much money must three senior citizens pay?

4. Ken leaves Nicetown at 12:30. He arrives at his uncle's farm at 2:00. Would it have taken less time to go to a friend's farm in Joyland?

5. Gates at Joyland Lake open at 6 A.M. and close at sunset. How many hours does the gate attendant work each day?

## Make Up Your Own

6. Write a problem with missing information. Have a classmate solve it.

**Problem-Solving Applications**

**Solve each problem and explain the method you used.**

1. Keri rides her bicycle around the park a dozen times each week. Which unit of measure should she use to record the distance she rides: inches, feet, or miles?

Imagine

2. Ms. Walls measured the length of the floor in the family room. Which is the most reasonable length: 12 feet, 12 inches, or 12 yards?

Name

3. Robin's test tube is $3\frac{1}{4}$ in. tall. Calvin's beaker is $3\frac{1}{2}$ in. tall. Whose glassware is taller?

Think

4. Joy's pet gained an average of 3 oz each week. In 6 weeks about how many pounds did her pet gain?

Compute

5. Matt used 7 pt of salt water and 3 qt of fresh water for an experiment. Did he use more fresh water or more salt water?

Check

**Use the map for problems 6−8.**

6. How many centimeters is the ant from the food? from the anthill?

7. How much closer is the ant to the flower than to the anthill?

8. An ant walks from the anthill to the flower. Then it walks to the food and back to the anthill. Does the ant walk more or less than 1 decimeter?

**Choose a strategy from the list or use another strategy you know to solve each problem.**

Use these strategies:

Missing Information
Multi-Step Problem
Draw a Picture
Extra Information
Logical Reasoning
Make a Table

9. Every 5 min a scientist added 200 mL of water to the fish tank. How long will it take her to add a liter of water?

10. Gil got up at 7 o'clock. He also went to the movies at 7 o'clock. How is this possible?

11. Lenore began her science experiment on February 22. How many days did she work on her experiment?

12. In science class the students found that each test tube weighs 32 g and is 12 cm long. Will ten test tubes weigh more or less than one kilogram?

13. At 8:45 the outdoor temperature was 56 °F. When Molly measured the temperature 50 minutes later, it was 5° warmer. What were the time and temperature when Molly measured again?

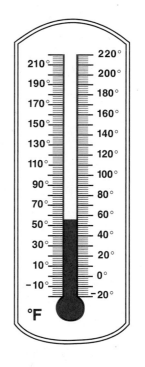

14. Every quarter hour two postal trucks leave the dock. How many trucks leave each hour?

15. Mother drove 10,560 ft to the mall. How many miles did she drive to the mall?

16. From a starting point Tran drew a line down 2 in. From there he drew a line to the right 1 in. Name the letter he drew.

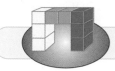

## Make Up Your Own

17. Use a centimeter ruler to draw a letter. Write a problem modeled on problem 16. Have a classmate solve it.

# Chapter Review and Practice

**Write the letter of the best estimate.**  *(See pp. 210–217.)*

**1.** length of a bed          **a.** 7 ft          **b.** 7 in.          **c.** 7 yd

**2.** length of a river          **a.** 100 ft          **b.** 50 yd          **c.** 3 mi

**Compare. Write <, =, or >.**  *(See pp. 218–221.)*

**3.** 3 pt _?_ 6 c          **4.** 1 gal _?_ 3 qt          **5.** 100 oz _?_ 3 lb

**Which unit is used to measure each: cm, dm, m, or km?**  *(See pp. 222–227.)*

**6.** width of a student desk          **7.** distance to the moon

**8.** height of a bookshelf          **9.** width of a toothbrush

**Which unit is used to measure each: mL or L?**  *(See pp. 228–229.)*

**10.** dose of medicine          **11.** juice in a glass

**12.** soda in a large bottle          **13.** water in a bathtub

**Which unit is used to measure each: g or kg?**  *(See pp. 230–231.)*

**14.** bag of sugar          **15.** teaspoon of salt          **16.** dinner roll

## PROBLEM SOLVING  *(See pp. 234–247.)*

**17.** Margaret is going on vacation. The temperature where she is going is 15°F. Do you think she is going to the beach or to a ski resort?

**18.** John left for lunch at a quarter to twelve. Write this number in standard form. How long was he gone?

**19.** How much time is between 9:30 A.M. and 11:15 A.M.?

**20.** Name the months that have 31 days.

*(See Still More Practice, pp. 448–449.)*

## AREA MODELS FOR MULTIPLICATION

On the planet Zitron all the children multiply by covering rectangles with squares.

Predict how many squares will cover the large rectangle. Write the multiplication fact.

Think: 5 × 3 = 15

5

3

## Write the multiplication fact for each.

**1.**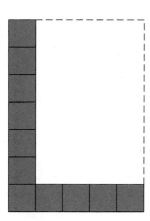

**2.**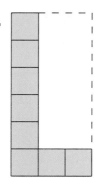

**3.**

**4.**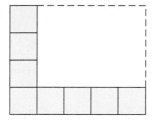

**5.**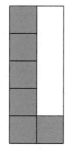

**6.**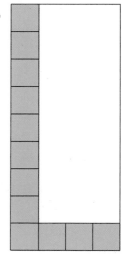

**7.** If the Zitron children and the Earth children had a contest to see who could mulitply the fastest, who do you think would win? Why?

*Communicate* ✓

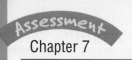
## Performance Assessment

**Draw a line $2\frac{1}{2}$ inches long.**

1. Explain how to tell how many $\frac{1}{4}$ in. units you drew.

**Write the letter of the best estimate.**

2. length of a car          **a.** 6 yd       **b.** 30 ft       **c.** 50 in.

3. height of a mountain     **a.** 2 mi       **b.** 20 yd       **c.** 60 ft

**Compare. Write <, =, or >.**

4. 2 pt  ?  3 c          5. 2 gal  ?  9 qt          6. 56 oz  ?  4 lb

**Which unit is used to measure each: cm, dm, m, or km?**

7. length of a thumb                8. distance from Denver to Chicago

9. height of a tree                 10. width of a penny

**Which unit is used to measure each: mL or L?**

11. milk in a baby bottle           12. gas in a car tank

13. water in a fish tank            14. tea in a cup

**Which unit is used to measure each: g or kg?**

15. man                             16. goldfish

**Write the time in standard form.**

17. quarter to four                 18. 17 minutes after 11

**PROBLEM SOLVING**    *Use a strategy you have learned.*

19. How much time is between         20. Name the months that
    10:45 P.M. and 11:00 P.M.?           have only 30 days.

# Cumulative Test I

**Choose the best answer.**

1. Choose the standard form of the number.

   9 thousands 2 hundreds

   **a.** 2,900
   **b.** 9,020
   **c.** 9,200
   **d.** 90,200

2. Round to the nearest dollar.

   $7.09

   **a.** $8.00
   **b.** $7.00
   **c.** $7.10
   **d.** $6.00

3.  8562
   + 658

   **a.** 8116
   **b.** 9120
   **c.** 9220
   **d.** 9110

4. 340 + 406 + 95 + 204

   **a.** 975
   **b.** 980
   **c.** 995
   **d.** 1045

5. 700 − 198

   **a.** 502
   **b.** 602
   **c.** 698
   **d.** 618

6.   $77.49
    − 16.98

   **a.** $60.50
   **b.** $60.51
   **c.** $61.45
   **d.** $61.51

7. 8 × 3

   **a.** 11  **b.** 20
   **c.** 24  **d.** not given

8.   5
   × 9

   **a.** 35  **b.** 40
   **c.** 45  **d.** not given

9. 4)32

   **a.** 6  **b.** 7
   **c.** 9  **d.** not given

10. 8¢ ÷ 2

    **a.** 4¢  **b.** 16¢
    **c.** 6¢  **d.** not given

11. _?_ = 5 × 5

    **a.** 1  **b.** 15
    **c.** 25  **d.** 10

12. _?_ = 3 ÷ 1

    **a.** 0  **b.** 1
    **c.** 3  **d.** 2

13. Estimate.

    weight of an apple

    **a.** 5 oz
    **b.** 5 lb
    **c.** 50 oz
    **d.** 50 lb

14. Estimate.

    height of a man

    **a.** 3 yd
    **b.** 6 ft
    **c.** 36 in.
    **d.** 10 ft

15. Which unit is used to measure the height of a flagpole?

    **a.** cm  **b.** dm
    **c.** m  **d.** km

16. What time will it be in 1 hour, 20 minutes, if it is now 7:45?

    **a.** 8:05  **b.** 8:45
    **c.** 9:05  **d.** 9:10

**Use a centimeter ruler. Measure each length to the nearest centimeter.**

17. |———————————————————————|

18. |————————————————————|

**PROBLEM SOLVING**

19. Copy and complete the tally chart. Then make a bar graph of the data.

20. Make a pictograph of the data.

21. How many more gray cars than blue cars were there?

22. If the gray cars were parked in 2 equal rows, how many gray cars were there in each row?

| Color of Car | Tally | Total |
|---|---|---|
| Red | ЖЖ ЖЖ ЖЖ I | ? |
| Gray | ЖЖ ЖЖ ЖЖ III | ? |
| Blue | ЖЖ ЖЖ II | ? |
| White | ЖЖ IIII | ? |

**Use the spinner at the right for problem 23.**

23. What is the probability of landing on

    **a.** red?    **b.** white?    **c.** yellow?

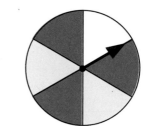

---

# For Rubric Scoring

**Listen for information on how your work will be scored.**

24. Explain how you would subtract 45 from 100. What is the difference? Draw a picture to show your work.

25. Describe how you would add the addends 376 and 124. What is the sum? Draw a picture to show how you solved the addition.

26. Explain how you can prove that 2 fives equals 5 twos. What is the product? Draw a picture to show your work.

27. Explain how you would divide 36 by 4. What is the quotient? Draw a picture to show your work.

# More Multiplication and Division Facts

## In this chapter you will:

Explore multiplication of 6 through 9
Multiply three factors
Explore dividing by 6 through 9
Learn about number patterns and
 fact families
Learn about the FOR-NEXT Loop
Solve problems by guess and test

**8**

## Critical Thinking/Finding Together

A tricycle has 3 wheels. A bicycle has
2 wheels. Suppose the mice in the poem
rode on bicycles. How many fewer wheels
would be riding on the ice?

## Nine Mice

Nine mice on tiny tricycles
went riding on the ice,
they rode in spite of warning signs,
they rode despite advice.

The signs were right, the ice was thin,
in half a trice, the mice fell in,
and from their chins down to their toes,
those mice entirely froze.

Nine mindless mice, who paid the price,
are thawing slowly by the ice,
still sitting on their tricycles
. . . nine white and shiny *micicles!*

*Jack Prelutsky*

253

# 8-1 Factors and Products

This multiplication machine multiplies 4 by each factor put in it. Bob put 3 into the machine. The product, 12, came out of the machine.

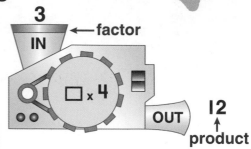

3 ← factor

IN

□ x 4

OUT 12

↑
product

$3 \times 4 = 12$   or

factors product

$$\begin{array}{r} 4 \leftarrow \text{factors} \\ \times 3 \leftarrow \\ \hline 12 \leftarrow \text{product} \end{array}$$

Remember:
Factors are the numbers you multiply.
Product is the answer when you multiply.

## Copy and complete each table.

**1.**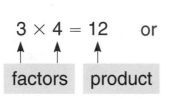

| IN | 0 | 1 | 2 | 3 | 4 | 5 | 6 | 7 | 8 | 9 |
|---|---|---|---|---|---|---|---|---|---|---|
| OUT | 0 | 2 | 4 | ? | ? | ? | ? | ? | ? | ? |

**2.**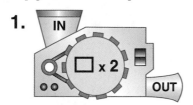

| IN | 0 | 1 | 2 | 3 | 4 | 5 | 6 | 7 | 8 | 9 |
|---|---|---|---|---|---|---|---|---|---|---|
| OUT | ? | ? | ? | ? | ? | ? | ? | ? | ? | ? |

**3.**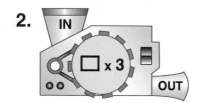

| IN | 3 | 7 | 2 | 8 | 9 | 5 | 0 | 4 | 1 | 6 |
|---|---|---|---|---|---|---|---|---|---|---|
| OUT | 12 | ? | ? | ? | ? | ? | ? | ? | ? | ? |

**4.**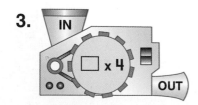

| IN | 6 | 4 | 2 | 8 | 7 | 5 | 9 | 0 | 1 | 3 |
|---|---|---|---|---|---|---|---|---|---|---|
| OUT | 30 | ? | ? | ? | ? | ? | ? | ? | ? | ? |

# Multiplying Sixes

Algebra

Find the product of $9 \times 6$ or $\begin{array}{r} 6 \\ \times\,9 \end{array}$ .

You can use counters or skip count to find the product.

## Hands-On Understanding

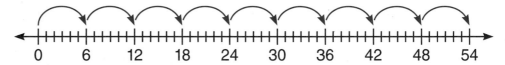

**You Will Need:** counters

**Step 1**  Model 9 groups of 6 counters.

How many groups are there in all?

How many counters are in each group?

How many counters are there altogether?

Write an addition sentence to represent your model.

Write a multiplication sentence to represent your model.

**Step 2**  Skip count by 6 to find the product of $9 \times 6$.

0   6   12   18   24   30   36   42   48   54

**Step 3**  Repeat Steps 1–2 for these groups.

| Number of groups | 0 | 1 | 2 | 3 | 4 | 5 | 6 | 7 | 8 | 9 |
|---|---|---|---|---|---|---|---|---|---|---|
| Counters in each group | 6 | 6 | 6 | 6 | 6 | 6 | 6 | 6 | 6 | 6 |
| Counters in all | ? | ? | ? | ? | ? | ? | ? | ? | ? | 54 |

## Communicate

Discuss

**1.** Look at the products for multiplying by 6.
Describe any pattern you see.

255

## 8-3 Multiplying Sevens

Mel's class went camping. They set up 6 rows of tents with 7 tents in each row. How many tents did the troop set up?

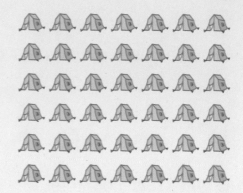

To find how many tents in all, multiply: $6 \times 7 = \underline{?}$  or  $\begin{array}{r} 7 \\ \times 6 \\ \hline \end{array}$

You can use an **array** or skip count to find the product of $6 \times 7$.

An **array** is a number of objects arranged in rows and columns.

 Hands-On Understanding

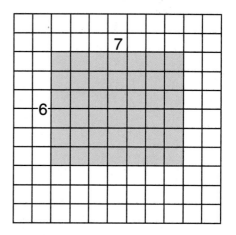

**You Will Need:** grid paper

**Step 1** Shade 6 rows of 7 squares each.

How many rows are there in all?

How many squares are shaded in each row?

How many small squares are shaded altogether?

Write an addition sentence to represent your model.

Write a multiplication sentence to represent your model.

**Step 2** Skip count by 7 to find the product of $6 \times 7$.

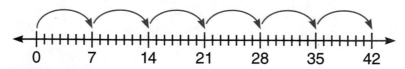

What is the product of $6 \times 7$?

How many tents did the troops set up?

Repeat Steps 1–2 for these models:

| Number of rows | 0 | 1 | 2 | 3 | 4 | 5 | 6 | 7 | 8 | 9 |
|---|---|---|---|---|---|---|---|---|---|---|
| Squares in each row | 7 | 7 | 7 | 7 | 7 | 7 | 7 | 7 | 7 | 7 |
| Small squares altogether | ? | ? | ? | ? | ? | ? | 42 | ? | ? | ? |

## Communicate

Discuss

1. Look at the products for multiplying by 7.
   Describe any pattern you see.

**Use counters or draw dots to show each multiplication. Complete.**

2. 3 sevens = _?_         7
   3 × 7   = _?_       ×3

3. 5 sevens = _?_         7
   5 × 7   = _?_       ×5

**Multiply. You may use counters or skip count to help.**

4.   7        5.   7        6.   7        7.   7        8.   7        9.   7
    ×5           ×2           ×7           ×0           ×4           ×8

10.   7       11.   7       12.   6       13.   7       14.   5       15.   7
     ×1           ×3           ×7           ×6           ×7           ×9

## Mental Math

**Study this example.**                                Communicate

$3 × 2 = 6$      What do you notice about the factors?

$6 × 2 = 12$     What do you notice about the products?

**Complete.**

16. $4 × 2 = 8$          17. $2 × 4 = 8$          18. $3 × 3 = ?$

    $8 × 2 = $ _?_            $4 × 4 = $ _?_           $6 × 3 = $ _?_

257

# Multiplying Eights

Find the product of $9 \times 8$ or $\begin{array}{r} 8 \\ \times\,9 \\ \hline \end{array}$.

You can use connecting cubes or skip count to find the product.

## Hands-On Understanding

**You Will Need:** connecting cubes

**Step 1**    Use your connecting cubes to model 1 group of 8 cubes.

**Step 2**    Now model 8 more groups of 8 cubes.

How many groups are there in all?

How many cubes are in each group?

How many cubes are there altogether?

Write an addition sentence to represent your model.

Write a multiplication sentence to represent your model.

**Step 3**    Skip count by 8 to show 9 groups of 8.

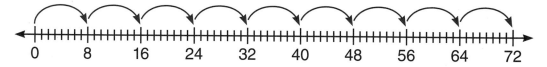

0     8    16    24    32    40    48    56    64    72

What is the product of $9 \times 8$?

**Step 4**    Repeat Steps 1–3 for these groups.

| Number of groups | 0 | 1 | 2 | 3 | 4 | 5 | 6 | 7 | 8 | 9 |
|---|---|---|---|---|---|---|---|---|---|---|
| Cubes in each group | 8 | 8 | 8 | 8 | 8 | 8 | 8 | 8 | 8 | 8 |
| Cubes in all | ? | ? | ? | ? | ? | ? | ? | ? | ? | 72 |

# Communicate

**1.** Look at the products for multiplying by 8.
Describe any pattern you see.

## Use counters or draw dots to show each multiplication. Complete.

**2.** 3 eights = _?_
$\quad\quad$ 3 × 8 = _?_

$$\begin{array}{r} 8 \\ \times 3 \\ \hline \end{array}$$

**3.** 5 eights = _?_
$\quad\quad$ 5 × 8 = _?_

$$\begin{array}{r} 8 \\ \times 5 \\ \hline \end{array}$$

**4.** 7 eights = _?_
$\quad\quad$ 7 × 8 = _?_

$$\begin{array}{r} 8 \\ \times 7 \\ \hline \end{array}$$

## Multiply. You may use counters or skip count to help.

**5.** $\begin{array}{r} 8 \\ \times 4 \\ \hline \end{array}$
**6.** $\begin{array}{r} 8 \\ \times 7 \\ \hline \end{array}$
**7.** $\begin{array}{r} 8 \\ \times 2 \\ \hline \end{array}$
**8.** $\begin{array}{r} 8 \\ \times 6 \\ \hline \end{array}$
**9.** $\begin{array}{r} 8 \\ \times 0 \\ \hline \end{array}$
**10.** $\begin{array}{r} 8 \\ \times 9 \\ \hline \end{array}$

**11.** $\begin{array}{r} 8 \\ \times 1 \\ \hline \end{array}$
**12.** $\begin{array}{r} 8 \\ \times 5 \\ \hline \end{array}$
**13.** $\begin{array}{r} 6 \\ \times 8 \\ \hline \end{array}$
**14.** $\begin{array}{r} 8 \\ \times 8 \\ \hline \end{array}$
**15.** $\begin{array}{r} 7 \\ \times 8 \\ \hline \end{array}$
**16.** $\begin{array}{r} 8 \\ \times 3 \\ \hline \end{array}$

## PROBLEM SOLVING

**17.** Kim found 8 fossils. Sara found 4 times as many fossils. How many fossils did Sara find?

**18.** Beverly found 8 shells. Jim found 6 times as many shells. How many shells did Jim find?

**19.** Barbara has 5 boxes of rocks. Each box has 8 rocks. How many rocks does Barbara have in all?

**20.** Josh found 5 minerals. Walt found 2 more than Josh. Larry found 7 times as many minerals as Walt. How many minerals did Larry find?

## Share Your Thinking

**21.** Explain how you can use doubles to find the product of 3 × 8. Can you can use doubles to find the products for all the multiplication facts of 8? Explain your answer.

## 8-5 Multiplying Nines

There are 5 rows of chairs in the school auditorium. Each row has 9 chairs. How many chairs are there in all?

To find how many chairs in all, multiply:  $5 \times 9 = \underline{\ ?\ }$  or

$$\begin{array}{r} 9 \\ \times\ 5 \\ \hline \end{array}$$

To find the product of $5 \times 9$ use counters or skip count to make 5 groups of 9.

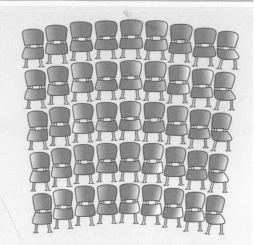

 Hands-On Understanding

**You Will Need:** counters, one dozen egg carton, marker or crayon

**Step 1**   Number 9 of the sections on your egg carton 1–9.

**Step 2**   Put the number of counters indicated by $5 \times 9$ in each section of your egg carton.

How many groups are there in all?

How many counters are in each group?

How many counters are there altogether?

Write an addition sentence to represent your model.

Write a multiplication sentence to represent your model.

**Step 3**   Skip count by 9 to show the product of $5 \times 9$.

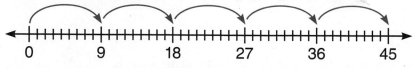

What is the product of $5 \times 9$?

How many chairs are there in all?

| Number of groups | 0 | 1 | 2 | 3 | 4 | 5 | 6 | 7 | 8 | 9 |
|---|---|---|---|---|---|---|---|---|---|---|
| Counters in each group | 9 | 9 | 9 | 9 | 9 | 9 | 9 | 9 | 9 | 9 |
| Counters altogether | ? | ? | ? | ? | ? | 45 | ? | ? | ? | ? |

# Communicate

Discuss ✓

1. Look at the products for multiplying by 9. Describe any pattern you see.

**Multiply. You may use counters or skip count to help.**

2.  $\begin{array}{r} 9 \\ \times 2 \\ \hline \end{array}$
3.  $\begin{array}{r} 9 \\ \times 5 \\ \hline \end{array}$
4.  $\begin{array}{r} 9 \\ \times 3 \\ \hline \end{array}$
5.  $\begin{array}{r} 9 \\ \times 7 \\ \hline \end{array}$
6.  $\begin{array}{r} 9 \\ \times 4 \\ \hline \end{array}$
7.  $\begin{array}{r} 9 \\ \times 6 \\ \hline \end{array}$

8.  $\begin{array}{r} 9 \\ \times 0 \\ \hline \end{array}$
9.  $\begin{array}{r} 9 \\ \times 9 \\ \hline \end{array}$
10. $\begin{array}{r} 6 \\ \times 9 \\ \hline \end{array}$
11. $\begin{array}{r} 9 \\ \times 1 \\ \hline \end{array}$
12. $\begin{array}{r} 7 \\ \times 9 \\ \hline \end{array}$
13. $\begin{array}{r} 9 \\ \times 8 \\ \hline \end{array}$

## Project

14. Make a multiplication table. Fill in all the facts to $9 \times 9$.

15. Draw a straight line from the product of $9 \times 0$ to the product of $0 \times 9$. What pattern do you see?

16. Use a blue crayon to mark all the products in the row and column for 5. What is the pattern in the ones places? the tens places?

| X | 0 | 1 | 2 | 3 | 4 | 5 |
|---|---|---|---|---|---|---|
| 0 | 0 | 0 | 0 | 0 | 0 | 0 |
| 1 | 0 | 1 | 2 | 3 | 4 | 5 |
| 2 | 0 | 2 | 4 | 6 | 8 | 10 |
| 3 | 0 | 3 | 6 | 9 | 12 | 15 |
| 4 | 0 | 4 | 8 | 12 | 16 | 20 |
| 5 | 0 | 5 | 10 | 15 | 20 | 25 |

17. Color all the even products red. What can you say about the product when both factors are odd? are even? one is odd and one is even?

Communicate ✓

**Multiplying Three Numbers**            *Algebra* ✓

There are 3 pairs of tennis socks in
a package. There are 3 packages
in each box. How many pairs
of socks are there in 2 boxes?

To find how many, multiply: $(3 \times 3 \times 2) = \underline{?}$

$3 \times 6 = 18$
$9 \times 2 = 18$

**To multiply three factors:**

- Group the first two factors
  using this symbol (  ).
- Multiply these factors first.
- Complete the multiplication.

$(3 \times 3) \times 2 = \underline{?}$

$9 \quad \times 2 = \underline{?}$
$9 \quad \times 2 = 18$

**or**

- Group the last two factors
  using this symbol (  ).
- Multiply these factors first.
- Complete the multiplication.

$3 \times (3 \times 2) = \underline{?}$

$3 \times \quad 6 \quad = \underline{?}$
$3 \times \quad 6 \quad = 18$

There are 18 pairs of socks
in 2 boxes.

Changing the **grouping** of the factors
does not change the product.

---

**Multiply.** Use the grouping shown.

**1.** $(4 \times 2) \times 3 = \underline{?}$
$\quad 8 \quad \times 3 = \underline{?}$

**2.** $4 \times (2 \times 3) = \underline{?}$
$\quad 4 \times \quad 6 \quad = \underline{?}$

**3.** $(2 \times 3) \times 2 = \underline{?}$
$\quad 6 \quad \times 2 = \underline{?}$

**4.** $2 \times (3 \times 2) = \underline{?}$
$\quad 2 \times \quad 6 \quad = \underline{?}$

**5.** $(3 \times 3) \times 2 = \underline{?}$
$\quad \underline{?} \quad \times 2 = \underline{?}$

**6.** $3 \times (3 \times 2) = \underline{?}$
$\quad 3 \times \quad \underline{?} \quad = \underline{?}$

**7.** $(9 \times 6) \times 1 = \underline{?}$
$\quad \underline{?} \quad \times 1 = \underline{?}$

**8.** $9 \times (6 \times 1) = \underline{?}$
$\quad 9 \times \quad \underline{?} \quad = \underline{?}$

**9.** $3 \times (3 \times 3) = \underline{?}$
$\quad 3 \times \quad \underline{?} \quad = \underline{?}$

**10.** $(2 \times 2) \times 2 = \underline{?}$
$\quad \underline{?} \quad \times 2 = \underline{?}$

**11.** $(1 \times 7) \times 0 = \underline{?}$
$\quad \underline{?} \quad \times 0 = \underline{?}$

**12.** $1 \times (7 \times 0) = \underline{?}$
$\quad 1 \times \quad \underline{?} \quad = \underline{?}$

**Find the product.** Use the grouping shown.

**13.** $2 \times (3 \times 3) = \underline{\ ?\ }$
$\phantom{2}2 \times \quad \underline{9} \quad = \underline{18}$

**14.** $(2 \times 2) \times 4 = \underline{\ ?\ }$
$\phantom{(2 \times 2)}\underline{\ ?\ } \times 4 = \underline{\ ?\ }$

**15.** $9 \times (1 \times 8) = \underline{\ ?\ }$
$\phantom{9 \times }9 \times \underline{\ ?\ } = \underline{\ ?\ }$

**16.** $5 \times (1 \times 2) = \underline{\ ?\ }$
$\phantom{5}\underline{\ ?\ } \times \underline{\ ?\ } = \underline{\ ?\ }$

**17.** $(3 \times 2) \times 4 = \underline{\ ?\ }$
$\phantom{(3 \times 2)}\underline{\ ?\ } \times \underline{\ ?\ } = \underline{\ ?\ }$

**18.** $2 \times (4 \times 2) = \underline{\ ?\ }$
$\phantom{2}\underline{\ ?\ } \times \underline{\ ?\ } = \underline{\ ?\ }$

**Find the product.** Use any grouping.

**19.** $3 \times 2 \times 2 = \underline{\ ?\ }$

**20.** $6 \times 1 \times 9 = \underline{\ ?\ }$

**21.** $6 \times 8 \times 0 = \underline{\ ?\ }$

**22.** $2 \times 0 \times 2 = \underline{\ ?\ }$

**23.** $4 \times 2 \times 2 = \underline{\ ?\ }$

**24.** $4 \times 1 \times 2 = \underline{\ ?\ }$

## PROBLEM SOLVING

**25.** A tennis player gives away 4 boxes of T-shirts. Each box has 2 packages. Each package has 3 shirts. How many shirts does she give away?

**26.** A company buys 2 sections of seats for a tennis match. Each section has 2 rows. Each row has 4 seats. How many seats does the company buy?

 **Skills to Remember**

**Find the missing factor.**

**27.** $\underline{4} \times 6 = 24$

**28.** $\underline{\ ?\ } \times 8 = 24$

**29.** $\underline{\ ?\ } \times 7 = 49$

**30.** $\underline{\ ?\ } \times 9 = 36$

**31.** $\underline{\ ?\ } \times 6 = 30$

**32.** $\underline{\ ?\ } \times 8 = 40$

**33.** $7 \times \underline{\ ?\ } = 42$

**34.** $9 \times \underline{\ ?\ } = 45$

**35.** $6 \times \underline{\ ?\ } = 48$

Tasha has 18 cassettes. She puts 3 cassettes in each group. How many groups of cassettes are there?

To find how many groups, divide: $18 \div 3 = \underline{?}$

$18 \div 3 = \underline{?}$    or    $3\overline{)18}^{?}$

↑ in all    ↑ in each group    ↑ groups

___ ? sets of 3 are 18.
$18 \div 3 = \underline{?}$

Division undoes multiplication.

**Every division fact has a related multiplication fact.**

Think:  $18 \div 3 = \underline{?}$
$\underline{?} \times 3 = 18$
$6 \times 3 = 18$ ⟶ $18 \div 3 = 6$    or    $3\overline{)18}^{6}$

quotient
divisor
dividend

There are 6 groups of cassettes.

Remember:  Divide to find how many groups or how many in each group.

**Study this example.**

Think:  $0 \div 6 = \underline{?}$
$\underline{?} \times 6 = 0$
$0 \times 6 = 0$
So  $0 \div 6 = 0.$

Zero divided by any number is zero.

**Divide.**

**1.** $6 \div 2$    **2.** $12 \div 3$    **3.** $8 \div 4$    **4.** $15 \div 5$    **5.** $4 \div 4$

**Use counters or draw dots to find each quotient.**

**6.** $18 \div 2$  **7.** $18 \div 3$  **8.** $20 \div 4$  **9.** $12 \div 4$  **10.** $0 \div 5$

**11.** $3\overline{)21}$  **12.** $2\overline{)2}$  **13.** $4\overline{)24}$  **14.** $5\overline{)10}$  **15.** $3\overline{)24}$

**16.** $2\overline{)12}$  **17.** $4\overline{)16}$  **18.** $3\overline{)0}$  **19.** $5\overline{)35}$  **20.** $3\overline{)9}$

**21.** $4\overline{)28}$  **22.** $3\overline{)27}$  **23.** $2\overline{)14}$  **24.** $4\overline{)0}$  **25.** $5\overline{)40}$

## Dividing by Zero

Remember that every division fact has a related multiplication fact.

$6 \div 3 = 2 \longrightarrow 2 \times 3 = 6$
$6 \div 2 = 3 \longrightarrow 3 \times 2 = 6$
$6 \div 1 = 6 \longrightarrow 6 \times 1 = 6$
$6 \div 0 = \underline{\phantom{?}} \qquad \underline{\phantom{?}} \times 0 = 6$

— No number works.

Remember:
You cannot divide
a number by zero.

**Divide.**

**26.** $3 \div 3$  **27.** $0 \div 4$  **28.** $45 \div 5$  **29.** $0 \div 5$  **30.** $4 \div 1$

**31.** $10 \div 2$  **32.** $0 \div 3$  **33.** $20 \div 5$  **34.** $0 \div 2$  **35.** $8 \div 4$

## PROBLEM SOLVING

**36.** Nancy has a tape case that holds 32 tapes. The case has 4 sections. How many tapes does each section hold?

**37.** Mr. Polski wraps 45 tapes in packages. Each package holds 5 tapes. How many packages does he wrap?

**38.** Franco buys 2 tapes a month. How many months will it take him to buy 16 tapes?

# 8-8 Dividing by 6

*Algebra* ✓

Find the quotient of 18 ÷ 6 or 6)‾18‾ with ? above

You can use counters or skip count to help you divide by 6.

## Hands-On Understanding

**You Will Need:** counters, record sheet

**Step 1**    Label your record sheet with these headings:

| Counters in all | 18 | | | |
|---|---|---|---|---|
| Counters in each group | 6 | | | |
| Number of groups | 3 | | | |
| Division sentence | | | | |
| Multiplication sentence | | | | |

**Step 2**    Model 18 counters.

How many counters do you have in all?

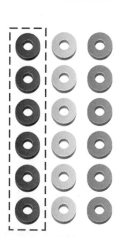

**Step 3**    Now make as many groups of 6 as you can.

How many counters are in each group?

How many groups did you make?

Write a division sentence to represent your model.

Now write a multiplication sentence to represent your model.

**Step 4**    Repeat Steps 2–3 for these counters:

| Counters in all | 54 | 48 | 42 | 36 | 30 | 24 | 12 | 6 | 0 |
|---|---|---|---|---|---|---|---|---|---|

You can also skip count to help you divide by 6.

Step 5 Count back by 6s. Start at 18 and stop at 0.

How many 6s are in 18?

Write a division sentence to show how many 6s are in 18.

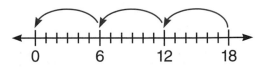

## Communicate

1. Look at the quotients on your record sheet. Describe any pattern you see.

2. Which meaning of division, separating or sharing, did you use to find each quotient? Explain your answer.

*Discuss* ✓

**Write a division sentence for each model.**

3.

4.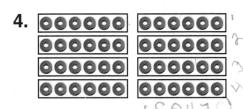

5.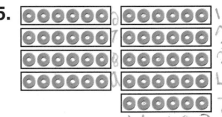

**Find the quotient.** Use counters or skip count to help.

6. $24 \div 6 = \underline{\ ?\ }$    7. $6 \div 6 = \underline{\ ?\ }$    8. $36 \div 6 = \underline{\ ?\ }$    9. $48 \div 6 = \underline{\ ?\ }$

10. $12 \div 6 = \underline{\ ?\ }$   11. $0 \div 6 = \underline{\ ?\ }$   12. $18 \div 6 = \underline{\ ?\ }$   13. $30 \div 6 = \underline{\ ?\ }$

14. $6\overline{)36}$     15. $6\overline{)54}$     16. $6\overline{)0}$     17. $6\overline{)42}$     18. $6\overline{)12}$

19. $6\overline{)48}$     20. $6\overline{)30}$     21. $6\overline{)18}$     22. $6\overline{)6}$     23. $6\overline{)24}$

## Critical Thinking

24. Sandy has fewer than 16 crayons. If she puts them in equal rows of 5, none are left over. If she puts them in equal rows of 6, four are left over. How many crayons are there?

# Dividing by 7

Mr. Johnson plants 35 tulips. He plants 7 equal groups of tulips. At most, how many tulips are in each group?

To find how many are in each group, divide:  $35 \div 7 = \underline{\ ?\ }$  or  $7\overline{)35}$

You can use counters or skip count to help you divide by 7.

## Hands-On Understanding

**You Will Need:** counters, 7 blank sheets of paper, record sheet

**Step 1**  Label your record sheet with these headings:

| Counters in all | 35 | | | |
| --- | --- | --- | --- | --- |
| Number of groups | 7 | | | |
| Counters in each group | 5 | | | |
| Division sentence | | | | |
| Multiplication sentence | | | | |

**Step 2**  Model 35 counters.

How many counters do you have in all?

**Step 3**  Place 7 blank sheets of paper on your worktable.

What do the 7 sheets of paper stand for?

**Step 4**  Now place one counter on each sheet of paper.

Do you have enough counters to put more counters on each sheet of paper?

**Step 5**   Continue to place counters on each sheet of paper until you can no longer give them out equally.

How many groups did you make?

How many counters are in each group?

How many tulips did Mr. Johnson plant in each group?

Write a division sentence to represent your model.

Write a multiplication sentence to represent your model.

**Step 6**   Repeat Steps 2–5 for these counters:

| Counters in all | 63 | 56 | 49 | 42 | 28 | 21 | 14 | 7 | 0 |
|---|---|---|---|---|---|---|---|---|---|

You can also skip count to help you divide by 7.

**Step 7**   Count back by 7s. Start at 35 and stop at 0.

How many 7s are there in 35?

So $35 \div 7 =$ _?_ .

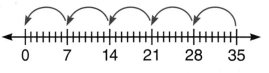

## Communicate

1. Look at the quotients on your record sheet. Describe any pattern you see.

   *Discuss* ✓

2. Which meaning of division, separating or sharing, did you use to find each quotient? Explain your answer.

**Find the quotient.** You may use counters or skip count to help.

3. $63 \div 7 =$ _?_   4. $42 \div 7 =$ _?_   5. $21 \div 7 =$ _?_   6. $14 \div 7 =$ _?_

7. $28 \div 7 =$ _?_   8. $7 \div 7 =$ _?_   9. $35 \div 7 =$ _?_   10. $0 \div 7 =$ _?_

11. $7)\overline{56}$     12. $7)\overline{14}$     13. $7)\overline{42}$     14. $7)\overline{63}$     15. $7)\overline{49}$

269

## 8-10 Dividing by 8

In the school library there are 48 chairs at tables. There are 8 chairs at each table. How many tables are there?

To find how many tables, divide: $48 \div 8 =$ ? or $8\overline{)48}$

You can use counters or skip count to help you divide by 8.

 **Hands-On Understanding**

**You Will Need:** counters, record sheet

| Step 1 | Label your record sheet with these headings: |

| Counters in all | 48 | | | |
|---|---|---|---|---|
| Counters in each group | 8 | | | |
| Number of groups | 6 | | | |
| Division sentence | | | | |
| Multiplication sentence | | | | |

**Step 2** Model 48 counters.

How many counters do you have in all?

**Step 3** Now make as many groups of 8 as you can.

How many counters are in each group?

How many groups did you make?

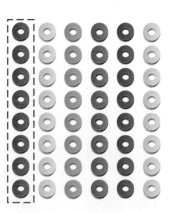

How many tables are there in the library?

Write a division sentence to represent your model.

Now write a multiplication sentence to represent your model.

**Step 4**   Repeat Steps 2 and 3. Choose a meaning of division, separating or sharing, to model these groups.

| Counters in all | 72 | 64 | 56 | 40 | 32 | 24 | 16 | 8 | 0 |
|---|---|---|---|---|---|---|---|---|---|

You can also skip count to help you divide by 8.

**Step 5**   Count back by 8s. Start at 48 and stop at 0.

How many 8s are there in 48?

So $48 \div 8 =$ _?_ .

0   8   16   24   32   40   48

# Communicate

Discuss

1. Look at the quotients on your record sheet. Describe any pattern you see.

**Divide. Then write the related multiplication fact.**

**2.** $16 \div 8$      **3.** $48 \div 8$      **4.** $8 \div 8$      **5.** $0 \div 8$      **6.** $56 \div 8$

**Find the quotient.** You may use counters or skip count.

**7.** $40 \div 8 =$ _?_      **8.** $24 \div 8 =$ _?_      **9.** $64 \div 8 =$ _?_     **10.** $0 \div 8 =$ _?_

**11.** $56 \div 8 =$ _?_     **12.** $8 \div 8 =$ _?_      **13.** $72 \div 8 =$ _?_     **14.** $16 \div 8 =$ _?_

**15.** $8\overline{)64}$        **16.** $8\overline{)32}$        **17.** $8\overline{)16}$        **18.** $8\overline{)24}$        **19.** $8\overline{)0}$

**20.** $8\overline{)48}$        **21.** $8\overline{)8}$        **22.** $8\overline{)40}$        **23.** $8\overline{)56}$        **24.** $8\overline{)72}$

# Dividing by 9

Miguel buys 54 stickers. There are 9 stickers on each sheet. How many sheets does Miguel buy?

To find how many sheets, divide:   $54 \div 9 = \underline{\phantom{?}}$   or   $9\overline{)54}$

You can use counters or skip count to help you divide by 9.

## Hands-On Understanding

**You Will Need:** counters, record sheet

**Step 1**  Label your record sheet with these headings:

| | | | | |
|---|---|---|---|---|
| Counters in all | 54 | | | |
| Counters in each group | 6 | | | |
| Number of groups | 9 | | | |
| Division sentence | | | | |
| Multiplication sentence | | | | |

**Step 2**  Model 54 counters.
How many counters do you have in all?

**Step 3**  Now make as many groups of 9 as you can.
How many counters are in each group?
How many groups did you make?
How many sheets does Miguel buy?
Write a division sentence to represent your model.
Write a multiplication sentence to represent your model.

**Step 4** Repeat Steps 2 and 3. Choose a meaning of division, separating or sharing, to model these groups:

| Counters in all | 81 | 72 | 63 | 45 | 36 | 27 | 18 | 9 | 0 |
|---|---|---|---|---|---|---|---|---|---|

You can also skip count to help you divide by 9.

**Step 5** Count back by 9s. Start at 54 and stop at 0.

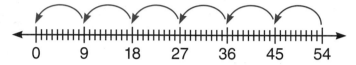

How many 9s are in 54?

So 54 ÷ 9 = _?_ .

## Communicate

Discuss

1. Look at the quotients on your record sheet. Describe any pattern you see.

**Find the quotient.** You may use counters or skip count.

2. 27 ÷ 9 = _?_    3. 81 ÷ 9 = _?_    4. 36 ÷ 9 = _?_    5. 72 ÷ 9 = _?_

6. 9 ÷ 9 = _?_    7. 45 ÷ 9 = _?_    8. 18 ÷ 9 = _?_    9. 63 ÷ 9 = _?_

10. 9)36    11. 9)63    12. 9)18    13. 9)27    14. 9)45

15. 9)0    16. 9)54    17. 9)81    18. 9)72    19. 9)9

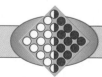 **Finding Together**    Algebra

20. List all the division facts for 12. How many facts did you find? (*Hint:* 12 = _?_ × _?_ )

21. List all the division facts for 24. How many facts did you find?

273

**Operation Patterns**

What number comes next in this pattern?     27, 9, 3, _?_

Think:  27,   9,   3,   ?          Rule:  Divide by 3.

$\div 3 \quad \div 3 \quad \div 3$

Now complete the pattern.

27, 9, 3, _?_

27, 9, 3, 1

Think:  $3 \div 3 = 1$

**Study these examples.**

1,   2,   4,   8,   ?

$\times 2 \quad \times 2 \quad \times 2 \quad \times 2$

Rule:
Multiply by 2.

Think:
$8 \times 2 = 16$

1, 2, 4, 8, 16

2,   6,   3,   9,   ?

$\times 3 \quad -3 \quad \times 3 \quad -3$

Rule:
Multiply by 3,
subtract 3.

Think:
$9 - 3 = 6$

2, 6, 3, 9, 6

**Write the rule. Then complete the pattern.**

**1.** 5, 10, 15, _?_

**2.** 1, 3, 5, 7, _?_

**3.** 45, 40, 35, 30, _?_

**4.** 16, 8, 4, _?_

**5.** 1, 3, 9, _?_

**6.** 18, 24, 30, 36, _?_

**7.** 2, 4, 3, 6, 5, _?_

**8.** 16, 8, 12, 6, 10, _?_

**9.** 2, 3, 6, 7, 14, _?_

**10.** 9, 18, 6, 12, 4, _?_

**Copy and complete each pattern.**

11.
$$\begin{array}{cccc} 6 & 7 & 8 & 9 \\ \times 3 & \times 3 & \times 3 & \times 3 \\ \hline ? & ? & ? & ? \end{array}$$

12.
$$\begin{array}{cccc} 9 & 9 & 9 & ? \\ \times 6 & \times 7 & \times 8 & \times ? \\ \hline ? & ? & ? & ? \end{array}$$

13.
$$\overset{?}{4\overline{)16}} \quad \overset{?}{4\overline{)20}} \quad \overset{?}{4\overline{)24}} \quad \overset{?}{?\overline{)?}}$$

14.
$$\overset{?}{7\overline{)35}} \quad \overset{?}{7\overline{)42}} \quad \overset{?}{7\overline{)49}} \quad \overset{?}{?\overline{)?}}$$

## PROBLEM SOLVING

15. May 2 is on a Wednesday. What are the dates of the next four Wednesdays?

16. June 1 is on a Friday. What date is the following Monday? What are the dates of the next three Mondays?

17. August 31 is on a Thursday. What are the dates of the four Thursdays before August 31?

18. February 28 is on a Tuesday. What is the date of the Tuesday two weeks before February 28? three weeks before it?

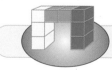 **Make Up Your Own**

19. Make up your own patterns like the ones on page 274. Write a rule for each of your patterns.

# Fact Families

Gary's teacher asked him to write two multiplication
and two division facts using 3, 4, and 12.
This is what Gary wrote:

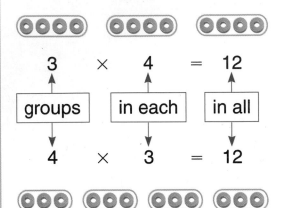

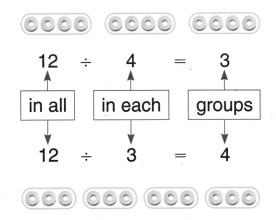

These sentences form a fact family for multiplication
and division.

## Write the fact family for each group.

1.
2.
3.

4.
5.
6.

**Copy and complete each fact family.**

**7.** $\underline{?} \times 3 = 12$
$12 \div 3 = \underline{?}$
$3 \times \underline{?} = 12$
$12 \div \underline{?} = 3$

**8.** $5 \times \underline{?} = 15$
$15 \div \underline{?} = 5$
$\underline{?} \times 5 = 15$
$15 \div 5 = \underline{?}$

**9.** $\underline{?} \times 4 = 20$
$20 \div 4 = \underline{?}$
$4 \times \underline{?} = 20$
$20 \div \underline{?} = 4$

**10.** $7 \times \underline{?} = 63$
$63 \div \underline{?} = 7$
$\underline{?} \times 7 = 63$
$63 \div 7 = \underline{?}$

**11.** $\underline{?} \times 5 = 10$
$10 \div 5 = \underline{?}$
$5 \times \underline{?} = 10$
$10 \div \underline{?} = 5$

**12.** $5 \times \underline{?} = 35$
$35 \div \underline{?} = 5$
$\underline{?} \times 5 = 35$
$35 \div 5 = \underline{?}$

**13.** $\underline{?} \times 3 = 18$
$18 \div 3 = \underline{?}$
$3 \times \underline{?} = 18$
$18 \div \underline{?} = 3$

**14.** $7 \times \underline{?} = 14$
$14 \div \underline{?} = 7$
$\underline{?} \times 7 = 14$
$14 \div 7 = \underline{?}$

**15.** $\underline{?} \times 6 = 24$
$24 \div 6 = \underline{?}$
$6 \times \underline{?} = 24$
$24 \div \underline{?} = 6$

**Write the complete fact family for each.**

**16.** 2, 8, 16

**17.** 4, 8, 32

**18.** 8, 9, 72

**19.** 3, 9, 27

**20.** 5, 6, 30

**21.** 3, 8, 24

**22.** 6, 8, 48

**23.** 5, 9, 45

**24.** 6, 7, 42

**25.** 5, 8, 40

**26.** 4, 7, 28

**27.** 4, 9, 36

**28.** 3, 7, 21

**29.** 6, 9, 54

**30.** 7, 8, 56

## Critical Thinking

**Write the fact family for each set of numbers.**

**31.** 3, 3, 9

**32.** 5, 5, 25

**33.** 9, 9, 81

**34.** How are these fact families like the ones on page 276 and above? How are they different?

**35.** How many facts can you write for 0, 0, 7? (*Hint:* you *cannot* divide by 0.)

# Applying Facts

**Use the skills and strategies you have learned to solve each problem.**

1. Jose has 8 party bags. Each party bag has 6 small cars. How many cars are there in all?

2. Taryn planted 3 plants in each of 6 flower boxes. How many plants did she plant?

3. How many ways can you divide the cubes above into equal groups? Write the facts for each model.

4. Thermal socks are on sale for $5 a pair. Mary buys 6 pairs. How much money does she spend?

5. Sarah spends 72 cents. She buys 9 pencils. Each pencil is the same price. How much does 1 pencil cost?

6. Robert read a book with 72 pages. He read 8 pages a day. How many days did it take to read the book?

7. There were 36 songs played during a parade. Each band played 4 songs. How many bands were in the parade?

8. Enrique baked 24 cookies for his birthday party. Each friend received 3 cookies. At most, how many friends were at the birthday party?

9. Nathan puts 7 shoe boxes in his closet. There are a pair of shoes in each box. How many shoes does Nathan have in his closet?

10. Daniel has 8 badges. Frank has 3 times as many badges. Does Frank have more or less than 30 badges? Explain.

11. Jo has 5 stacks of 7 pennies. Jerry has 8 stacks of 4 pennies. Who has more money? Explain.

**Read each problem carefully before you solve it.**

12. The art club has 9 members. The computer club has 3 times as many members. How many members does the computer club have?

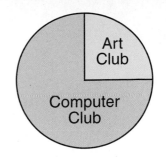

13. A band has 54 instruments. They are on 6 shelves. Each shelf holds the same number of instruments. How many instruments are on each shelf?

14. There are 63 trees. They are planted in rows of 7. How many rows of trees are there?

15. Mandy pays a library fine of 40¢. Her book was 5 days overdue. How much did she pay for each day?

## Challenge

**Use counters to act out each problem.**

16. Jack is packing 43 books. He can fit 5 books into each box. What is the least number of boxes he needs to pack all the books?

17. Jill has 33 beads. She puts exactly 5 beads on each bracelet she makes. How many more beads does she need to finish the last bracelet?

18. Ms. Levine gets 2 bunches of grapes from each of 3 children. There are 5 grapes in each bunch. How many grapes did she get in all?

# TECHNOLOGY

## FOR-NEXT loop

You can write a BASIC program to list multiplication and division facts.

Use a **LET** statement and a **FOR-NEXT** loop.

A LET statement assigns a value to a variable.

A FOR-NEXT loop tells the computer how many times the statements within the loop should be executed. It also assigns a value to the given variable.

```
10  PRINT "List the division facts for 2."
20  LET y = 2
30  FOR x = 0 to 18 step 2
40  PRINT x; "÷"; y; "="; x/y
50  NEXT x
60  END
```

Assigns the value 2 to *y*.

Step 2 tells the computer to skip count by 2s and store the even numbers 0–18 in *x*. The first number stored is 0. The second number stored is 2. The last number stored is 18.

Tells the computer to loop back to line 30 and assign the next number to *x*.

Semicolons print the output items right next to each other. Anything inside quotation marks will be printed exactly.

### Program Output

List the division facts for 2.

| | |
|---|---|
| $0 \div 2 = 0$ | $10 \div 2 = 5$ |
| $2 \div 2 = 1$ | $12 \div 2 = 6$ |
| $4 \div 2 = 2$ | $14 \div 2 = 7$ |
| $6 \div 2 = 3$ | $16 \div 2 = 8$ |
| $8 \div 2 = 4$ | $18 \div 2 = 9$ |

**Write the output for each program.**

1. 10 PRINT "List the multiplication facts for 10."
   20 LET $y = 10$
   30 FOR $x = 0$ to 9
   40 PRINT $x$; "×"; $y$; "="; $x*y$
   50 NEXT $x$
   60 END

2. 10 PRINT "List the multiplication facts for 11."
   20 LET $y = 11$
   30 FOR $x = 0$ to 9
   40 PRINT $x$; "×"; $y$; "="; $x*y$
   50 NEXT $x$
   60 END

3. 10 PRINT "List the division facts for 3."
   20 LET $y = 3$
   30 FOR $x = 0$ to 27 step 3
   40 PRINT $x$; "÷"; $y$; "="; $x/y$
   50 NEXT $x$
   60 END

**Use the programs above to answer each question.**

4. What does the $x$ represent in the multiplication programs? What does the $y$ represent?

5. What does the $x$ represent in the division program? What does the $y$ represent?

6. What does *step 3* tell the computer to do?

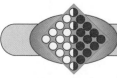

**Finding Together**

7. Write a multiplication program and a division program to list the facts for 12.

# 8-16 | Problem Solving: Guess and Test   *Algebra* ✓

**Problem:**   During vacation Gary and Robin collected
bumper stickers from different towns.
The number of stickers Gary collected
is 6 times the number Robin collected.
Together they have 28 bumper stickers.
How many did each child collect?

**1 IMAGINE**   Look at the picture.
Think about the problem.

**2 NAME**   *Facts:*   Gary — 6 times the number
Robin collected
Together — 28 bumper stickers

*Question:*   How many bumper stickers
did each child collect?

**3 THINK**   First *guess.*   | Robin's stickers |   | Gary's stickers |
2          12 ◄— 6 × 2

Then *test* by adding. 2 + 12 = 14 too low

**4 COMPUTE**   Guess and test again. Use a table.

| Robin's Stickers | Gary's Stickers | Test 28 stickers in all |
|---|---|---|
| 2 | 6 × 2 = 12 | 2 + 12 = 14  too low |
| 5 | 6 × 5 = 30 | 5 + 30 = 35  too high |
| 4 | 6 × 4 = 24 | 4 + 24 = 28 |

**5 CHECK**   Robin collected 4 stickers and
Gary collected 24.
Does 24 equal 6 times 4? Yes.
Does 24 + 4 = 28? Yes.

## Solve by Guess and Test.

**1.** Lydia spent 55¢ for a souvenir. She paid the cashier with 4 coins. What coins did she give the cashier?

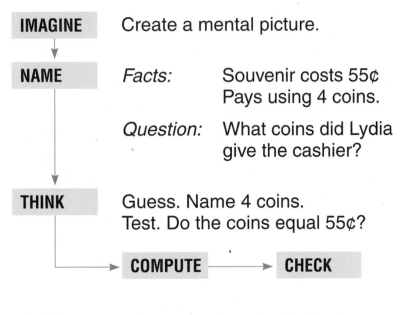

**IMAGINE**    Create a mental picture.

**NAME**

*Facts:*    Souvenir costs 55¢
Pays using 4 coins.

*Question:*    What coins did Lydia give the cashier?

**THINK**    Guess. Name 4 coins.
Test. Do the coins equal 55¢?

**COMPUTE** ⟶ **CHECK**

**2.** The sum of two numbers is 11. Their product is 24. What are the two numbers?

**3.** The quotient of two numbers is 9. The sum of the same two numbers is 70. What are the two numbers?

**4.** Eight friends went to the lake. Some paid $4 each for tickets to the water slide. The rest paid $7 each for boat rentals. The group paid $38 in all. How many friends rented boats?

**5.** Sal bought some postcards. If he sends the same number to each of 6 friends, he will have none left over. If he sends the same number to each of 7 friends, he will have 3 left over. How many postcards did Sal buy?

# Problem-Solving Applications

**Solve each problem and explain the method you used.**

1. Sue bought 4 bunches of bananas. Each bunch had 6 bananas. How many bananas did Sue have in all?

2. Adam bought 6 packs of 9 apples. Mark bought 7 packs of 8 plums. Who bought more fruit? How much more?

3. Each silk flower costs 9¢. How much money does Judy need to buy 7 silk flowers?

Imagine

4. Lee uses 8 oranges in each fruit salad. How many salads can she make with 32 oranges?

Name

5. There are 4 rows of 2 papayas in each tray. How many papayas are in 4 trays?

Think

6. Luis packs grapefruits in 4 packages. Each package has 2 rows of 3 grapefruits. How many grapefruits does Luis pack?

Compute

7. The fruit-of-the-month club offers 18 different kinds of fruit. The Heltz family receives 2 kinds of fruit each month. How many months will it take the family to get all 18 fruits?

Check

8. Ms. Price spent $36 on four identical fruit baskets. How much did each fruit basket cost?

9. A chef uses 40 cherries to decorate 8 cakes in the same way. How many cherries does the chef use on one cake?

**Choose a strategy from the list or use another strategy you know to solve each problem.**

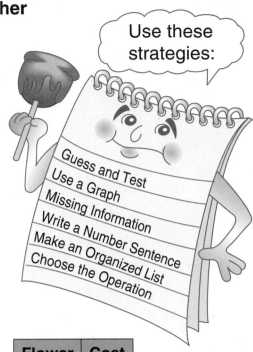

Use these strategies:

Guess and Test
Use a Graph
Missing Information
Write a Number Sentence
Make an Organized List
Choose the Operation

10. The fruit stand used 54 pieces of fruit to make 6 baskets. There is an equal number of fruit in each basket. How many pieces of fruit are in each basket?

11. There are 8 caramel apples in each row on a tray. How many caramel apples are there in all?

12. Cantaloupes cost 85¢ each. Ross used 4 coins to pay for one. What coins might he have used?

13. The fruit stand sells silk flowers. Julie spent 31¢ for some daisies and violets. How many of each did she buy?

14. Mr. Lee bought a total of 10 mums and daisies. He bought more mums than daisies. He spent 84¢. How many of each did he buy?

| Flower | Cost |
|--------|------|
| mum    | 9¢   |
| daisy  | 7¢   |
| violet | 5¢   |

**Use the graph for problems 15–18.**

15. How many Macintosh apple trees does the orchard have?

16. How many more Yellow Delicious trees than Red Delicious trees are there?

17. The gardener plants a dozen more Red Delicious trees. How many trees of that kind are there now?

**Apple Orchard**

| Kind | Number of Trees |
|------|-----------------|
| Macintosh | 🍎🍎🍎🍎🍎 |
| Red Delicious | 🍎🍎🍎 |
| Yellow Delicious | 🍎🍎🍎🍎🍎🍎🍎 |
| Key: Each 🍎 = 6 apple trees. | |

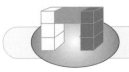

**Make Up Your Own**

18. Write a problem of your own using the graph. Then solve it.

# Chapter 8

# Chapter Review and Practice

**Find the product.** (See pp. 254–261.)

**1.** 7 × 7 = _?_    **2.** 5 × 8 = _?_    **3.** _?_ = 3 × 6    **4.** _?_ = 4 × 8

**5.**   9    **6.**   6    **7.**   9    **8.**   8    **9.**   6    **10.**   8
   ×5         ×8         ×2         ×9         ×1            ×3

**11.**   0    **12.**   6    **13.**   7    **14.**   8    **15.**   9    **16.**   7
   ×5         ×7         ×3         ×8         ×4            ×5

**Multiply.** (See pp. 262–263.)

**17.** 4 × 1 × 6 = _?_                **18.** 8 × (4 × 1) = _?_

**19.** 6 × (2 × 4) = _?_              **20.** 6 × 6 × 0 = _?_

**Find the quotient.** (See pp. 264–273.)

**21.** 32 ÷ 8        **22.** 63 ÷ 7        **23.** 45 ÷ 9        **24.** 64 ÷ 8

**25.** 6)‾42‾        **26.** 7)‾35‾        **27.** 6)‾54‾        **28.** 7)‾56‾        **29.** 8)‾72‾

**Write the rule. Then complete the pattern.** (See pp. 274–275.)

**30.** 1, 6, 11, 16, _?_                **31.** 3, 6, 4, 8, 6, _?_

**Write the complete fact family for each.** (See pp. 276–277.)

**32.** 7, 8, 56                **33.** 5, 6, 30                **34.** 9, 7, 63

## PROBLEM SOLVING
(See pp. 278–279, 282–285.)

**35.** The difference of two numbers is 3. Their product is 54. What are the two numbers?

**36.** Mr. Rodriguez spent $72 on 8 identical cassettes for his nieces and nephews. How much did each cassette cost?

(See *Still More Practice*, pp. 449–450.)

# BUILDING SQUARE NUMBERS

**Study the grids below.**

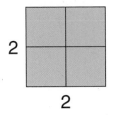

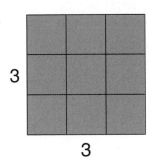

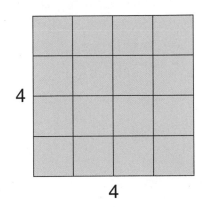

| 2 rows | 3 rows | 4 rows |
|---|---|---|
| 2 columns | 3 columns | 4 columns |
| 4 squares in all | 9 squares in all | 16 squares in all |
| $2 \times 2 = 4$ | $3 \times 3 = 9$ | $4 \times 4 = 16$ |

Each grid has the same number of squares in each row and in each column. When both factors are the same, the product is called a **square number**. So the numbers 4, 9, and 16 are square numbers.

**Use grid paper to find the square numbers.
Follow the pattern above.**

**1.** $7 \times 7 = $ _?_

**2.** $6 \times 6 = $ _?_

**3.** $5 \times 5 = $ _?_

**4.** $8 \times 8 = $ _?_

## PROBLEM SOLVING

**5.** I am a number between 60 and 70. I am a square number. The difference between my digits is 2. What number am I?

**6.** I am a number between 40 and 50. I am a square number. The sum of my digits is 13. What number am I?

# Check Your Mastery

## Performance Assessment

**Use a number line.**

1. Show how many 6s are in 30. Then write the related division and multiplication sentences.

**Find the product.**

2. $9 \times 8$    3. $6 \times 6$    4. $7 \times 8$    5. $6 \times 9$

6. $\begin{array}{r} 9 \\ \times 7 \\ \hline \end{array}$    7. $\begin{array}{r} 0 \\ \times 8 \\ \hline \end{array}$    8. $\begin{array}{r} 9 \\ \times 9 \\ \hline \end{array}$    9. $\begin{array}{r} 7 \\ \times 4 \\ \hline \end{array}$    10. $\begin{array}{r} 8 \\ \times 8 \\ \hline \end{array}$    11. $\begin{array}{r} 6 \\ \times 1 \\ \hline \end{array}$

12. $\begin{array}{r} 9 \\ \times 5 \\ \hline \end{array}$    13. $\begin{array}{r} 8 \\ \times 6 \\ \hline \end{array}$    14. $\begin{array}{r} 6 \\ \times 7 \\ \hline \end{array}$    15. $\begin{array}{r} 9 \\ \times 4 \\ \hline \end{array}$    16. $\begin{array}{r} 8 \\ \times 3 \\ \hline \end{array}$    17. $\begin{array}{r} 7 \\ \times 7 \\ \hline \end{array}$

18. $4 \times (6 \times 1) = \underline{\ ?\ }$    19. $3 \times 3 \times 7 = \underline{\ ?\ }$

**Find the quotient.**

20. $48 \div 8$    21. $49 \div 7$    22. $36 \div 9$    23. $42 \div 6$

24. $7\overline{)63}$    25. $7\overline{)35}$    26. $8\overline{)56}$    27. $9\overline{)81}$    28. $8\overline{)40}$

**Write the rule. Then complete the pattern.**

29. 1, 8, 15, 22, $\underline{\ ?\ }$    30. 18, 9, 10, 5, 6, $\underline{\ ?\ }$

**Write the complete fact family for each.**

31. 9, 8, 72    32. 6, 7, 42    33. 6, 9, 54

**PROBLEM SOLVING**    *Use a strategy you have learned.*

34. Posters are on sale for $6 each. Chad buys 8 posters. How much money does he need?

35. Mr. Hunt spent $42 on identical gifts for his 6 children. How much did each gift cost?

# Geometry

9

## Sunflakes

If sunlight fell like snowflakes,
gleaming yellow and so bright,
we could build a sunman,
we could have a sunball fight,
we could watch the sunflakes
drifting in the sky.
We could go sleighing
in the middle of July
through sundrifts and sunbanks,
we could ride a sunmobile,
and we could touch sunflakes—
I wonder how they'd feel.

*Frank Asch*

**In this chapter you will:**

Explore plane and space figures,
  perimeter, and volume
Learn about congruent and similar figures
  and symmetry
Study how figures move
Investigate ordered pairs and area
Solve problems using logical reasoning

**Critical Thinking/Finding Together**

Look at the sunflakes. How are they alike?
How are they different? Trace a sunflake
and fold the tracing in half. Tell what happens.

# 9-1 Polygons and Circles

A **polygon** is any closed flat figure with straight sides.
Polygons also have special names.

triangle

square

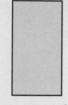
rectangle

pentagon

hexagon

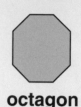
octagon

Any polygon with 4 straight sides is called a **quadrilateral**.
A quadrilateral that has opposite sides, parallel and equal,
is a **parallelogram**.

## Hands-On Understanding

**You Will Need:** dot paper, record sheet

**Step 1** Label the columns on your record sheet
with the following headings:

| Figure | Number of Sides | Number of Corners |
|--------|-----------------|-------------------|
|        |                 |                   |

**Step 2** Copy the names of the following figures
onto your record sheet: triangle, square,
rectangle, pentagon, hexagon and octagon.

**Step 3** Draw each figure listed on your record sheet
on dot paper.

How many sides does each figure have?

How many corners does each figure have?

**Step 4**    Draw a three-sided figure that is not a polygon.

Explain why this figure is not classified as a polygon.

**Step 5**    Use a coin to trace a circle.

How many sides does a circle have?

How many corners does a circle have?

Is a circle a polygon? Explain your answer.

# Communicate

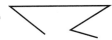
Discuss

1. Look at the data on your record sheet.
   What pattern do you notice?

2. Explain why squares and rectangles are parallelograms.
   How are they alike? How are they different?

**Is each figure a polygon? Write *Yes* or *No*.**

3.

4.

5.

6.

7.

8.

9.

10.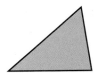

## Project

**Cut four 6-in. strips and four 4-in. strips from drinking straws.**
From these 8 strips how many of each of these figures can you form?

11. triangles      12. squares      13. pentagons

14. rectangles      15. hexagons      16. octagons

A **line** is straight. It goes on forever in both directions.

A **line segment** is part of a line with two endpoints.

A **ray** is part of a line with one endpoint. It goes on forever in one direction.

**Name each: line, line segment, ray, or none of these.**

1.

2. ⟷

3.

4.

5.

6.

7.

8.

292

## Special Lines

**Parallel lines** are lines on a flat surface that never meet.

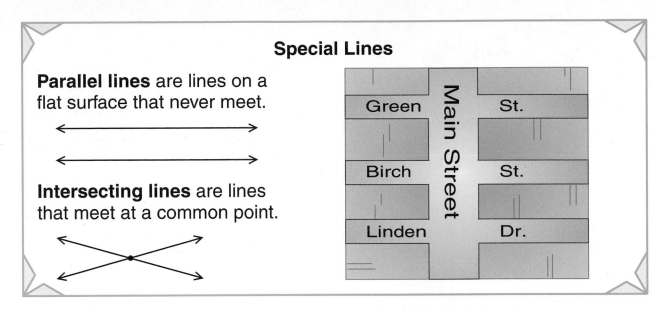

**Intersecting lines** are lines that meet at a common point.

**Using the street map above, answer these questions.**

9. Name two streets that are parallel.

10. Name two streets that intersect.

**Copy these lines onto dot paper.**

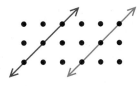

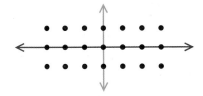

**Draw a line that:**

11. is parallel to the red line.

12. is parallel to the green line.

13. intersects the red line and the blue line.

14. intersects the green line and the purple line.

## Critical Thinking

Communicate

15. Which picture is most like a ray? Explain.

# Angles

**Angles** are formed by two rays
that start out from the same point.

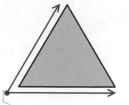

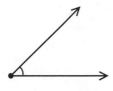 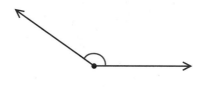 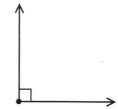 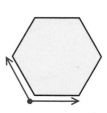

A **right angle** is an angle that
forms a square corner.

  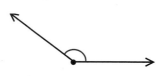 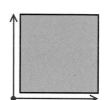

| right angle | less than a right angle | greater than a right angle |
|---|---|---|

**How many angles does each figure have?**

**1.**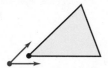
4

**2.**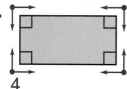

**3.**

**4.**

**Tell if the angle is a right angle, greater than
a right angle, or less than a right angle.**

**5.**

**6.**

**7.**

**8.**

**Tell if the angle formed by the hands of the clock is a right angle, greater than a right angle, or less than a right angle.**

9.

10.

11.

**Use Figures A and B for exercises 12–14.**

12. Which figure has 1 right angle?

13. Which figure has 2 right angles?

14. Which figure has 2 angles that are less than a right angle?

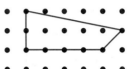

Figure A

**Use grid paper to make block capital letters.**

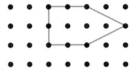

Figure B

15. How many right angles do the letters **L** and **F** each have?

16. Which letters of the alphabet have more than three right angles?

 **Critical Thinking**

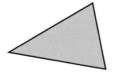 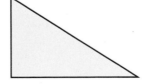 

**Use the figures above.**

17. A triangle that has one right angle is a **right triangle**. Identify the right triangle.

295

**Congruent Figures**

**Congruent figures** have the same size and the same shape.

The triangles are congruent figures.

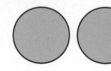

The circles are congruent figures.

The rectangles are *not* congruent figures.

**Copy the figures onto dot paper.**
**Then draw figures that are congruent to each.**

1.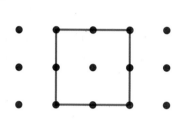

2.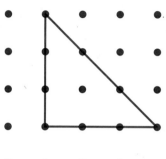

3.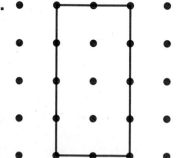

4.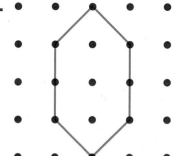

5. Explain in your Math Journal how you can tell whether two figures are congruent.

Math Journal

**Do the pairs look congruent? Write *Yes* or *No*.**

6.

7.

8.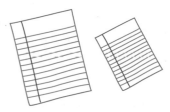

9.

10.

11.

**Find the two congruent figures. Write their letters.**

12. a.    b.    c.    d.

13. a.    b.    c.    d.

 **Challenge**

14. Laura writes a letter to her cousin. She puts the letter in an envelope. Then she puts a stamp on the envelope. Identify and draw the figures that represent Laura's letter, envelope, and stamp. Are any of these figures congruent? Explain your answer.

**Similar figures** have the same shape.
They may or may not be the same size.

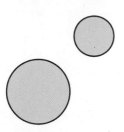

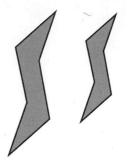

same shape
different size

same shape
same size

same shape
different size

**Do the figures look similar? Write *Yes* or *No*.**

**1.**

**2.**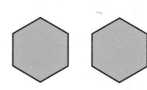

**3.**

**4.**

**5.**

**6.**

**Find the two similar figures. Write their letters.**

**7.** **a.**      **b.** 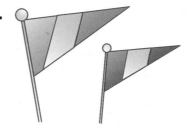     **c.**     **d.**

**8.** **a.**      **b.** 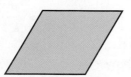 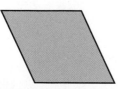     **c.**       **d.**

**Draw each figure on dot paper.**
**Then double each side to draw a similar figure.**

9.

10.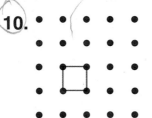

11.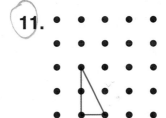

12. Explain in your Math Journal the difference between congruent and similar figures.

13. Can congruent figures be similar? Explain your answer.

14. Can similar figures be congruent? Explain your answer.

Math Journal ✓

Communicate ✓

## Connections: Literature

The figure below is called a **tangram**. A tangram is a geometric puzzle that originated in China over 4000 years ago. It starts out as a square and then is cut into seven pieces. Tangrams are traditionally used to tell stories. You might want to read *Grandfather Tang's Story* by Ann Tompert.

**Trace the tangram on a sheet of paper, and then cut along the lines to separate the shapes.**

15. Name the shapes of the tangram.

16. How are shapes 2 and 5 alike?

17. Is shape 1 similar to shapes 2 and 5? Explain your answer.

18. Is shape 4 congruent to any other shape?

19. Which shape(s) contain right angles?

20. Which shape(s) contain angles less than a right angle? more than a right angle?

21. Use all the tangram pieces to make a design or a shape. Write a few sentences describing your design.

**Ordered Pairs**

**Ordered pairs** locate points on a graph.

▶ Look at the graph. What animal is at point (5,1)?

To find out:

- Begin at 0.

- The **first number** tells you to move 5 spaces to the right.

- The **second number** tells you to move 1 space up.

The cat is located at point (5,1).

▶ Locate the mouse. Name the ordered pair for that point.

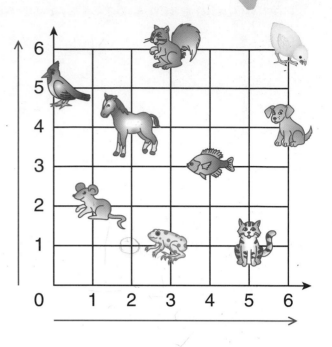

The mouse is at the point (1,2).    The mouse is 1 space to the *right* and 2 spaces *up*.

---

**Use the graph to answer each question.**

**1.** What animal is at (2,4)?    **2.** What animal is at (3,1)?

**Write the ordered pair for each animal.**

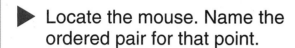

**3.** dog        **4.** bird        **5.** fish        **6.** squirrel

**PROBLEM SOLVING**

**7.** I am the animal two spaces to the right of the frog. What am I? Name the ordered pair of my position.

*Communicate* ✓

**8.** Tell how you would find the ordered pair to locate the chick.

**Write the letter for each ordered pair. Use the graph at the right.**

9. (1,2)   10. (4,4)   11. (3,3)

12. (2,1)   13. (1,5)   14. (4,1)

15. (2,3)   16. (3,2)   17. (4,5)

18. (4,0)   19. (1,6)   20. (5,5)

21. (5,3)   22. (0,1)   23. (6,0)

**Write the ordered pair for each letter.**

24. G     25. I     26. A

27. X     28. N     29. H

30. O     31. R     32. U

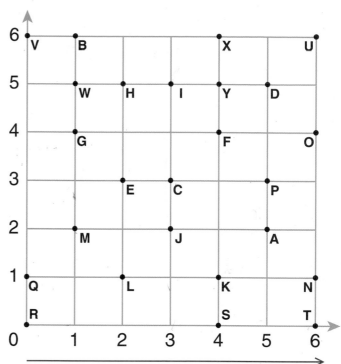

**Write the place for each ordered pair. Use the graph at the left to locate each.**

33. (1,5)          34. (2,3)

35. (1,1)          36. (3,6)

37. (3,0)          38. (1,2)

**Write the ordered pair for each place.**

39. Playground

40. Velma's House

41. Park

42. Susan's House

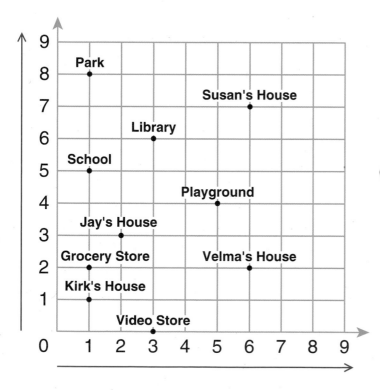

# Symmetry

Each of these figures can be folded along the dashed line so that the two parts match.

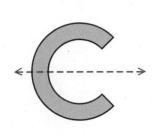  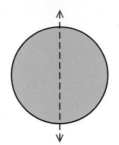  

The dashed line is a **line of symmetry**.

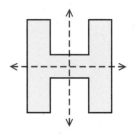

Some figures have two or more lines of symmetry.

---

**Is the dashed line a line of symmetry? Write *Yes* or *No*.**

1.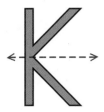

2.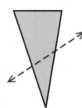

3.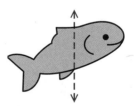

4.

**Use dot paper. Draw each figure and its matching part.**

5.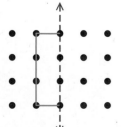

6.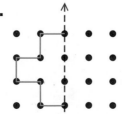

7.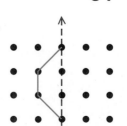

8.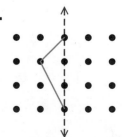

9. Take an 8-inch square piece of paper. Fold it to see how many lines of symmetry a square has.

# Slides, Flips, and Turns

You can move an object in any direction.

up

left | right

down

You can also move an object by sliding, flipping, or turning it.

- A **slide** is a move in a straight line without changing direction.

- A **flip** is a move over an imaginary line. A flip puts the object in the opposite direction or on the opposite side.

- A **turn** is a move around a point. A turn moves an object clockwise or counterclockwise.

**Write how the object was moved: slide, flip, or turn.**

1.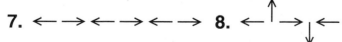

2.

3.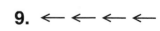

4.

5.

6.

**Tell if the patterns show slides, flips, or turns.**

7. ← → ← → ← →   8. ← ↑ → ↓ ←   9. ← ← ← ←

10. Name the two moves.   ← → →

303

# 9-9 Exploring Space Figures

corner

edge    face

**cube**

6 faces
12 edges
8 corners

**Space figures** are all around us.
Some space figures have faces,
edges, and corners.

A **face** is a flat surface surrounded by
   line segments.
Two faces meet at a line segment called
   an **edge**.
Three or more edges meet at a **corner**.

Some space figures do not have faces, edges, and
corners because they do *not* have line segments.

Remember: Space figures have special names.

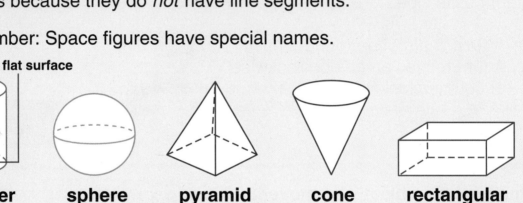

flat surface

curved
surface

**cylinder**     **sphere**     **pyramid**     **cone**     **rectangular prism**

## Hands-On Understanding

**You Will Need:** a net for a cube, a pyramid, and a rectangular prism;
            tape; record sheet

**Step 1**   Label the columns on your record sheet with the
         following headings:

| Space Figure | Number of Flat Surfaces | Number of Curved Surfaces | Number of Edges | Number of Corners |
|---|---|---|---|---|
|  |  |  |  |  |

| **Step 2** | Use the net like the one at the right. Fold it to form a space figure. |
|---|---|

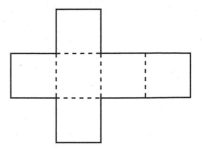

What space figure did you form?

How many flat surfaces, edges, and corners does the space figure have?

| **Step 3** | Repeat Step 2 for the following nets. |
|---|---|

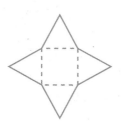   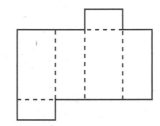

A net is the shape made by opening a solid figure and laying it flat.

| **Step 4** | Now study the pictures of the cylinder, sphere, and cone on page 304. |
|---|---|

How many flat surfaces, curved surfaces, edges, and corners does each have?

## Communicate

Discuss ✓

1. Compare the data in your record sheet for a cube and a rectangular prism. What do you notice?

2. Which space figures have only flat surfaces?

3. Which space figures have both flat surfaces and curved surfaces?

4. Which space figure has only a curved surface?

## Skills to Remember

**Add.**

**5.** 7 + 8 + 4 + 5

**6.** 52 + 43 + 52 + 43

**7.** 19 + 19 + 19 + 19

**8.** 108 + 46 + 62 + 387

*Update your skills. See page 13.*

# 9-10 Exploring Perimeter

**Perimeter** is the distance around a figure or an object.

 Hands-On Understanding

**You Will Need:** geoboard or dot paper, rubber bands, 36 inches of string or yarn, record sheet, pen, ruler

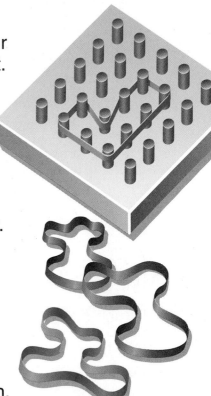

| **Step 1** | Stretch rubber bands around pegs of your geoboard to copy the polygon at the right. |
| --- | --- |
| **Step 2** | Count the units around the polygon. |

 is 1 unit

What is the perimeter of the polygon?

| **Step 3** | Now use string to measure the perimeter. Pick a starting point, and wrap the string around the polygon until you are back at the starting point. |
| --- | --- |
| **Step 4** | Mark this point on the string with a pen. |
| **Step 5** | Lay the string along the ruler to measure the length of the string to the nearest inch. |

What is the perimeter of the polygon measured to the nearest inch?

## Communicate

Discuss ✓

1. Which is a better method to measure perimeter? Explain.

306

**Use a string to measure the perimeter of these figures. Then measure the length of the string to the nearest inch.**

**2.** a closed folder

**3.** an open folder

**4.** a textbook

**5.** a flat desk top

**Use your metric ruler to find the perimeter.**

**6.**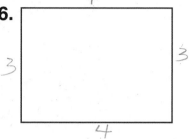

**7.**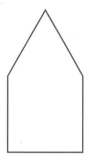

**8.**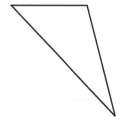

**9.**

**10.**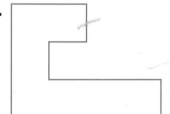

**11.**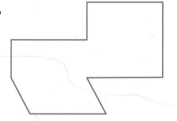

## PROBLEM SOLVING

**12.** Alice made a square frame for her needlepoint. Each side measured 8 in. What was the perimeter of the frame?

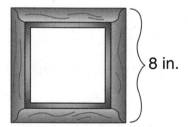

8 in.

**13.** Mr. Diaz uses wood to make a new frame around the bedroom window. One side is 67 cm, and another is 95 cm. What is the perimeter of the window frame?

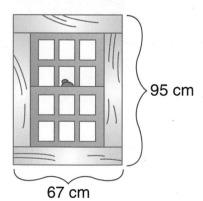

95 cm

67 cm

# Area

Jamie has a new picture. One way to describe its size is to find its **area**.

Area is the number of square units needed to cover a surface.

 1 square unit

You can count the number of square units to find the area.

The area of the picture is 36 square units.

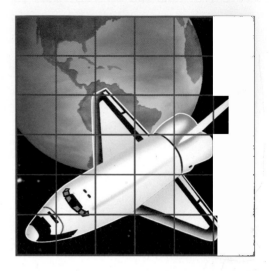

**Study these examples.**

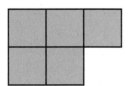

The area is 5 square units.

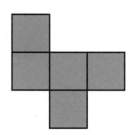

The area is 5 square units.

**Write how many square units.**

1.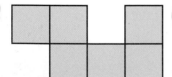

2.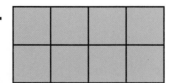

3.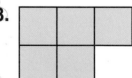

**Use grid paper. Draw a figure for each area.**

4. 12 square units

5. 4 square units

6. 25 square units

## PROBLEM SOLVING  Use grid paper to help.

7. Mark makes a picture that measures 5 square units across and 2 square units down. What is the area of his picture?

8. Charisse puts a puzzle together. The finished puzzle measures 6 square units across and 3 square units down. What is the area of the puzzle?

9. Maria opened a box and laid it flat. (See right.) She wants to find the number of square units of wrapping paper needed to cover the box. How much wrapping paper does Maria need?

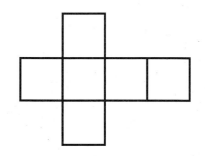

 **Share Your Thinking**

*Communicate* ✓

10. Estimate which figure has the greater area. Check your answer. Explain the method you used to determine your answer.

a.

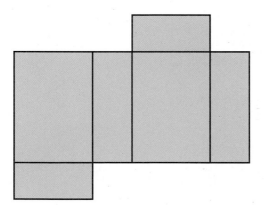

b.

11. Use grid paper to draw as many different figures as you can that have an area of 10 square units. How many figures did you draw? Do you think there are more figures that you can draw that have an area of 10 square units? Explain.

# Volume

The **volume** of a space figure is the number of cubic units the figure contains.

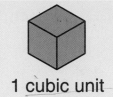

1 cubic unit

## Hands-On Understanding

**You Will Need:** connecting cubes, record sheet

**Step 1** Use your connecting cubes to model 3 groups of 4 cubes, forming straight lines or stacks.

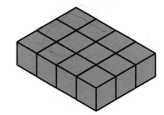

**Step 2** Place each stack of cubes next to the others.

How many cubes does your new model contain?

What space figure did you make?

What is the volume of your space figure?

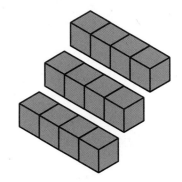

**Step 3** Using more cubes, repeat Steps 1 and 2.

How many cubes does this model contain?

**Step 4** Now place one of the models on top of the other.

What space figure did you make?

What is the volume of this space figure?

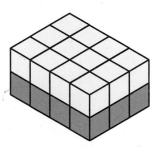

## Communicate

1. How did you find the volume of your space figure?

2. Compare your model from Step 2 with your model from Step 4. What do you notice?

**Find the volume. You may use connecting cubes.**

**3.**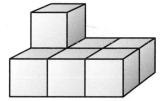

**4.**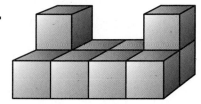

**5.**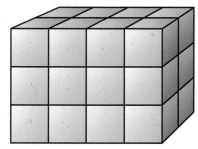

**6.**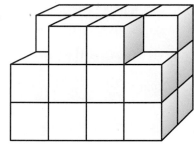

## PROBLEM SOLVING

**7.** Alice makes a pile of cubes. She makes one layer that has 3 rows of 3 cubes. She puts 2 more layers of the same size on top of the first layer. What is the volume of the pile?

**8.** Rick makes a box that is 2 cubes long, 3 cubes wide, and 2 cubes high. What is the volume of the box?

**9.** Look at the numbers for the length, width, and height of Rick's box in problem 8. Now look at your answer for the volume of the box. What do you notice?

*Communicate*

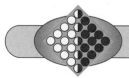 **Finding Together**

**Use connecting cubes to model:**

**10.** two different solid figures, each made of 8 cubic units.

**11.** two different solid figures, each made of 11 cubic units.

# Problem Solving: Logical Reasoning

**Problem:** Erin, Taylor, and Sabrina ran the 100-meter dash. Erin, wearing number 34, did *not* win. Taylor, wearing number 47, did *not* come in third. Taylor finished two seconds behind the runner wearing number 20. In what order did the three runners finish the race?

**1 IMAGINE** Create a mental picture.

**2 NAME**

*Facts:* Erin did not win.
Taylor did not come in third.
Taylor finished 2 seconds behind the runner wearing number 20.

*Question:* In what order did the three runners finish the race?

| Runner | Place | | |
|---|---|---|---|
| | 1st | 2nd | 3rd |
| Erin | no | no | yes |
| Taylor | no | yes | no |
| Sabrina | yes | no | no |

**3 THINK** Use the facts to complete the chart.
Write **yes** when the fact is true.
Write **no** when the fact is false.

**4 COMPUTE** Erin did *not* win. (Write no under 1st.)
Taylor did not finish 3rd. (Write no under 3rd.)
Since Taylor finished 2 seconds *behind* the runner wearing number 20, he was not 1st, but 2nd. (Write no under 1st; write yes under 2nd.)
Erin had to finish 3rd. (Write yes under 3rd.)
So Sabrina, wearing number 20, finished 1st.

**5 CHECK** Compare the completed chart to the facts given.

312

**Use Logical Reasoning to solve each problem.**

1. Each of the six cars on this Ferris wheel is of a different color and shape. Find the purple car and match each car color and shape, given that:

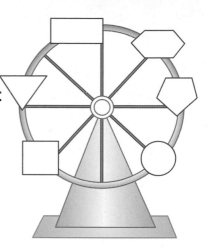

   • The red car, nearest the top, and the blue car both have only right angles.

   • The green car is between the yellow car and the brown car, which is a hexagon.

| **IMAGINE** | Create a mental picture. |
|---|---|
| **NAME** | *Facts:*   Six cars of different colors and shapes |
| | red car — nearest the top; has only right angles |
| | blue car — has only right angles |
| | green car — between a yellow and a brown car |
| | brown car — is a hexagon |
| | *Question:*   Find the purple car. |
| **THINK** | Use the facts and logical reasoning to fill in one color at a time. |

**COMPUTE** ⟶ **CHECK**

2. Jody's birthday is between the 9th and the 17th of April. It is before the 16th but after the 12th, and it is on an even-numbered date. When is Jody's birthday?

3. Only Bob, Lyn, John, and Vicky sit in the 3rd row. Neither Bob nor Lyn is first. Vicky sits between two boys. In what order do the children sit in the row?

4. Tina and Len made emblems. Tina's emblem had 4 triangles and a circle. Len's had 4 triangles, a circle, and 2 quadrilaterals. Which emblem did each child make?

**A**
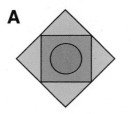

**B**
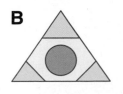

**Solve each problem and explain the method you used.**

1. Yvonne used space figures to make these 3 drawings. Name the space figures.

   **a.**    **b.**    **c.**

Imagine

Name

Think

Compute

Check

2. Jamal drew this kite.
   **a.** Is the kite a quadrilateral?
   **b.** Is the kite a rectangle?
   **c.** How many angles less than a right angle does the kite have?
   **d.** What colors are the sides that are parallel?
   **e.** How many lines of symmetry does Jamal's kite have?

3. Ben wants to make a square using the points on this grid.
   **a.** Name the ordered pairs of 4 points he should connect.
   **b.** What is the area of Ben's square?
   **c.** Tess connected (0,3); (4,3); (4,1); (2,1); (1,0). What polygon did she make?

**Write *always* or *never* to make each sentence true.**

4. A square __?__ has 4 right angles.

5. A circle is __?__ a polygon.

6. A line __?__ has 2 endpoints.

7. Intersecting lines __?__ meet.

8. When two squares are the same size, they are __?__ congruent figures.

9. When Leanne traces the shape of a dime and of a quarter, she __?__ makes similar figures.

**Choose a strategy from the list or use another strategy you know to solve each problem.**

Use these strategies:

Logical Reasoning
Guess and Test
Write a Number Sentence
Make a Table
Find a Pattern
Draw a Picture

**10.** The letter **H** looks the same after 1 half turn. Name 3 other letters that look the same after 1 half turn.

**11.** Bryan wrote the numbers between 4 and 10 once in these figures. He wrote even numbers in the square and odd numbers in the triangle. The difference in the products of the numbers inside the triangle and the square is 15. The sum of the numbers inside the circle is 20. In which figure did Bryan write each number?

**12.** Four people can sit around a square table. When two tables are pushed together, 6 people can sit around them. How many people can sit around five tables that are pushed together?

8 yards

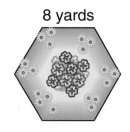

**13.** Tawny put a fence around her garden. Each side measured 8 yards. Find its perimeter.

**14.** Ms. Glenic made a quilt from squares of fabric. The quilt is 8 squares across and 5 squares down. How many squares did she use to make the quilt?

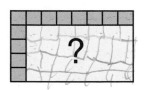

**15.** Koyi made a model building using 27 cubic units. The building is shaped like a cube. How many cubic units were in each row?

**16.** Jamal built the sand castle pictured at the right. Which space figures did he use in building his castle? How many of each space figure did he use?

**Name the figure. Then tell how many sides** *(See pp. 290–291, 304–305.)*
**or edges, faces, and corners each has.**

**1.**

**2.**

**3.**

**4.**

**Draw each.** *(See pp. 292–295.)*

**5.** line          **6.** parallel lines          **7.** right angle

**Do the figures look congruent?**
**Write *Yes* or *No*.**

**Do the figures look similar?**
**Write *Yes* or *No*.** *(See pp. 296–299.)*

**8.**

**9.**

**Write the ordered pairs.**

**Write slide, flip, or turn.**
*(See pp. 300–301, 303.)*

**10.**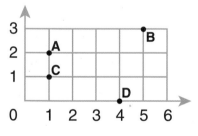

$$
\begin{array}{cc}
A & \underline{?} \\
B & \underline{?} \\
C & \underline{?} \\
D & \underline{?}
\end{array}
$$

**11.**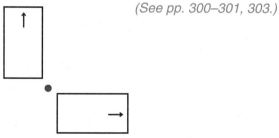

**Is the dashed line a line of**
**symmetry? Write *Yes* or *No*.**

**Find the perimeter.** *(See pp. 302, 306–307.)*

**12.**

**13.**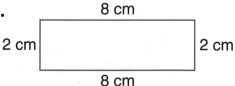

**Find the area in**
**square units.**

**Find the volume**
**in cubic units.** *(See pp. 308–311.)*

**14.**

**15.**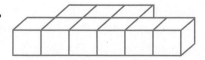

*(See Still More Practice, pp. 450–451.)*

# TRIANGLES

Triangles are polygons
with three sides.

- A triangle with one right
  angle is a **right triangle**.

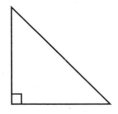

**right triangle**

- A triangle with at least
  two sides that are equal
  is an **isosceles triangle**.

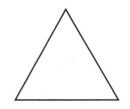

**isosceles triangle**

- A triangle with all the
  sides equal is an
  **equilateral triangle**.

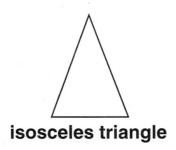

**equilateral triangle**

## Name each triangle.

1.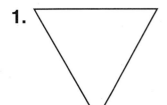

2.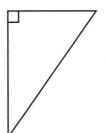

3.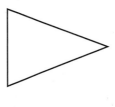

4.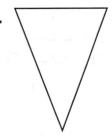

**5.** On dot paper draw your own right triangle, isosceles
triangle, and equilateral triangle. Then look at the 3
angles in each triangle you drew. Color the angles
less than a right angle blue.

Math
Journal

## Performance Assessment

Use dot paper.

**1.** Draw a rectangle and a square. Describe the figures.

**2a.** Draw 2 congruent triangles.

**2b.** Double each side of one triangle. Is the new triangle similar or congruent?

**Name the figure. Then tell how many sides or edges, faces, and corners each has.**

**3.**   **4.**   **5.**   **6.**

**Draw each.**

**7.** ray  **8.** intersecting lines  **9.** right angle

**Write the ordered pairs.**

**10.**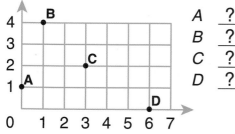

A  ?
B  ?
C  ?
D  ?

**Write slide, flip, or turn.**

**11.**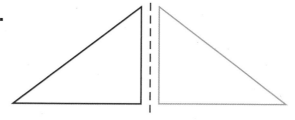

**Is the dashed line a line of symmetry? Write Yes or No.**

**12.**

**Find the perimeter.**

**13.** 8 cm  6 cm  10 cm

**Find the area in square units.**

**14.**

**Find the volume in cubic units.**

**15.**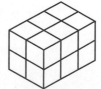

# Cumulative Review III

**Choose the best answer.**

**1.** What is the amount?

2 five-dollar bills,
2 nickels,
2 pennies

  **a.** $10.12
  **b.** $10.17
  **c.** $10.22
  **d.** $10.52

**2.** Choose the order from greatest to least.

  **a.** 4997; 5013; 4896
  **b.** 4997; 4896; 5013
  **c.** 5013; 4896; 4997
  **d.** 5013; 4997; 4896

**3.** Which is 2550 more than 3680?

  **a.** 1130    **b.** 5230
  **c.** 6230    **d.** 610

**4.** Which gives an answer that is less than 400?

  **a.** 500 − 46    **b.** 800 − 350
  **c.** 900 − 279    **d.** 700 − 302

**5.** $7 \times 6$

  **a.** 13    **b.** 36
  **c.** 42    **d.** not given

**6.** $4 \times \underline{\ ?\ } = 36$

  **a.** 7    **b.** 8
  **c.** 9    **d.** not given

**7.** $\underline{\ ?\ } = 9 \div 9$

  **a.** 0    **b.** 1
  **c.** 81    **d.** not given

**8.** $(3 \times 3) \times 8 = \underline{\ ?\ }$

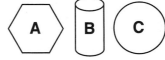

  **a.** 24    **b.** 64
  **c.** 72    **d.** 48

**9.** $8\overline{)40}$

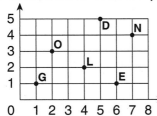

  **a.** 5    **b.** 6
  **c.** 20    **d.** not given

**10.** 54¢ ÷ 6

  **a.** 7¢    **b.** 8¢
  **c.** 9¢    **d.** not given

**11.** Choose the ordered pair for *L*.

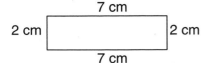

  **a.** (4,2)
  **b.** (4,3)
  **c.** (5,2)
  **d.** (5,5)

**12.** Which is a polygon?

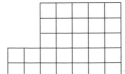

  **a.** A
  **b.** B
  **c.** C
  **d.** none of these

**13.** Find the perimeter.

7 cm
2 cm    2 cm
7 cm

  **a.** 9 cm
  **b.** 14 cm
  **c.** 16 cm
  **d.** 18 cm

**14.** How many square units are there?

  **a.** 24
  **b.** 25
  **c.** 29
  **d.** 34

# Ongoing Assessment III

## For Your Portfolio

Solve each problem. Explain the steps and the strategy or strategies you used for each. Then choose one from problems 1–4 for your Portfolio.

1. Ana used 4 cups of milk in a recipe, and Leo used 2 quarts of milk. Who used more?

2. The distance from A to B is 100 cm and the distance from C to D is 2 m. Which is longer?

**Use the tally chart for exercises 3–5.**

3. Which vacation place received double the number of votes of mountains?

4. In a pictograph, each ✗ = 2 votes. How many ✗ would you need to represent each vacation place?

| Favorite Vacation Place | |
| --- | --- |
| Beach | JHT JHT JHT I |
| Theme park | JHT JHT JHT IIII |
| Campgrounds | JHT JHT IIII |
| Mountains | JHT II |

5. Make a pictograph or bar graph to represent the data.

**Tell about it.**

6. Estimate the total number of people who chose a favorite vacation place. Explain how you made your estimate.

Communicate ✓

## For Rubric Scoring

**Listen for information on how your work will be scored.**

7. In the figure for exercise 14 on page 319, how many more squares are needed to make a rectangle? What is the area of the rectangle?

8. Copy the figure in exercise 14 on page 319 on centimeter grid paper. Make a figure to cover it, using 3 layers of cubes. What is the volume of the space figure? Explain.

# Multiply by One Digit

# 10

## Things You Don't Need to Know

Don't test a rattlesnake's rattle.
Don't count the teeth of a shark.
Don't stick your head in the mouth of a bulldog
to find out what's making him bark.

Don't count the stripes on a tiger.
Don't squeeze an elephant's trunk.
Don't pet the scales of a boa constrictor
and don't lift the tail of a skunk.

Don't inspect spots on a leopard.
Don't check the charge of an eel.
Don't count the claws on a grizzly bear's paws
regardless of how brave you feel.

Try to learn all that you'd like to.
Study wherever you go.
Take my advice, though, and never forget
there are some things you don't need to know.

*Kenn Nesbitt*

### In this chapter you will:

Estimate products
Explore multiplication with regrouping
Learn about LOGO
Solve problems by working backwards

### Critical Thinking/Finding Together

Carlos needs to know about how
many beans are in the jar to win a
contest. Explain a strategy that
Carlos can learn to estimate the
number of beans in the jar.

321

# Estimating Products

Nhan buys 4 binders for his brothers. About how much money does he spend?

An estimate tells *about* how much or *about* how many.

$.89

To find about how much Nhan spends, estimate:  4 × $5.75 = _?_

Here are two ways to estimate the product.

$5.75

▶ **Rounding:** Round the greater factor to its greatest place value. Then multiply.

$5.75 ⟶ $6.00
×    4        ×    4
         about $24.00

Remember: Write the dollar sign and the decimal point.

$6.09

▶ **Front-End Estimation:** Use the value of the front digit of the greater factor. Then multiply.

$5.75 ⟶ $5.00
×    4        ×    4
         about $20.00

Nhan spends between $20.00 and $24.00.

## Study these examples.

| Round to the nearest hundred. | 328 ⟶ 300<br>× 5 × 5<br>about 1500 | Round to the nearest ten. | 65 ⟶ 70<br>× 3 × 3<br>about 210 |
|---|---|---|---|

## Multiply mentally. Describe any pattern you see.

*Communicate* ✓

**1.** 2 × 20    **2.** 2 × 30    **3.** 2 × 40    **4.** 2 × 50    **5.** 2 × 60

**6.** 6 × 100   **7.** 6 × 200   **8.** 6 × 300   **9.** 6 × 400   **10.** 6 × 500

# Estimating to the Nearest Ten Cents

Estimate:  3 × 28¢

- Use rounding:

$$28¢ \longrightarrow 30¢$$
$$\underline{× \ 3} \qquad \underline{× \ 3}$$
$$\text{about} \quad 90¢$$

Remember:  90¢ = $.90

Estimate:  8 × $.22

- Use front-end estimation:

$$\$.22 \longrightarrow \$.20$$
$$\underline{× \ 8} \qquad \underline{× \ 8}$$
$$\text{about} \quad \$1.60$$

Remember:  $1.60 = 160¢

**Estimate. First round, then use front-end estimation.**

11.  17
   × 4

12.  73
   × 8

13.  11¢
   × 5

14.  $.26
   ×   2

15.  $.88
   ×   3

16. 3 × 35

17. 2 × 14

18. 3 × $.39

19. 5 × $.48

20.  135
   × 6

21.  274
   × 3

22.  $3.48
   ×    2

23.  $8.09
   ×    4

24.  $6.77
   ×    4

25. 7 × 757

26. 3 × 522

27. 5 × $5.98

28. 4 × $5.79

**PROBLEM SOLVING** Use the sale items on page 322.

29. Ms. Li bought 3 gym bags at the sale. About how many dollars did she spend?

30. Jacinta bought a half dozen rulers. About how much money did she spend?

 **Share Your Thinking**

The exact product for exercise 28 is $23.16.

31. Look at your estimates for exercise 28. Which method resulted in an estimate that was greater than the exact product? Explain why.

32. Which method resulted in an estimate that was less than the exact product? Why?

## 10-2 Multiplying Two Digits

Kindra has 2 sheets of stamps. Each sheet has 24 stamps. How many stamps does Kindra have?

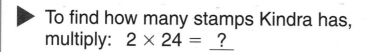

▶ First estimate the product:

$$
\begin{array}{r} 24 \\ \times\ 2 \end{array} \longrightarrow \begin{array}{r} 20 \\ \times\ 2 \\ \hline \end{array}
$$
about 40

▶ To find how many stamps Kindra has, multiply:  $2 \times 24 = $ _?_

Multiply the ones by 2. ⟶ Multiply the tens by 2.

| tens | ones |
|---|---|
| 2 | 4 |
| × | 2 |
| | 8 |

$2 \times 4$ ones = 8 ones
Write 8 in the ones place.

| tens | ones |
|---|---|
| 2 | 4 |
| × | 2 |
| 4 | 8 |

$2 \times 2$ tens = 4 tens
Write 4 in the tens place.

$$
\begin{array}{r} 24 \\ \times\ 2 \\ \hline 48 \end{array}
$$

Kindra has 48 stamps.

Think: 48 is close to 40.
The answer is reasonable.

## Multiply mentally.

1. Multiply 10 by:  3, 7, 8, 2, 9, 4, 5, 6

2. Multiply 20 by:  1, 3, 6, 2, 5, 8, 7, 4, 9

## Estimate. Then multiply.

3. $\begin{array}{r} 11 \\ \times\ 6 \\ \hline \end{array}$
4. $\begin{array}{r} 12 \\ \times\ 2 \\ \hline \end{array}$
5. $\begin{array}{r} 21 \\ \times\ 4 \\ \hline \end{array}$
6. $\begin{array}{r} 11 \\ \times\ 8 \\ \hline \end{array}$
7. $\begin{array}{r} 12 \\ \times\ 3 \\ \hline \end{array}$

**Multiply.**

**8.** 33
× 3

**9.** 23
× 2

**10.** 23
× 3

**11.** 40
× 2

**12.** 13
× 2

**13.** 22
× 4

**14.** 32
× 3

**15.** 34
× 2

**16.** 13
× 3

**17.** 31
× 2

**18.** 41
× 2

**19.** 21
× 3

**20.** 32
× 2

**21.** 43
× 2

**22.** 22
× 3

**Align. Then find the product.**

**23.** 3 × 31      **24.** 2 × 33      **25.** 2 × 42      **26.** 3 × 30

**27.** 2 × 22      **28.** 9 × 11      **29.** 2 × 30      **30.** 2 × 44

**PROBLEM SOLVING**

**31.** Liam mounts stamps on pages. Each page has 12 stamps. How many stamps are on 4 pages?

**32.** Chau collects coins. She has 2 sheets. Each sheet has 24 coins. How many coins does she have in all?

 **Skills to Remember**

**Regroup.**

**33.** 17 ones =
\_?\_ ten \_?\_ ones

**34.** 36 ones =
\_?\_ tens \_?\_ ones

**35.** 25 tens =
\_?\_ hundreds \_?\_ tens

**36.** 81 tens =
\_?\_ hundreds \_?\_ ten

# 10-3 Multiplication Models

Algebra

You can use place-value models to show multiplication.

Multiply: $3 \times 14 = \underline{\ ?\ }$

## Hands-On Understanding

**You Will Need:** base ten blocks, record sheet

**Step 1** First estimate the product.

**Step 2** Model 3 groups of 14.

**Step 3** Combine the groups.

How many ones and tens do you have altogether?

**Step 4** Regroup ones as tens.

How many ones did you regroup as tens?

How many ones and tens do you have now?

Write the number in standard form that represents your model.

What is the product of $3 \times 14$?

Is your answer reasonable? Explain why or why not.

124

**Step 5** Use models to find the product of $4 \times 31$.

First estimate the product.

Then model 4 groups of 31.

| Step 6 | Regroup tens as hundreds. |
|---|---|

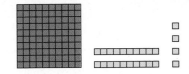

How many ones, tens, and hundreds do you have now?

Write the number in standard form that represents your model.

What is the product of 4 × 31?

Is your answer reasonable? Explain why or why not?

**Which multiplication sentence goes with each model?**
Choose the letter of the correct answer.

1.

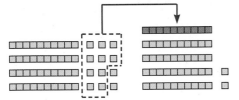

2.

**a.** 4 × 12 = 48    **b.** 4 × 13 = 52    **c.** 3 × 16 = 48    **d.** 3 × 13 = 39

---

### Area Model of Multiplication

Multiply:  3 × 14 = ?

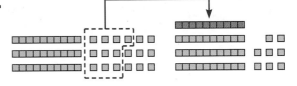

←—— 14 ——→ Factor
10  +  4

Factor 3

—Product

3 × 14 = 42

3 tens—→ 30
12 ones—→ +12
        42

---

**Use your base ten blocks to show each model. Then complete.**

3.

? tens —→ ?
? ones —→ + ?
            ?

6 × 12 = ?

4.

? tens —→ ?
? ones —→ + ?
            ?

8 × 14 = ?

## Communicate

5. Which method of multiplication is easier for you? Explain your answer.

## 10-4 Multiplying with Regrouping

Jamal buys 6 packs of postcards. Each pack has 12 postcards. How many postcards does he buy?

▶ First estimate the product: $6 \times 10 = 60$

▶ Then to find the number of postcards, multiply: $6 \times 12 = \underline{?}$

| Multiply the ones. Regroup. | Multiply the tens. Then add the 1 ten. |
|---|---|

$$\begin{array}{r} \overset{1}{\phantom{0}} \\ 1\,2 \\ \times\ \ 6 \\ \hline 2 \end{array}$$

$$\begin{array}{r} \overset{1}{\phantom{0}} \\ 1\,2 \\ \times\ \ 6 \\ \hline 7\,2 \end{array}$$

Think: 72 is close to 60. The answer is reasonable.

$6 \times 2$ ones $= 12$ ones
12 ones $= 1$ ten 2 ones

$6 \times 1$ ten $= 6$ tens
6 tens plus 1 ten $= 7$ tens

Jamal buys 72 postcards.

### Study these examples.

$$\begin{array}{r} \overset{2}{\phantom{0}} \\ 29\cent \\ \times\ \ 3 \\ \hline 87\cent \end{array}$$

Remember: Write the cent sign.

$$\begin{array}{r} \overset{3}{\phantom{0}} \\ \$19 \\ \times\ \ 4 \\ \hline \$76 \end{array}$$

Remember: Write the dollar sign.

---

**Multiply.**

1.  $\begin{array}{r} 16\cent \\ \times\ 4 \\ \hline \end{array}$

2.  $\begin{array}{r} 27\cent \\ \times\ 2 \\ \hline \end{array}$

3.  $\begin{array}{r} \$14 \\ \times\ 6 \\ \hline \end{array}$

4.  $\begin{array}{r} \$13 \\ \times\ 7 \\ \hline \end{array}$

5.  $\begin{array}{r} \$25 \\ \times\ 3 \\ \hline \end{array}$

6. $7 \times 14$

7. $4 \times 19$

8. $3 \times 24$

9. $2 \times 28$

328

**Estimate. Then multiply.**

10.  15
    × 5

11.  18
    × 4

12.  35
    × 2

13.  29
    × 2

14.  13
    × 6

15.  19
    × 3

16.  38
    × 2

17.  47
    × 2

18.  17
    × 5

19.  12
    × 7

20.  17¢
    × 4

21.  26¢
    × 3

22.  $13
    × 5

23.  $37
    × 2

24.  $15
    × 4

**Find the product.**

25. 3 × 18

26. 2 × 45

27. 4 × 23

28. 4 × 14

29. 4 × 13¢

30. 4 × 16¢

31. 2 × 39¢

32. 3 × 28¢

33. 4 × $21

34. 3 × $27

35. 5 × $18

36. 2 × $49

**PROBLEM SOLVING** Use the chart.

37. Chad buys 5 small packs. How many postcards does he buy?

38. Leah buys 2 large packs. How many postcards does she buy?

39. Sara buys 4 medium packs. About how much money does she spend?

**Postcard Packs**

| Size of Pack | Number in the Pack | Price of Pack |
|---|---|---|
| Small | 12 | $1.29 |
| Medium | 24 | $1.99 |
| Large | 36 | $2.59 |

**Finding Together**

*Algebra* ✓

40. How many ways are there to buy 48 postcards? Write a multiplication or an addition sentence for each way. Share your ways with classmates.

*Hint:* Make a List
Use a Table

41. Which is the cheapest way to buy 48 postcards?

# More Multiplying with Regrouping

A truck travels 43 miles on the same route each day. How many miles does the truck travel in 3 days?

▶ First estimate the product: $3 \times 40 = 120$

▶ Then to find the number of miles, multiply: $3 \times 43 = \underline{?}$

| Multiply the ones. ⟶ | Multiply the tens. Regroup. |
|---|---|

$$
\begin{array}{r} 4\ 3 \\ \times\ \ \ 3 \\ \hline 9 \end{array}
\qquad
\begin{array}{r} 4\ 3 \\ \times\ \ \ \ 3 \\ \hline 1\ 2\ 9 \end{array}
\qquad
\begin{array}{r} 43 \\ \times\ \ 3 \\ \hline 129 \end{array}
$$

$3 \times 3$ ones = 9 ones

$3 \times 4$ tens = 12 tens
12 tens = 1 hundred 2 tens

The truck travels 129 miles in 3 days.

Sometimes you need to regroup twice.

▶ Multiply: $5 \times 37 = \underline{?}$

| Multiply the ones. Regroup. ⟶ | Multiply the tens. Then add the 3 tens. Regroup again. |
|---|---|

$$
\begin{array}{r} \overset{3}{3}\ 7 \\ \times\ \ \ 5 \\ \hline 5 \end{array}
\qquad
\begin{array}{r} \overset{3}{3}\ 7 \\ \times\ \ \ \ 5 \\ \hline 1\ 8\ 5 \end{array}
\qquad
\begin{array}{r} 37 \\ \times\ \ 5 \\ \hline 185 \end{array}
$$

$5 \times 7$ ones = 35 ones
35 ones = 3 tens 5 ones

$5 \times 3$ tens = 15 tens
15 tens + 3 tens = 18 tens
18 tens = 1 hundred 8 tens

**Estimate. Then multiply.**

| 1. | 41<br>× 3 | 2. | 52<br>× 4 | 3. | 40<br>× 7 | 4. | 63<br>× 3 | 5. | 54<br>× 2 |
|---|---|---|---|---|---|---|---|---|---|

| 6. | 21<br>× 6 | 7. | 31<br>× 5 | 8. | 62<br>× 3 | 9. | 71<br>× 8 | 10. | 51<br>× 4 |

| 11. | 47<br>× 5 | 12. | 23<br>× 4 | 13. | 74<br>× 8 | 14. | 37<br>× 6 | 15. | 64<br>× 7 |

| 16. | 23<br>× 9 | 17. | 37<br>× 4 | 18. | 75<br>× 3 | 19. | 78<br>× 2 | 20. | 94<br>× 5 |

| 21. | 27<br>× 3 | 22. | 42<br>× 7 | 23. | 43<br>× 8 | 24. | 35<br>× 6 | 25. | 86<br>× 4 |

**PROBLEM SOLVING**  Use the chart.

| Miles Per Gallon of Gas | |
|---|---|
| Compact car | 28 miles |
| Mid-size car | 24 miles |
| Sports car | 22 miles |
| Truck | 18 miles |

26. How far can a compact car travel on 5 gallons of gas?

27. How far can a mid-size car travel on 8 gallons of gas?

28. A sports car has 5 gallons of gas in its tank. A truck has 7 gallons. Which vehicle can travel farther? How much farther?

 **Choose a Computation Method**   *Communicate* ✓

**Solve. Use paper and pencil or mental math.
Tell why you chose the method you used.**

**29.** $8 \times 30$    **30.** $8 \times 24$    **31.** $2 \times 70$    **32.** $4 \times 22$

**33.** $5 \times 73$    **34.** $9 \times 50$    **35.** $6 \times 67$    **36.** $3 \times 49$

# 10-6 Multiplying Three Digits

Terrell and Tom collect seashells.
Terrell has 123 shells. Tom has twice
that many. How many shells does
Tom have?

▶ First estimate the product:  $2 \times 100 = 200$

▶ Then to find the number of shells,
multiply:  $2 \times 123 = \underline{?}$

| Multiply the ones. → | Multiply the tens. → | Multiply the hundreds. |
|---|---|---|

$$
\begin{array}{r} 1\ 2\ 3 \\ \times\ \ \ \ 2 \\ \hline 6 \end{array}
\qquad
\begin{array}{r} 1\ 2\ 3 \\ \times\ \ \ \ 2 \\ \hline 4\ 6 \end{array}
\qquad
\begin{array}{r} 1\ 2\ 3 \\ \times\ \ \ \ 2 \\ \hline 2\ 4\ 6 \end{array}
$$

| $2 \times 3$ ones = 6 ones | $2 \times 2$ tens = 4 tens | $2 \times 1$ hundred = 2 hundreds |
|---|---|---|

Tom has 246 shells.

Think:  246 is close to 200.
The answer is reasonable.

## Estimate. Then multiply.

| | | | | | | | | | |
|---|---|---|---|---|---|---|---|---|---|
| **1.** | 422 × 2 | **2.** | 132 × 3 | **3.** | 102 × 4 | **4.** | 343 × 2 | **5.** | 122 × 4 |
| **6.** | 401 × 2 | **7.** | 111 × 5 | **8.** | 231 × 3 | **9.** | 144 × 2 | **10.** | 320 × 2 |
| **11.** | 110 × 6 | **12.** | 121 × 4 | **13.** | 101 × 7 | **14.** | 431 × 2 | **15.** | 332 × 3 |
| **16.** | 202 × 3 | **17.** | 123 × 2 | **18.** | 333 × 3 | **19.** | 111 × 2 | **20.** | 444 × 2 |

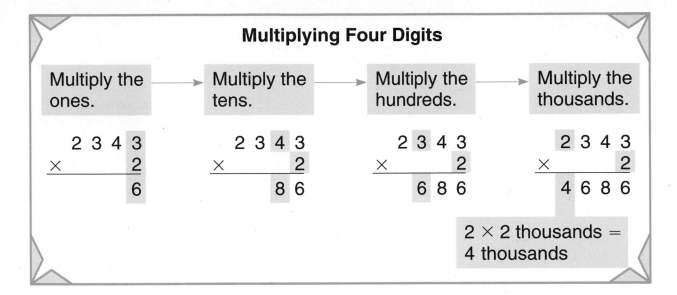

## Multiplying Four Digits

| Multiply the ones. | Multiply the tens. | Multiply the hundreds. | Multiply the thousands. |
|---|---|---|---|

```
   2 3 4 3          2 3 4 3          2 3 4 3          2 3 4 3
 ×       2        ×       2        ×       2        ×       2
 ─────────        ─────────        ─────────        ─────────
         6              8 6          6 8 6          4 6 8 6
```

2 × 2 thousands =
4 thousands

## Multiply.

**21.**  1313
×    3

**22.**  3424
×    2

**23.**  1111
×    9

**24.**  2323
×    3

**25.**  2120
×    4

**26.**  3321
×    3

**27.**  4242
×    2

**28.**  3302
×    3

**29.**  1110
×    7

**30.**  2102
×    3

## PROBLEM SOLVING

**31.** Jason made 212 free-throw shots. Craig made 3 times as many shots. How many shots did Craig make?

**32.** Dong made 131 free throws on each of three days. How many throws did Dong make in all?

 ## Critical Thinking  Algebra

**Complete. Write <, =, >.**

**33.** 2 × 122 _?_ 2 × 102

**34.** 3 × 131 _?_ 3 × 231

**35.** 4 × 111 _?_ 3 × 111

**36.** 2 × 2212 _?_ 2 × 2214

**37.** 3 × 3303 _?_ 3 × 2223

**38.** 3 × 2203 _?_ 2 × 2202

# Regrouping in Multiplication

A certain type of plane has 224 seats.
How many seats do 3 planes have?

▶ First estimate the product:  $3 \times 200 = 600$

▶ Then to find the number of seats,
multiply:  $3 \times 224 =$  ?

| Multiply the ones. Regroup. | → | Multiply the tens. Add the regrouped ten. | → | Multiply the hundreds. |
|---|---|---|---|---|

$$\begin{array}{r} \overset{1}{2}\,2\,4 \\ \times\ \ \ \ 3 \\ \hline 2 \end{array}$$

$$\begin{array}{r} \overset{1}{2}\,2\,4 \\ \times\ \ \ \ 3 \\ \hline 7\,2 \end{array}$$

$$\begin{array}{r} \overset{1}{2}\,2\,4 \\ \times\ \ \ \ 3 \\ \hline 6\,7\,2 \end{array}$$

| $3 \times 4$ ones $= 12$ ones | $3 \times 2$ tens $= 6$ tens | $3 \times 2$ hundreds |
| 12 ones $= 1$ ten 2 ones | 6 tens $+ 1$ ten $= 7$ tens | $= 6$ hundreds |

Three planes have 672 seats.

**Study this example.**

Sometimes you need to regroup tens.

Multiply:  $3 \times 273 =$  ?

| Multiply the ones. | → | Multiply the tens. Regroup. | → | Multiply the hundreds. Add the regrouped hundreds. |
|---|---|---|---|---|

$$\begin{array}{r} 2\,7\,3 \\ \times\ \ \ \ 3 \\ \hline 9 \end{array}$$

$$\begin{array}{r} \overset{2}{2}\,7\,3 \\ \times\ \ \ \ 3 \\ \hline 1\,9 \end{array}$$

$$\begin{array}{r} \overset{2}{2}\,7\,3 \\ \times\ \ \ \ 3 \\ \hline 8\,1\,9 \end{array}$$

| $3 \times 7$ tens $= 21$ tens | $3 \times 2$ hundreds $=$ |
| 21 tens $= 2$ hundreds 1 ten | 6 hundreds |
| | 6 hundreds $+ 2$ hundreds $=$ |
| | 8 hundreds |

## Estimate. Then multiply.

| | | | | |
|---|---|---|---|---|
| **1.** 238 × 2 | **2.** 112 × 5 | **3.** 102 × 8 | **4.** 218 × 4 | **5.** 325 × 3 |
| **6.** 105 × 7 | **7.** 127 × 2 | **8.** 107 × 9 | **9.** 437 × 2 | **10.** 102 × 6 |
| **11.** 242 × 3 | **12.** 131 × 6 | **13.** 192 × 4 | **14.** 253 × 3 | **15.** 121 × 8 |
| **16.** 182 × 4 | **17.** 121 × 8 | **18.** 150 × 6 | **19.** 142 × 3 | **20.** 493 × 2 |

## Regrouping with Four Digits

| Multiply the ones. Regroup. | → | Multiply the tens. Add the regrouped ones. | → | Multiply the hundreds. Then multiply thousands. |
|---|---|---|---|---|

$$\begin{array}{r} \overset{1}{1\ 2\ 1\ 3} \\ \times\ \ \ \ 4 \\ \hline 2 \end{array}$$

4 × 3 ones =
12 ones =
1 ten 2 ones

$$\begin{array}{r} 1\ 2\ \overset{1}{1}\ 3 \\ \times\ \ \ \ 4 \\ \hline 5\ 2 \end{array}$$

$$\begin{array}{r} \overset{1}{1\ 2\ 1\ 3} \\ \times\ \ \ \ 4 \\ \hline 4\ 8\ 5\ 2 \end{array}$$

4 × 1213 = 4852

## Find the product.

| | | | |
|---|---|---|---|
| **21.** 1113 × 5 | **22.** 3126 × 2 | **23.** 1025 × 3 | **24.** 1417 × 2 |
| **25.** 3251 × 3 | **26.** 2263 × 2 | **27.** 1070 × 4 | **28.** 2042 × 4 |

**More Regrouping in Multiplication**

Ted sells 5 couches. How much money does he get?

| TED'S FURNITURE STORE | |
|---|---|
| Item | Price |
| Bookcase | $ 278 |
| Cabinet | $ 469 |
| Desk | $ 129 |
| Table | $ 257 |
| Couch | $1129 |

▶ First estimate:  $5 \times \$1000 = \$5000$

▶ Then to find how much money he gets, multiply:  $5 \times \$1129 = \underline{\ ?\ }$

Multiply the ones. Regroup.

$$
\begin{array}{r}
\overset{4}{\phantom{0}} \\
\$\,1\,1\,2\,9 \\
\times \qquad 5 \\
\hline
5
\end{array}
$$

$5 \times 9$ ones $= 45$ ones
$\phantom{5 \times 9 \text{ ones}} = 4$ tens $5$ ones

Multiply the tens. Add the regrouped ten. Regroup again.

$$
\begin{array}{r}
\overset{1\ 4}{\phantom{0}} \\
\$\,1\,1\,2\,9 \\
\times \qquad 5 \\
\hline
4\,5
\end{array}
$$

$5 \times 2$ tens $= 10$ tens
$10$ tens $+ 4$ tens
$= 14$ tens
$= 1$ hundred $4$ tens

Multiply the hundreds. Add the regrouped hundred.

$$
\begin{array}{r}
\overset{1\ 4}{\phantom{0}} \\
\$\,1\,1\,2\,9 \\
\times \qquad 5 \\
\hline
6\,4\,5
\end{array}
$$

$5 \times 1$ hundred $= 5$ hundreds
$5$ hundreds $+ 1$ hundred
$= 6$ hundreds

Ted gets $5645.

Multiply the thousands.

$$
\begin{array}{r}
\overset{1\ 4}{\phantom{0}} \\
\$\,1\,1\,2\,9 \\
\times \qquad 5 \\
\hline
5\,6\,4\,5
\end{array}
$$

$5 \times 1$ thousand $= 5$ thousands

Think:  $5645 is close to $5000. The answer is reasonable.

**Estimate. Then multiply.**

| 1. | 2. | 3. | 4. | 5. |
|---|---|---|---|---|
| 355 | 247 | $153 | $499 | 122 |
| × 2 | × 4 | × 5 | × 2 | × 6 |

| 6. | 7. | 8. | 9. | 10. |
|---|---|---|---|---|
| 2248 | 1128 | 2224 | 1113 | 1134 |
| × 3 | × 7 | × 4 | × 8 | × 5 |

| 11. | 12. | 13. | 14. | 15. |
|---|---|---|---|---|
| $135 | $158 | $1172 | $3386 | $2243 |
| × 7 | × 6 | × 5 | × 2 | × 4 |

**Find the product.**

| 16. | 17. | 18. | 19. | 20. |
|---|---|---|---|---|
| 235 | 141 | 3239 | 1106 | 303 |
| × 4 | × 7 | × 3 | × 8 | × 2 |

| 21. | 22. | 23. | 24. | 25. |
|---|---|---|---|---|
| 256 | 2164 | 117 | $137 | $1105 |
| × 3 | × 4 | × 8 | × 6 | × 5 |

**PROBLEM SOLVING** Use the chart on page 336.

**26.** Kari buys 3 desks at Ted's. How much money does she spend?

**27.** Ted sells 2 cabinets. How much money does he get?

**28.** Bryant buys 2 bookcases. How much money does he spend?

**29.** Ted sells 3 couches. How much money does he get?

## Challenge

**30.** Colin has $900. Does he have enough money for 3 tables? If so, how much extra money does he have? If not, how much more money does he need?

**31.** Wayne has $1600. Does he have enough money for 2 tables and 1 couch? If so, how much extra money does he have? If not, how much more money does he need?

# TECHNOLOGY

## LOGO

**LOGO** is a computer language that lets you draw figures by moving a turtle around the computer screen.

Commands are used to tell the turtle in which direction it should move.

The LOGO turtle always starts out in the center of the screen, heading in the north direction.

LOGO turtle

Here are some commands used to move the turtle.

| Command | What you enter | How the turtle moves |
|---------|---------------|---------------------|
| FORWARD | FD 50 | Forward 5 steps |
| BACK | BK 50 | Back 5 steps |
| RIGHT | RT 90 | Makes a right angle |
| LEFT | LT 45 | Makes an angle less than a right angle |

FD
↓

Tells the direction.

90 ←
↓

Tells how far.

Numbers tell the turtle how far to move. Since the turtle takes very small steps, add a 0 to the end of each number when moving forward or back.

These commands tell the turtle to draw each figure.

FD 50
RT 90
FD 50
RT 90
RT 45
FD 70

FD 30
RT 90
FD 50
RT 90
FD 30
RT 90
FD 50

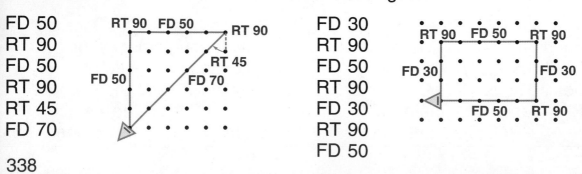

## Match each movement with the correct LOGO command.

a.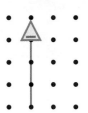

b.

c.

d.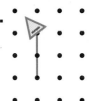

**1.** FD 20
RT 90

**2.** FD 30

**3.** FD 20
LT 45

**4.** BK 30

## Use each set of commands to draw each figure. Name the figure.

**5.** FD 40
RT 90
FD 40
RT 90
FD 40
RT 90
FD 40

**6.** FD 20
RT 45
FD 20
RT 90
FD 20
RT 45
FD 20
RT 90
FD 30

**7.** LT 90
FD 50
RT 90
FD 30
BK 20
FD 40
RT 90
RT 45
FD 70

**8.** LT 90
FD 30
RT 45
FD 30
RT 90
FD 30
RT 45
FD 30
RT 45
FD 30
RT 90
FD 30

## Write the commands to draw each letter.

**9.**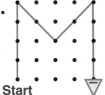

Start

**10.**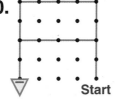

Start

**11.**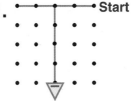
Start

**12.**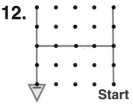

Start

**9.** FD 40 → FD 20 → RT 90
RT 90   LT 90   RT 45
RT 45 ⌐ FD 20 ⌐ FD 40

339

# 10-10 | Problem Solving: Working Backwards Algebra ✓

**Problem:** The librarian got a box full of new books. Gina gave her 4 more books. They put 9 books on the shelf. There are now 12 books left in the box. How many new books were in the box?

**1 IMAGINE** Think about what happened.

**2 NAME**
*Facts:* box of _?_ new books
4 books from Gina
9 books on the shelf
12 books left

*Question:* How many new books were in the box?

**3 THINK** To find how many books were in the box at the beginning, start at the end and work backwards.

| 12 | + | 9 | − | 4 | = | ? |
|----|---|---|---|---|---|---|
| books left | | books on shelf | | books *not* in box | | new books in box |

**4 COMPUTE**
12  +  9  =  21  ⟵ total number of books
21  −  4  =  17  ⟵ number of new books in the box

There were 17 new books in the box.

**5 CHECK** Work from the beginning to check.

17  +  4  =  21 and 21  −  9  =  12
number  number                         number  number
in box  from Gina                      on shelf  left

## Work backwards to solve each problem.

**1.** Phil finished his research paper at 4:00 after working on it for 3 hours. What time did he begin?

| IMAGINE | Think about the problem. |
|---------|--------------------------|

| NAME | *Facts:* | Phil worked until 4:00. He worked for 3 hours. |
|------|----------|-----------------------------------------------|
|      | *Question:* | What time did he begin? |

| THINK | You can count backwards to find the time Phil began working. |
|-------|-------------------------------------------------------------|

→ **COMPUTE** → **CHECK**

**2.** Maria worked in the library until 3:30. She spent 1 hour on a research paper, 45 minutes returning books to shelves, and 15 minutes making copies. What time did she begin?

**3.** The library closes at 8:00 P.M. If it is open for nine hours each day, what time does the library open?

**4.** Kyle's 3-day book sale made a total of $684. Today he made $252. Yesterday he made $197. How much money did he make the first day?

**5.** Right now Karen has 15 books. When she first began a library, her friend gave her 8 books, she bought 1 book, and she lost 2 books. How many books did she have at first in her library?

Solve each problem and explain the method you used.

1. A toy store gets 3 cartons of stuffed bears. Each carton has 20 stuffed bears. How many stuffed bears did the store get?

2. A salesperson sold 4 bicycles. Each bicycle sold for $123. Did the salesperson make more or less than $800?

Imagine

3. Adam has 9 pieces of train track. Each piece is 32 centimeters long. How long will the entire track be when he connects all 9 pieces?

Name

4. Ms. Shaw bought 5 games for her family. She spent $12.59 for each game. About how much money did she spend for the games?

Think

5. Brent goes a distance of 23 inches with each hop on his pogo stick. At this rate about how many inches can Brent go in 6 hops? Will he beat Ellen, who can go 120 inches in 6 hops?

Compute

6. Lori made 7 bracelets. She used 22 beads on each. Faith used 31 beads on each of the 5 bracelets that she made. Which girl used more beads? How many more beads did she use?

Check

7. A toy factory makes 38 electronic games every hour. At this rate how many games will the factory make in 7 hours?

8. Sue and each of the twins buy roller skates. Each pair costs $25.39. About how much money do the girls pay altogether?

**Choose a strategy from the list or use another strategy you know to solve each problem.**

Use these strategies:

9. Ms. Leone makes toy keyboards. Each of them has a dozen black keys and 17 white keys. How many keys of each color does she need for 5 keyboards?

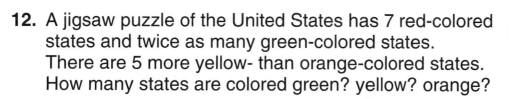

Guess and Test
Logical Reasoning
Working Backwards
Multi-Step Problem
Hidden Information
Write a Number Sentence

10. Raisa and her two friends want to buy the same jewelry kits. About how much money will each girl pay if the total cost will be $23.67?

11. Chris arranged his models on a shelf in left-to-right order. The spaceship is to the left of the truck. The truck is not next to the boat. The jet is the fourth model. Which model is second?

12. A jigsaw puzzle of the United States has 7 red-colored states and twice as many green-colored states. There are 5 more yellow- than orange-colored states. How many states are colored green? yellow? orange?

13. Marsha bought a pack of colored felt. She loaned 5 pieces to Russ. Then Laura gave Marsha 3 pieces. After Marsha used 4 pieces, she had 6 pieces left. How many pieces of felt were in the pack Marsha bought?

14. Altogether four children collected 54 toys for the hospital. Liz collected a dozen. Randy, Carla, and Angie collected an equal number of toys. How many toys did Carla collect?

 **Make Up Your Own**

15. Write a problem. Have a classmate solve it.

**Estimate. First round, then use front-end estimation.**    *(See pp. 322–323.)*

| 1. | 59 | 2. | 62 | 3. | 273 | 4. | 38¢ | 5. | $6.04 |
|----|----|----|----|----|----|----|----|----|----|
|    | × 3 |    | × 2 |    | × 6 |    | × 8 |    | × 9 |

**Multiply.**    *(See pp. 324–325, 328–337.)*

| 6. | 21 | 7. | 32 | 8. | 14¢ | 9. | 53 | 10. | 65 |
|----|----|----|----|----|----|----|----|----|----|
|    | × 4 |    | × 3 |    | × 2¢ |    | × 8 |    | × 6 |

| 11. | $29 | 12. | $87 | 13. | 224 | 14. | 113 | 15. | 240 |
|----|----|----|----|----|----|----|----|----|----|
|    | × 2 |    | × 7 |    | × 2 |    | × 4 |    | × 3 |

| 16. | $164 | 17. | $106 | 18. | 1322 | 19. | 2143 | 20. | 3256 |
|----|----|----|----|----|----|----|----|----|----|
|    | × 5 |    | × 9 |    | × 2 |    | × 3 |    | × 2 |

**Align. Then find the product.**

**21.** $2 \times 40$     **22.** $9 \times 16$     **23.** $3 \times 138$     **24.** $4 \times \$1209$

**Compare. Write <, =, or >.**

**25.** $4 \times 111$ _?_ $3 \times 121$     **26.** $2 \times 404$ _?_ $3 \times 303$

**PROBLEM SOLVING**    *(See pp. 340–343.)*

**27.** Joseph has 15 coins in a box. Andrew has 4 times as many. How many coins does Andrew have?

**28.** A movie theater ticket costs $7.75. About how much does it cost for 3 people to go to the movies?

**29.** Dad bought a puzzle for Mona and a puzzle for Jim. The puzzles cost $9.89 each. About how much did Dad spend?

**30.** Mona and Jim finished one of the puzzles at 9:30, after working on it for 4 hours. What time did they begin?

*(See Still More Practice, pp. 451–452.)*

## LOGIC AND MISSING DIGITS

*Algebra* ✓

Someone ate Lady Snoot's cookies. Lady Snoot thinks the culprit was the maid, the cook, or the butler. Can you help her solve this mystery?

Be a detective. Follow the steps below and you will find the culprit.

**Step 1** Find the missing digits.

**Step 2** Look at the code at the bottom of the page. Each missing digit has a matching letter.

**Step 3** Write the letters in the order of the exercises to find "whodunit."

| 1. | 2. | 3. |
|---|---|---|
| $\boxed{2}6$ <br> $\times\ \ 5$ <br> $130$ | $3\square8$ <br> $\times\ \ \ \ 2$ <br> $616$ | $129$ <br> $\times\ \ \ \square$ <br> $516$ |

| 4. | 5. | 6. |
|---|---|---|
| $15\square$ <br> $\times\ \ \ \ 3$ <br> $471$ | $\square2$ <br> $\times\ \ 9$ <br> $738$ | $279$ <br> $\times\ \ \square$ <br> $837$ |

| 0 | 7 | 2 | 4 | 3 | 8 |
|---|---|---|---|---|---|
| E | D | H | R | G | O |

**7.** Write a mystery. Then make up exercises and a code to solve the mystery.

**Performance Assessment**

**Estimate.**

1. You have $200 to spend on science materials for your classroom. Use estimation to decide how to spend the money.

| Software | $34.59 each |
|---|---|
| Video | $19.95 each |
| Magazine | $4.37 for 1 copy |
| Fossil Kit | $44.95 each |
| Model | $12.95 each |

**Multiply.**

2. $\begin{array}{r} 44 \\ \times\ 2 \\ \hline \end{array}$
3. $\begin{array}{r} 13 \\ \times\ 3 \\ \hline \end{array}$
4. $\begin{array}{r} 10 \\ \times\ 9 \\ \hline \end{array}$
5. $\begin{array}{r} 12¢ \\ \times\ 4 \\ \hline \end{array}$
6. $\begin{array}{r} 24¢ \\ \times\ 2 \\ \hline \end{array}$

7. $\begin{array}{r} 19 \\ \times\ 5 \\ \hline \end{array}$
8. $\begin{array}{r} 26 \\ \times\ 3 \\ \hline \end{array}$
9. $\begin{array}{r} 74 \\ \times\ 4 \\ \hline \end{array}$
10. $\begin{array}{r} 47 \\ \times\ 6 \\ \hline \end{array}$
11. $\begin{array}{r} \$63 \\ \times\ 8 \\ \hline \end{array}$

12. $\begin{array}{r} \$123 \\ \times\ \ \ 3 \\ \hline \end{array}$
13. $\begin{array}{r} \$327 \\ \times\ \ \ 2 \\ \hline \end{array}$
14. $\begin{array}{r} \$190 \\ \times\ \ \ 3 \\ \hline \end{array}$
15. $\begin{array}{r} \$179 \\ \times\ \ \ 4 \\ \hline \end{array}$
16. $\begin{array}{r} \$207 \\ \times\ \ \ 4 \\ \hline \end{array}$

**Align. Then find the product.**

17. $3 \times 30$

18. $2 \times 171$

19. $4 \times \$203$

20. $2 \times 2443$

21. $2 \times 4306$

22. $3 \times 3197$

**PROBLEM SOLVING**  *Use a strategy you have learned.*

23. There are 7 students in a group. There are 14 groups. How many students are there in all?

24. Brian rides his bike 12 kilometers a day. How many kilometers does he ride in 9 days?

25. Ryan bought 8 rolls of film. The film cost $3.98 a roll. About how much money did he pay for the film?

26. Min bought 5 coloring books. Each book cost $1.75. About how much money did Min pay for the books?

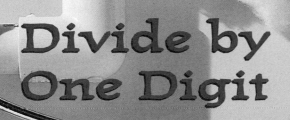

# Divide by One Digit

## SOS

Sammy's head is pounding—
Sammy's in pain—
A long division's got
Stuck in his brain—
Call for the locksmith
Call the engineer
Call for the plumber
To suck out his ear,
Call the brain surgeon
To pry out the mess,
Call out the Coast Guard
SOS,
Because—
Sammy's head is pounding—
Sammy's in pain—
A long division's got
Stuck in his brain.

*Beverly McLoughland*

**In this chapter you will:**

Find 1- and 2-digit
  quotients
Explore division with
  remainders
Estimate quotients
Solve problems by
  interpreting the remainder

**Critical Thinking/
Finding Together**

List all the division facts
that you know. What do
you think long division is?

347

# Division Sense

Estimate quotients before you divide.

▶ **Estimate**: $34 \div 8 = \underline{?}$

When the divisor is *greater than* the tens digit in the dividend, use multiplication facts to help find the answer.

Think: About how many 8s in 34?

3 ones × 8 = 24 ones ←——too small

4 ones × 8 = 32 ones ←—— 34 is between 32 and 40.

5 ones × 8 = 40 ones ←——too large

So $34 \div 8$ is about 4.

▶ **Estimate**: $76 \div 3 = \underline{?}$

When the divisor is *less than* the tens digit in the dividend, use tens to help find the answer.

Think: $\underline{?}$ tens × 3 = 7 tens

1 ten × 3 = 3 tens ←—— too small

2 tens × 3 = 6 tens ←—— 7 tens is between 6 tens and 9 tens.

3 tens × 3 = 9 tens ←——too large

Try 2. Write zeros for the other digits. $3\overline{)76}$ $\overset{20}{\phantom{)}}$

So $76 \div 3$ is about 20.

---

**Use facts to estimate.**

**1.** $7\overline{)25}$      **2.** $3\overline{)14}$      **3.** $2\overline{)17}$      **4.** $3\overline{)28}$

**5.** $6\overline{)46}$      **6.** $9\overline{)22}$      **7.** $8\overline{)36}$      **8.** $4\overline{)29}$

**Use tens to estimate.**

9. $2\overline{)27}$        10. $5\overline{)71}$        11. $4\overline{)92}$        12. $7\overline{)85}$

13. $8\overline{)94}$        14. $6\overline{)79}$        15. $3\overline{)82}$        16. $4\overline{)66}$

**Estimate. Explain how you estimated each quotient.**                    *Communicate* ✓

17. $5\overline{)43}$        18. $4\overline{)51}$        19. $6\overline{)49}$        20. $7\overline{)83}$

---

## Patterns in Division
*Algebra* ✓

**Look at the pattern.**

$4 \div 2 = 2$   or   $2\overline{)4}^{\,2}$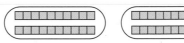

4 ones $\div$ 2 = 2 ones

$40 \div 2 = 20$ or $2\overline{)40}^{\,20}$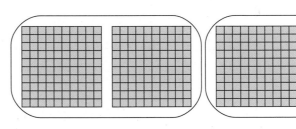

4 tens $\div$ 2 = 2 tens

$400 \div 2 = 200$

or

$2\overline{)400}^{\,200}$

4 hundreds $\div$ 2 = 2 hundreds

---

**Divide mentally.**

21. $6 \div 3 = \underline{\ ?\ }$     22. $8 \div 4 = \underline{\ ?\ }$     23. $9 \div 3 = \underline{\ ?\ }$     24. $8 \div 8 = \underline{\ ?\ }$

$60 \div 3 = \underline{\ ?\ }$          $80 \div 4 = \underline{\ ?\ }$          $90 \div 3 = \underline{\ ?\ }$          $80 \div 8 = \underline{\ ?\ }$

$600 \div 3 = \underline{\ ?\ }$          $800 \div 4 = \underline{\ ?\ }$          $900 \div 3 = \underline{\ ?\ }$          $800 \div 8 = \underline{\ ?\ }$

**Find the quotient.**

25. $2\overline{)80}$        26. $7\overline{)70}$        27. $3\overline{)600}$        28. $2\overline{)800}$        29. $5\overline{)50}$

## 11-2 | Division with Remainders

Mrs. Ming has 15 tickets for rides at the park. They are to be shared equally among her 4 children. At most, how many tickets will each child receive? How many tickets will be left over?

To find how many tickets, divide: $15 \div 4 = \underline{\ ?\ }$ or $4\overline{)15}$

You can use counters to model division.

## Hands-On Understanding

**You Will Need:** counters, record sheet, 4 blank sheets of paper

 **Step 1** Label the columns on your record sheet with these headings:

| Dividend (Number of tickets) | Divisor (Number of children) | Quotient (Tickets per child) | Remainder (Tickets left over) |
|---|---|---|---|

**Step 2** Model 15 counters.

What do the 15 counters represent?

**Step 3** Place 4 blank sheets of paper on your worktable.

What do the 4 sheets of paper stand for?

**Step 4** Now place one counter on each sheet of paper.

Do you have enough counters to put more counters on each sheet of paper?

**Step 5** Continue to place one counter on each sheet of paper until you can no longer give them out equally.

How many counters are on each sheet of paper?

Do you have any counters left over?

How many counters do you have left over?

How many tickets will each child receive?

How many tickets will be left over?

> The number left over after dividing is called the **remainder**.

**Step 6** Use counters to find out how Mrs. Ming would share the 15 tickets among 2 children, 3 children, and 5 children.

# Communicate

Discuss ✓

1. Look at the size of the divisors and the remainders on your record sheet. What do you notice?

2. Explain why there are no remainders when you divide the 15 tickets among 3 and 5 children.

3. Can the remainder in a division sentence ever be larger than the divisor? Explain.

**Find the quotient and remainder. Act it out, use counters, or draw dots to show each division.**

4. $7 \div 2 = \underline{\ ?\ }$    5. $10 \div 3 = \underline{\ ?\ }$    6. $18 \div 5 = \underline{\ ?\ }$    7. $21 \div 4 = \underline{\ ?\ }$

8. $17 \div 6 = \underline{\ ?\ }$    9. $13 \div 4 = \underline{\ ?\ }$    10. $11 \div 5 = \underline{\ ?\ }$    11. $15 \div 2 = \underline{\ ?\ }$

## PROBLEM SOLVING

12. Tim shares 16 cards equally among 6 boys. At most, how many cards will each boy get? How many will be left over?

13. Mike has 19 tickets. Each ride costs 4 tickets. At most, how many rides can he go on? How many tickets will he have left over?

# 11-3 | One-Digit Quotients

The teacher has 17 stickers. He wants to divide the stickers equally among 3 children. At most, how many stickers does each child get? How many stickers are left over?

To find the number of stickers, divide:  $17 \div 3 = \underline{\ ?\ }$  or  $3\overline{)17}^{\ ?}$

- Estimate:  About how many 3s in 17?

$4 \times 3 = 12$ ◄——too small

$5 \times 3 = 15$ ◄—— 17 is between 15 and 18. Try 5.

$6 \times 3 = 18$ ◄——too large

- Write 5 in the quotient above the 7 ones in the dividend.

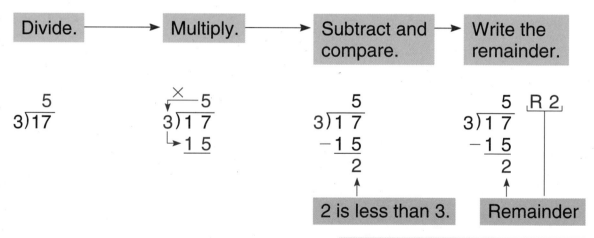

| Divide. | → | Multiply. | → | Subtract and compare. | → | Write the remainder. |

$$3\overline{)17}^{\,5}$$

$$\begin{array}{r} \times\ 5 \\ 3\overline{)1\ 7} \\ 1\ 5 \end{array}$$

$$\begin{array}{r} 5 \\ 3\overline{)1\ 7} \\ -1\ 5 \\ \hline 2 \end{array}$$

2 is less than 3.

$$\begin{array}{r} 5\ \text{R 2} \\ 3\overline{)1\ 7} \\ -1\ 5 \\ \hline 2 \end{array}$$

Remainder

Each child gets 5 stickers. Two stickers are left over.

The remainder must always be less than the divisor.

## Estimate. About how many:

**1.** 4s in 10?     **2.** 3s in 26?     **3.** 6s in 39?     **4.** 5s in 23?

**5.** 2s in 11?     **6.** 8s in 15?     **7.** 9s in 39?     **8.** 7s in 24?

352

**Copy and complete.**

$$\begin{array}{r} 8\ R\ \underline{\ ?\ } \\ 9.\ 2\overline{)1\ 7} \\ -1\ 6 \\ \hline 1 \end{array}$$

$$\begin{array}{r} 5\ R\ \underline{\ ?\ } \\ 10.\ 4\overline{)2\ 2} \\ -2\ 0 \\ \hline ? \end{array}$$

$$\begin{array}{r} ?\ R\ \underline{\ ?\ } \\ 11.\ 5\overline{)3\ 3} \\ -?\ ? \\ \hline ? \end{array}$$

$$\begin{array}{r} ?\ R\ \underline{\ ?\ } \\ 12.\ 6\overline{)3\ 5} \\ -?\ ? \\ \hline ? \end{array}$$

**Find the quotient and the remainder.**

**13.** $2\overline{)15}$    **14.** $3\overline{)14}$    **15.** $4\overline{)38}$    **16.** $5\overline{)32}$    **17.** $6\overline{)56}$

**18.** $9\overline{)18}$    **19.** $4\overline{)17}$    **20.** $5\overline{)29}$    **21.** $8\overline{)26}$    **22.** $6\overline{)28}$

---

## Checking Division

**To check division, use multiplication.**

- Multiply the quotient and the divisor.
- Add the remainder.
- The answer should equal the dividend.

$$\begin{array}{r} 6\ R\ 3 \\ 4\overline{)2\ 7} \\ -2\ 4 \\ \hline 3 \end{array}$$

$$\begin{array}{r} 6 \text{ Quotient} \\ \times 4 \text{ Divisor} \\ \hline 2\ 4 \\ +\ 3 \text{ Remainder} \\ \hline 2\ 7 \text{ Dividend} \end{array}$$

The answer checks.

---

**Divide and check.**

**23.** $5\overline{)42}$    **24.** $9\overline{)30}$    **25.** $4\overline{)25}$    **26.** $5\overline{)37}$    **27.** $4\overline{)33}$

**28.** $9 \div 5 = \underline{\ ?\ }$    **29.** $15 \div 6 = \underline{\ ?\ }$    **30.** $41 \div 6 = \underline{\ ?\ }$    **31.** $37 \div 4 = \underline{\ ?\ }$

**32.** $19 \div 2 = \underline{\ ?\ }$    **33.** $18 \div 5 = \underline{\ ?\ }$    **34.** $22 \div 3 = \underline{\ ?\ }$    **35.** $89 \div 9 = \underline{\ ?\ }$

**36.** $30 \div 7 = \underline{\ ?\ }$    **37.** $56 \div 7 = \underline{\ ?\ }$    **38.** $75 \div 9 = \underline{\ ?\ }$    **39.** $42 \div 8 = \underline{\ ?\ }$

**40.** $52 \div 8 = \underline{\ ?\ }$    **41.** $45 \div 6 = \underline{\ ?\ }$    **42.** $26 \div 4 = \underline{\ ?\ }$    **43.** $51 \div 7 = \underline{\ ?\ }$

## PROBLEM SOLVING

**44.** Lou has 13 cards. He sends the same number of cards to each of 6 friends. What is the greatest number of cards he can send to each friend? How many cards will be left over?

# Two-Digit Quotients

Divide: $96 \div 4 = \underline{\ ?\ }$ or $4\overline{)96}^{??}$

To find the first digit in the quotient:

- Estimate: $\underline{\ ?\ }$ tens $\times 4 = 9$ tens

  1 ten $\times 4 = 4$ tens ←—too small

  2 tens $\times 4 = 8$ tens ←— 9 tens is between 8 tens and 12 tens. Try 2.

  3 tens $\times 4 = 12$ tens ←—too large

Write 2 in the quotient above the 9 tens in the dividend.

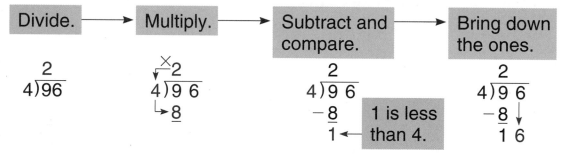

| Divide. | → | Multiply. | → | Subtract and compare. | → | Bring down the ones. |
|---|---|---|---|---|---|---|

$$\begin{array}{r} 2 \\ 4\overline{)96} \end{array} \qquad \begin{array}{r} \times 2 \\ 4\overline{)9\ 6} \\ 8 \end{array} \qquad \begin{array}{r} 2 \\ 4\overline{)9\ 6} \\ -8 \\ \hline 1 \end{array} \text{ 1 is less than 4.} \qquad \begin{array}{r} 2 \\ 4\overline{)9\ 6} \\ -8\downarrow \\ \hline 1\ 6 \end{array}$$

## Repeat the steps to divide the ones.

To find the second digit in the quotient:

- Estimate: $\underline{\ ?\ }$ ones $\times 4 = 16$ ones

  4 ones $\times 4 = 16$ ones ←— Try 4.

Write 4 in the quotient above the 6 ones in the dividend.

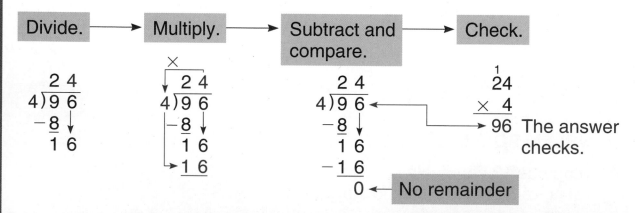

| Divide. | → | Multiply. | → | Subtract and compare. | → | Check. |
|---|---|---|---|---|---|---|

$$\begin{array}{r} 2\ 4 \\ 4\overline{)9\ 6} \\ -8\downarrow \\ \hline 1\ 6 \end{array} \qquad \begin{array}{r} \times \\ 2\ 4 \\ 4\overline{)9\ 6} \\ -8\downarrow \\ \hline 1\ 6 \\ 1\ 6 \end{array} \qquad \begin{array}{r} 2\ 4 \\ 4\overline{)9\ 6} \\ -8\downarrow \\ \hline 1\ 6 \\ -1\ 6 \\ \hline 0 \end{array} \text{← No remainder} \qquad \begin{array}{r} 1\\ 24 \\ \times\ 4 \\ \hline 96 \end{array} \text{ The answer checks.}$$

**Copy and complete.**

$$
\begin{array}{r}
1\ 6 \\
2\overline{)3\ 2} \\
-2\phantom{\ }\downarrow \\
\hline
1\ 2 \\
-1\ 2 \\
\hline
0
\end{array}
$$
1.

$$
\begin{array}{r}
1\ ? \\
3\overline{)3\ 9} \\
-3\phantom{\ }\downarrow \\
\hline
9 \\
-\ ? \\
\hline
?
\end{array}
$$
2.

$$
\begin{array}{r}
1\ ? \\
5\overline{)6\ 5} \\
-5\phantom{\ }\downarrow \\
\hline
1\ 5 \\
-?\ ? \\
\hline
?
\end{array}
$$
3.

$$
\begin{array}{r}
1\ ? \\
4\overline{)6\ 4} \\
-? \\
\hline
?\ ? \\
-?\ ? \\
\hline
?
\end{array}
$$
4.

**Divide and check.**

5. $2\overline{)26}$  6. $4\overline{)44}$  7. $3\overline{)72}$  8. $2\overline{)38}$  9. $4\overline{)76}$

10. $4\overline{)52}$  11. $6\overline{)84}$  12. $9\overline{)99}$  13. $6\overline{)78}$  14. $2\overline{)96}$

15. $4\overline{)56}$  16. $5\overline{)75}$  17. $6\overline{)96}$  18. $7\overline{)84}$  19. $7\overline{)91}$

---

## Three-Digit Dividends

Divide: $145 \div 5 = \underline{\ ?\ }$

Use the division steps to find the quotient.

Think: $5\overline{)145}$  $5<1$  Not enough hundreds
$5\overline{)145}$  $5>14$  Enough tens

| Divide the tens. | Divide the ones. | Check. |
|---|---|---|

$$
\begin{array}{r}
2 \\
5\overline{)1\ 4\ 5} \\
-1\ 0\phantom{\ }\downarrow \\
\hline
4
\end{array}
\qquad
\begin{array}{r}
2\ 9 \\
5\overline{)1\ 4\ 5} \\
-1\ 0 \\
\hline
4\ 5 \\
-4\ 5 \\
\hline
0
\end{array}
\qquad
\begin{array}{r}
^4\phantom{0} \\
29 \\
\times\ \ 5 \\
\hline
145
\end{array}
$$

**Division Steps**
- Divide.
- Multiply.
- Subtract.
- Compare.
- Bring down.
- Repeat the steps as necessary.
- Check.

---

**Divide and check.**

20. $3\overline{)138}$  21. $6\overline{)324}$  22. $7\overline{)147}$  23. $4\overline{)296}$  24. $5\overline{)375}$

# Quotients with Remainders

A camp counselor shares 47 marbles equally among 3 children. At most, how many marbles will each child get? How many will be left over?

To find the number of marbles, divide: $47 \div 3 = \underline{\ ?\ }$   or   $3)\overline{47}$   with ?? above

| Divide tens. → | Multiply. → | Subtract and compare. → | Bring down the ones. |
|---|---|---|---|

$$\begin{array}{r} 1 \\ 3)\overline{47} \end{array} \qquad \begin{array}{r} \times 1 \\ 3)\overline{4\ 7} \\ 3 \end{array} \qquad \begin{array}{r} 1 \\ 3)\overline{4\ 7} \\ -3 \\ \hline 1 \end{array} \qquad \begin{array}{r} 1 \\ 3)\overline{4\ 7} \\ -3 \downarrow \\ \hline 1\ 7 \end{array}$$

1 is less than 3.

## Repeat the steps to divide the ones.

| Divide ones. → | Multiply. → | Subtract and compare. → | Check. |
|---|---|---|---|

$$\begin{array}{r} 1\ 5 \\ 3)\overline{4\ 7} \\ -3 \downarrow \\ \hline 1\ 7 \end{array} \qquad \begin{array}{r} \times \\ 1\ 5 \\ 3)\overline{4\ 7} \\ -3 \downarrow \\ \hline 1\ 7 \\ 1\ 5 \end{array} \qquad \begin{array}{r} 1\ 5 \quad \text{R 2} \\ 3)\overline{4\ 7} \\ -3 \downarrow \\ \hline 1\ 7 \\ -1\ 5 \\ \hline 2 \end{array} \leftarrow \text{Remainder}$$

2 is less than 3.

$$\begin{array}{r} 1 \\ 15 \\ \times\ 3 \\ \hline 45 \\ +\ 2 \\ \hline 47 \end{array}$$ The answer checks.

The camp counselor gives each child 15 marbles. There are 2 marbles left over.

**Copy and complete.**

$$\begin{array}{r} 1\ 1\ \text{R}\ \underline{?} \\ 4\overline{)4\ 5} \\ -\underline{4}\downarrow \\ 5 \\ -\underline{?} \\ ? \end{array}$$

**1.**

$$\begin{array}{r} 1\ ?\ \text{R}\ \underline{?} \\ 3\overline{)4\ 6} \\ -\underline{3}\downarrow \\ 1\ 6 \\ -\underline{?\ ?} \\ ? \end{array}$$

**2.**

$$\begin{array}{r} 1\ ?\ \text{R}\ \underline{?} \\ 5\overline{)5\ 9} \\ -\underline{?}\downarrow \\ 9 \\ -\underline{?} \\ ? \end{array}$$

**3.**

$$\begin{array}{r} ?\ ?\ \text{R}\ \underline{?} \\ 3\overline{)7\ 0} \\ -\underline{?}\downarrow \\ ?\ 0 \\ -\underline{?\ ?} \\ ? \end{array}$$

**4.**

**Divide and check.**

**5.** $5\overline{)68}$     **6.** $2\overline{)83}$     **7.** $8\overline{)89}$     **8.** $4\overline{)63}$     **9.** $3\overline{)58}$

**10.** $3\overline{)82}$     **11.** $6\overline{)95}$     **12.** $5\overline{)81}$     **13.** $2\overline{)71}$     **14.** $7\overline{)92}$

**15.** $4\overline{)89}$     **16.** $3\overline{)40}$     **17.** $6\overline{)79}$     **18.** $7\overline{)79}$     **19.** $5\overline{)77}$

**Find the quotient and the remainder.**

**20.** $73 \div 7$     **21.** $45 \div 2$     **22.** $85 \div 4$     **23.** $77 \div 3$

**24.** $98 \div 3$     **25.** $92 \div 8$     **26.** $74 \div 5$     **27.** $94 \div 4$

**PROBLEM SOLVING**

**28.** Six campers share 75 balloons equally. At most, how many does each child get? How many are left over?

**29.** Two groups share 45 badges equally. What is the greatest number each group gets? How many are left over?

**30.** Chris has 38 marbles. He puts 3 marbles in each bag. What is the greatest number of bags he needs? How many marbles are left over?

**31.** Nick puts 57 books into equal stacks of 4 books each. At most, how many stacks are there? How many books are left over?

 **Skills to Remember**

**Round to the nearest ten.**

**32.** 78     **33.** 85     **34.** 33

**Round to the nearest dollar.**

**35.** $1.46     **36.** $27.51     **37.** $31.92

# Estimating Quotients

Ali, Juanita, and Sara earned $11.75 doing jobs in the neighborhood. They shared the money equally. About how much money did each girl receive?

To find about how much money, estimate: $11.75 ÷ 3 = _?_

**To estimate a quotient:**
- Round the money amount to the nearest dollar.
- Divide mentally.

$11.75 ⟶ $12

$12 ÷ 3 = $4

Each girl received about $4.

**Study these examples.**

| | | |
|---|---|---|
| 83 ÷ 2 = _?_ | $8.40 ÷ 8 = _?_ | $5.75 ÷ 3 = _?_ |
| 83 rounds to 80. | $8.40 rounds to $8. | $5.75 rounds to $6. |
| 80 ÷ 2 = 40 | $8 ÷ 8 = $1 | $6 ÷ 3 = $2 |

**Round to the nearest ten.**

1. 51    2. 29    3. 12    4. 83    5. 65

6. 39    7. 44    8. 78    9. 55    10. 56

11. 91    12. 16    13. 24    14. 62    15. 32

**Round to the nearest dollar.**

16. $4.60    17. $2.30    18. $1.75    19. $5.10    20. $1.55

21. $10.10    22. $20.90    23. $25.50    24. $17.99    25. $12.15

**Estimate the quotient.**

**26.** $11 \div 5$ **27.** $25 \div 6$ **28.** $39 \div 4$ **29.** $56 \div 3$

**30.** $\$4.60 \div 5$ **31.** $\$12.30 \div 4$ **32.** $\$1.75 \div 2$ **33.** $\$5.10 \div 5$

**34.** $\$12.10 \div 6$ **35.** $\$17.99 \div 9$ **36.** $\$24.50 \div 5$ **37.** $\$63.40 \div 3$

**38.** $3\overline{)31}$ **39.** $4\overline{)22}$ **40.** $5\overline{)37}$ **41.** $6\overline{)59}$ **42.** $9\overline{)91}$

**43.** $4\overline{)\$4.10}$ **44.** $2\overline{)\$7.65}$ **45.** $3\overline{)\$9.35}$ **46.** $5\overline{)\$4.95}$ **47.** $3\overline{)\$8.75}$

**48.** $6\overline{)\$17.60}$ **49.** $7\overline{)\$48.60}$ **50.** $8\overline{)\$56.25}$ **51.** $9\overline{)\$71.79}$ **52.** $4\overline{)\$24.25}$

**PROBLEM SOLVING** Use the pictures and estimate to solve.

3 for $11.58

4 for $35.50

1 for $19.88

**53.** Estimate the cost of 1 towel.

**54.** How many towels can you buy with $10?

**55.** Estimate the cost of 1 pail.

**56.** How many pails can you buy with $10?

**57.** Estimate the cost of 2 tape players.

**58.** How many tape players can you buy with $70?

## Finding Together

Roll a number cube two times and form a 2-digit dividend. Then roll the cube again to find a divisor.

**59.** Estimate the quotient. Then find the exact answer.

**60.** Compare the exact answer with the estimated answer, and write a conclusion about them.

Communicate

## 11-7 | Problem Solving: Interpret the Remainder

**Problem:**

There were 56 people in the diner. Only 6 people could be seated at the counter. How many counters would be needed to seat all the people at once? Would each counter be filled?

**1 IMAGINE**

Create a mental picture.

**2 NAME**

*Facts:*    56 people in the diner
Only 6 people at the counter

*Questions:*  How many counters would be needed?
Would each counter be filled?

**3 THINK**

To find how many counters would be filled, divide:  56 ÷ 6 = _?_

To find how many people would be at the last counter, find the remainder.

**4 COMPUTE**

**First estimate.**     **Then divide.**

$60 ÷ 6 = 10$
about 10 counters

$$\begin{array}{r} 9 \text{ R } 2 \\ 6\overline{)56} \\ -\underline{54} \\ 2 \end{array}$$

There would be 9 full counters.
There would be only 2 people at the last counter.
To seat 56 people, 10 counters would be needed.

**5 CHECK**

Multiply the quotient by the divisor.
Then add the remainder.
Does the sum equal the dividend?
Yes. Your answer is correct.

$$\begin{array}{r} 9 \\ \underline{\times 6} \\ 54 \\ \underline{+2} \\ 56 \end{array}$$

**Interpret the remainder to solve each problem.**

1. Pete has 45 slices of bread. He uses 2 slices to make each sandwich. What is the greatest number of sandwiches Pete can make?

**IMAGINE**    Create a mental picture.

**NAME**    *Facts:*    45 slices of bread in all
2 slices for each sandwich

*Question:*    How many sandwiches can he make?

**THINK**    To find the number of sandwiches, divide:   $45 \div 2 = \underline{?}$

**COMPUTE** → **CHECK**

2. Cans of juice are sold in packs of 6. The school needs 94 cans of juice. How many packs should the school buy?

3. Ramón has 23 bottles of spring water. If he puts the greatest number of bottles on each of 4 shelves, does he use all 23 bottles?

4. Elena had $19. She spent an equal amount on each of 4 sandwiches. Then she had $3 left. How much did each sandwich cost?

5. The deli packs 8 boxed lunches in each carton. How many full cartons should the scoutmaster get to feed 98 scouts at a picnic?

6. Beth received 75 orders. She gave the same number of orders to each of 9 delivery persons. At most, how many orders did each delivery person get? How many orders did Beth have to deliver herself?

**Solve each problem and explain the method you used.**

1. The bareback rider sews 8 sequins on each of the horses' hats. She has 62 sequins. How many hats can she sew? How many leftover sequins does she have?

Imagine

Name

Think

Compute

Check

2. A clown has 38 patches. He puts an equal number of patches on each of 3 jackets. At most, how many patches are on each jacket? How many extra patches does the clown have?

3. Each of the 9 clowns took the same number of balloons. There were 95 balloons. What is the greatest number of ballons each clown took? How many balloons were left over?

4. Mr. Oslow spent $34.50 for circus tickets for 5 children. About how much did each child's ticket cost?

5. The acrobat performs 97 stunts every week. She performs an equal number of stunts for each daily show during 1 week. At most, how many stunts does she perform at each show?

6. The divisor is 8. The dividend is 91. What are the quotient and remainder?

7. The product is 120. One factor is 4. What is the missing factor?

**Choose a strategy from the list or use another strategy to solve each problem.**

Use these strategies:

Draw a Picture
Make a Table
Logical Reasoning
Use Simpler Numbers
Missing Information

8. A circus parade had a bear, an elephant, a cyclist, and a ringmaster. What is their position in the parade if the elephant is not first, the ringmaster is last, and the cyclist is just before the bear?

9. Under the Big Top, 4568 people sat in section A, 3469 in section B, and 3087 in section C. How many people were sitting under the Big Top?

10. Using stilts to walk the inside of a circus ring, how far did Eddy walk in 12 steps if one step is 24 inches?

11. The highest high-wire act was by Philippe Petit at a height of 1350 feet between the towers of the World Trade Center. How far did Mr. Petit walk?

12. Kara jumps 6 feet high on the trampoline. Gus jumps double the 4-foot height that Yoko jumps. How much higher must each jump to reach 9 feet?

13. Kara jumped a total of 96 centimeters in 6 jumps. Each jump was the same height. How many centimeters high was each jump?

14. It takes 15 clowns to build a human tower. The base uses 2 more clowns than the third row. The second row uses 4 clowns, while the fourth uses 2 clowns, and the fifth has one clown. How many clowns are on the base and third rows?

# Chapter Review and Practice

**Find the quotient.** *(See p. 349.)*

**1.** $6 \div 3$      **2.** $60 \div 3$      **3.** $600 \div 3$

**4.** $4\overline{)160}$      **5.** $9\overline{)810}$      **6.** $7\overline{)700}$

**Find the quotient and the remainder.** *(See pp. 348–349, 352–357.)*

**7.** $5\overline{)21}$    **8.** $3\overline{)29}$    **9.** $6\overline{)38}$    **10.** $9\overline{)74}$

**11.** $2\overline{)25}$    **12.** $8\overline{)95}$    **13.** $5\overline{)66}$    **14.** $6\overline{)73}$

**Divide and check.**

**15.** $56 \div 7$    **16.** $43 \div 9$    **17.** $427 \div 7$    **18.** $8\overline{)328}$

**Estimate the quotient.** *(See pp. 358–359.)*

**19.** $6\overline{)58}$    **20.** $5\overline{)32}$    **21.** $4\overline{)19}$    **22.** $8\overline{)37}$

**23.** $7\overline{)\$13.90}$    **24.** $3\overline{)\$27.15}$    **25.** $9\overline{)\$80.72}$    **26.** $5\overline{)\$38.89}$

## PROBLEM SOLVING

*(See pp. 360–363.)*

**27.** Betty Sue arranges all of her tapes in 9 rows. She has 82 tapes. How many tapes are in each row? How many tapes are left over?

**28.** Kyle has 96 baseball cards. He divides them equally among 9 friends. At most, how many baseball cards does each friend receive? How many are left over?

**29.** Jamal arranged 67 chairs into rows. There is room for only 9 chairs in each row. How many rows of chairs will there be?

**30.** Jill has $10 to buy a tape. The tapes are on sale at 3 for $26.75. Does Jill have enough money to buy one tape? Explain.

*(See Still More Practice, pp. 452–453.)*

## COMPATIBLE NUMBERS IN DIVISION

**Compatible numbers** are numbers that are easy to work with mentally. You can use compatible numbers to help you estimate quotients.

| X | 0 | 1 | 2 | 3 | 4 | 5 |
|---|---|---|---|---|---|---|
| 0 | 0 | 0 | 0 | 0 | 0 | 0 |
| 1 | 0 | 1 | 2 | 3 | 4 | 5 |
| 2 | 0 | 2 | 4 | 6 | 8 | 10 |
| 3 | 0 | 3 | 6 | 9 | 12 | 15 |
| 4 | 0 | 4 | 8 | 12 | 16 | 20 |
| 5 | 0 | 5 | 10 | 15 | 20 | 25 |

Estimate:  $23 \div 7$

Think:  $3 \times 7 = 21$    21 is close to 23.

7 and 21 are compatible numbers.

$$7\overline{)21} = 3$$

So $23 \div 7$ is about 3.

**Write the compatible numbers you would use to estimate each quotient.**

**1.** $8\overline{)67}$    **2.** $9\overline{)52}$    **3.** $6\overline{)49}$    **4.** $3\overline{)192}$    **5.** $5\overline{)243}$

**Estimate the quotient.** Use compatible numbers.

**6.** $9\overline{)38}$    **7.** $2\overline{)17}$    **8.** $4\overline{)35}$    **9.** $6\overline{)331}$    **10.** $3\overline{)220}$

**11.** $65 \div 7$    **12.** $74 \div 8$    **13.** $47 \div 9$    **14.** $22 \div 4$    **15.** $800 \div 9$

## PROBLEM SOLVING

**16.** Grandmother Curry gave her 3 grandchildren a twenty-dollar bill for cleaning the garage. The children shared the money equally. About how much money did each child receive?

**17.** The Curry children collected 750 bottles for recycling. They put the bottles into boxes. Each box held 8 bottles. About how many boxes did the children fill?

## Performance Assessment

**How many digits will each quotient have? Use or draw base ten models to explain your thinking.**

1. dividend: 250
   divisor: 5

2. dividend: 400
   divisor: 2

3. dividend: 120
   divisor: 4

**Find the quotient and the remainder.**

4. $2\overline{)17}$

5. $7\overline{)39}$

6. $5\overline{)42}$

7. $9\overline{)61}$

8. $3\overline{)37}$

9. $4\overline{)57}$

10. $8\overline{)92}$

11. $7\overline{)85}$

**Divide and check.**

12. $72 \div 9$

13. $36 \div 7$

14. $97 \div 8$

15. $2\overline{)62}$

**Estimate the quotient.**

16. $3\overline{)29}$

17. $4\overline{)22}$

18. $8\overline{)39}$

19. $5\overline{)31}$

20. $4\overline{)\$11.99}$

21. $6\overline{)\$36.25}$

22. $9\overline{)\$63.40}$

23. $7\overline{)\$41.75}$

**PROBLEM SOLVING**     *Use a strategy you have learned.*

24. Chad received $25 for his birthday. He wants to share the money equally among his 4 nephews. At most, how much money will each nephew get? How much will be left?

25. Kate divided 50 muffins equally among her six nephews. What is the greatest number of muffins each nephew received? How many muffins were left?

26. There are 23 slices of bread in a loaf. If a mother uses 2 slices for each sandwich, how many sandwiches can she make?

27. Danika bought the same gift for each of her 3 friends. She spent $26.79. About how much did each gift cost?

# Cumulative Review IV

**Choose the best answer.**

**1.** What number comes next?

2988, 2992, 2996, __?__

    **a.** 2984
    **b.** 2998
    **c.** 3000
    **d.** 3002

**2.** Find the missing factor.

$1 \times 7 = 7 \times$ __?__

    **a.** 0
    **b.** 1
    **c.** 7
    **d.** 8

**3.** Mike has 8 canoes. Each canoe has 2 oars. How many oars are there altogether?

    **a.** 4    **b.** 10
    **c.** 16    **d.** 18

**4.** Which is greater:

300 mL or 30 L?

    **a.** 300 mL
    **b.** 30 L
    **c.** same amount
    **d.** cannot tell

**5.** $9\overline{)36}$

    **a.** 3    **b.** 4
    **c.** 6    **d.** 4 R2

**6.** $72 \div 8 =$ __?__

    **a.** 7    **b.** 8
    **c.** 9    **d.** 80

**7.** Which figure is a rectangle?

A    B    C

    **a.** Figure A    **b.** Figure B
    **c.** Figure C    **d.** none of these

**8.** Which describes the figure?

    **a.** rectanglular prism, 8 edges
    **b.** pyramid, 5 edges
    **c.** cube, 12 edges
    **d.** pyramid, 8 edges

**9.**
$$\begin{array}{r} 124 \\ \times\ 6 \\ \hline \end{array}$$

    **a.** 624    **b.** 644
    **c.** 744    **d.** 724

**10.** $3\overline{)38}$

    **a.** 11 R2    **b.** 12 R2
    **c.** 15 R2    **d.** 12 R1

**11.** Estimate. Round to the nearest ten.

$$\begin{array}{r} 18 \\ \times\ 4 \\ \hline \end{array}$$

    **a.** 70    **b.** 80
    **c.** 90    **d.** 100

**12.** Estimate. Round to the nearest ten cents.

$8 \times \$.36$

    **a.** $1.60    **b.** $3.20
    **c.** $8.00    **d.** $32.00

**13.** $600 \div 3$

    **a.** 20    **b.** 100
    **c.** 300    **d.** not given

**14.** Estimate.

$9\overline{)\$26.50}$

    **a.** $3.00    **b.** $5.00
    **c.** $6.00    **d.** $7.00

# Ongoing Assessment IV

## For Your Portfolio

Solve each problem. Explain the steps and the strategy
or strategies you used for each. Then choose one from
problems 1–4 for your Portfolio.

1. Vince bought 3 lamps at $9.05 each. How much change did he receive from $30.00?

2. The difference between two numbers is 214. One number is 86. What is the other number?

3. Each shelf of a bookcase holds 7 books. Marsha has 46 books. Marsha fills as many shelves as she can and then has some books left over. How many books are left over?

4. Mr. Kickingbird has black, blue, and gray slacks. He also has red and green sweatshirts. How many different outfits can he make?

### Tell about it.

5. Nine friends are going to the museum. Each ticket costs $7.49. About how much money do they need altogether for the tickets?

Communicate

6. For problem 5, did you use front-end estimation or rounding? Why?

7. What other strategy can you use to solve problem 4?

## For Rubric Scoring

Listen for information on how your work will be scored.

8. There are 22 students going to the lunchroom. Only 5 students can sit at each table. How many tables are needed to seat all of the students? Explain how you solved the problem. What do the quotient and the remainder represent?

# Fractions

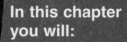

## 12

## QUEUE

The
life
of
this
queue
depends
only
on
you.
One
step
out
of
line

and
it
all
breaks
in
two.

*Sylvia Cassedy*

**In this chapter
you will:**

Explore equivalent
  fractions

Compare, add,
  and subtract
  fractions with
  models

Relate fractions
  and circle graphs

Solve problems by
  using fractions

Solve problems by
  using a drawing
  or model

**Critical Thinking/
Finding Together**

Relate the fraction
strip for thirds at the
bottom of the page
to the children.

$\frac{1}{3}$ $\frac{1}{3}$ $\frac{1}{3}$

# 12-1 Fractions

A **fraction** can name one or more equal parts
of a whole or of a set.

Each fraction strip in the table is divided into equal parts.

## Fraction Table

| | |
|---|---|
| 1 | 1 whole |
| $\frac{1}{2}$    $\frac{1}{2}$ | 2 halves |
| $\frac{1}{3}$    $\frac{1}{3}$    $\frac{1}{3}$ | 3 thirds |
| $\frac{1}{4}$    $\frac{1}{4}$    $\frac{1}{4}$    $\frac{1}{4}$ | 4 fourths |
| $\frac{1}{5}$    $\frac{1}{5}$    $\frac{1}{5}$    $\frac{1}{5}$    $\frac{1}{5}$ | 5 fifths |
| $\frac{1}{6}$    $\frac{1}{6}$    $\frac{1}{6}$    $\frac{1}{6}$    $\frac{1}{6}$    $\frac{1}{6}$ | 6 sixths |
| $\frac{1}{7}$    $\frac{1}{7}$    $\frac{1}{7}$    $\frac{1}{7}$    $\frac{1}{7}$    $\frac{1}{7}$    $\frac{1}{7}$ | 7 sevenths |
| $\frac{1}{8}$    $\frac{1}{8}$    $\frac{1}{8}$    $\frac{1}{8}$    $\frac{1}{8}$    $\frac{1}{8}$    $\frac{1}{8}$    $\frac{1}{8}$ | 8 eighths |
| $\frac{1}{9}$    $\frac{1}{9}$    $\frac{1}{9}$    $\frac{1}{9}$    $\frac{1}{9}$    $\frac{1}{9}$    $\frac{1}{9}$    $\frac{1}{9}$    $\frac{1}{9}$ | 9 ninths |
| $\frac{1}{10}$ (×10) | 10 tenths |
| $\frac{1}{12}$ (×12) | 12 twelfths |

$$1 = \frac{2}{2} = \frac{3}{3} = \frac{4}{4} = \frac{5}{5} = \frac{6}{6} = \frac{7}{7} = \frac{8}{8} = \frac{9}{9} = \frac{10}{10} = \frac{12}{12}$$

**numerator** ⟶ 5 ⟵ number of equal parts
**denominator** ⟶ 8 ⟵ total number of equal parts in
a whole or in the set

**Word name**: five eighths      **Write**: $\frac{5}{8}$

$\frac{5}{8}$ of this

figure is

colored purple.

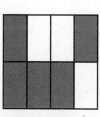

$\frac{5}{8}$ of the

set of hearts

is colored red.

**Write each as a fraction.**

**1.** three fourths **2.** one half **3.** five sixths

**4.** one eighth **5.** two ninths **6.** three thirds

**Write the word name for each fraction.**

**7.** $\frac{1}{4}$ **8.** $\frac{2}{7}$ **9.** $\frac{4}{5}$ **10.** $\frac{7}{9}$ **11.** $\frac{1}{12}$

**12.** $\frac{3}{10}$ **13.** $\frac{5}{8}$ **14.** $\frac{1}{6}$ **15.** $\frac{2}{3}$ **16.** $\frac{5}{9}$

## PROBLEM SOLVING

**17.** There are 12 months in a year. What part of the year is July?

**18.** What part of a year begins with the letter J?

**19.** Does the figure at the right show fourths? Why or why not?

*Communicate* ✓

**20.** Do the fractional parts get larger or smaller when you divide a whole into more and more equal parts? Use a model or draw a picture to explain your answer.

**21.** Joseph cut an apple pie into 10 equal pieces and a cherry pie into 8 equal pieces. Which pie is cut into larger pieces?

 **Connections: Art**

**Draw a picture to show each fraction.**

**22.** $\frac{1}{4}$ **23.** $\frac{1}{2}$ **24.** $\frac{5}{6}$ **25.** $\frac{3}{8}$ **26.** $\frac{2}{5}$

# Equivalent Fractions

Algebra

Different fractions can name the same amount. Fractions that name the same amount are called **equivalent fractions**.

## Hands-On Understanding

**You Will Need:** fraction strips for halves, thirds, fourths, fifths, sixths, eighths, ninths, tenths, and twelfths; record sheet; scissors

**Step 1**   Label your record sheet with the following headings:

| Fraction | Equivalent Fractions |
|----------|---------------------|
|          |                     |

**Step 2**   Cut each fraction strip into equal parts and put the pieces from each fraction strip into separate piles.

**Step 3**   Place a fraction piece for $\frac{1}{2}$ on your worktable.

Record this fraction on your record sheet.

**Step 4**   Now model $\frac{1}{2}$ using the pieces of the fraction strip that show fourths.

How many pieces of the fourths strip did you use to equal the length of the fraction piece for $\frac{1}{2}$?

What fraction does your fourths model represent?

Record this equivalent fraction on your record sheet.

| Step 5 | Use your fraction strips for sixths, eighths, tenths, and twelfths to model equivalent fractions for $\frac{1}{2}$. |
|---|---|

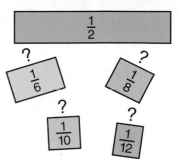

Record each equivalent fraction in the column to the right of $\frac{1}{2}$.

| Step 6 | Place a fraction piece for $\frac{1}{3}$ on your worktable. Record this fraction on your record sheet. |
|---|---|

| Step 7 | Now model equivalent fractions for $\frac{1}{3}$ using the pieces of the fraction strips that show sixths, ninths, and twelfths. |
|---|---|

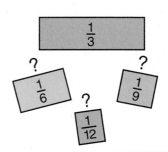

How many pieces of each fraction strip did you use to equal the length of the fraction piece for $\frac{1}{3}$?

What fraction does each model represent?

Record each equivalent fraction.

## Communicate

Discuss

1. Look on your record sheet at the equivalent fractions for $\frac{1}{2}$ and $\frac{1}{3}$. What do you notice?

**Use the second fraction strip to write a fraction that is equivalent to the first strip.**

2.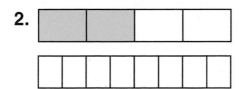

3.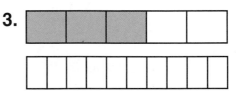

**Write the equivalent fraction for each. Use fraction strips.**

4. $\frac{2}{4} = \frac{?}{8}$    5. $\frac{3}{6} = \frac{?}{12}$    6. $\frac{3}{4} = \frac{?}{8}$    7. $\frac{3}{5} = \frac{?}{10}$

373

# 12-3 Estimating Fractions

You can estimate using fractions to tell *about* how much.

About how full is each cup?

Write *about* $\frac{1}{4}$, *about* $\frac{1}{2}$, or *about* a whole to estimate parts of a whole.

*about* $\frac{1}{4}$

*about* $\frac{1}{2}$

*about* a whole

## Is the estimate correct? Write *Yes* or *No*.

**1.** About what part is light blue?

Estimate: *about* a whole

**2.** About how much pizza is left?

Estimate: *about* $\frac{1}{2}$

**3.** About what part of the hour has passed?

Estimate: *about* $\frac{1}{2}$

**4.** About how much of the pie is left?

Estimate: *about* $\frac{1}{4}$

**5.** About how much sand is in the bucket?

Estimate: *about* a whole

**6.** About what part of the sandwich is left?

Estimate: *about* $\frac{1}{4}$

**Estimate the fraction for the part of each set that is shaded. Write *less than* $\frac{1}{2}$ or *more than* $\frac{1}{2}$ .**

**7.** ★★
★☆
☆☆
☆☆
☆☆
less than $\frac{1}{2}$

**8.** ■ ■ ■ ■
■ ■ ■ □
□ □ □ □

**9.** △ △ △ △
△ △ △ △

**10.** ⬠ ⬠ ⬠
⬠ ⬠ ⬠
⬠ ⬠ ⬠

**11.** ● ○ ○ ○ ○
○ ○ ○ ○ ○

**12.** ⬠ ⬠
⬠ ⬠
⬠ ⬠
⬠ ⬠
⬠ ⬠

**Estimate the time.**

**13.**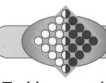
about $\frac{1}{4}$ to two

**14.**

**15.**

**16.**

### Finding Together

**17.** Have a classmate cut a piece of string equal to the length of your arm. First estimate, then measure the length of the piece of string. Choose objects about the same length, about $\frac{1}{4}$ of the length, and about $\frac{1}{2}$ of the length of the piece of string. Then find the actual measure of each object to see if the estimate is reasonable.

375

# Comparing Fractions

Sue has colored $\frac{5}{6}$ of the stars on her banner. Ben has colored $\frac{3}{6}$ of the stars on his banner. Who has colored the greater part?

To find who has colored the greater part, compare the fractions. Use fraction strips.

## Hands-On Understanding

**You Will Need:** 2 fractions strips that show sixths, 1 fraction strip that shows fifths, 1 fraction strip that shows fourths, crayons

Compare $\frac{5}{6}$ and $\frac{3}{6}$.

**Step 1**  Shade 5 equal parts of a fraction strip that show sixths.

| $\frac{1}{6}$ | $\frac{1}{6}$ | $\frac{1}{6}$ | $\frac{1}{6}$ | $\frac{1}{6}$ | $\frac{1}{6}$ |
|---|---|---|---|---|---|

What fraction does your model represent?

**Step 2**  Shade 3 equal parts of another fraction strip that show sixths.

| $\frac{1}{6}$ | $\frac{1}{6}$ | $\frac{1}{6}$ | $\frac{1}{6}$ | $\frac{1}{6}$ | $\frac{1}{6}$ |
|---|---|---|---|---|---|

What fraction does this model represent?

**Step 3**  Place your fraction models one below the other.

| $\frac{1}{6}$ | $\frac{1}{6}$ | $\frac{1}{6}$ | $\frac{1}{6}$ | $\frac{1}{6}$ | $\frac{1}{6}$ |
|---|---|---|---|---|---|

| $\frac{1}{6}$ | $\frac{1}{6}$ | $\frac{1}{6}$ | $\frac{1}{6}$ | $\frac{1}{6}$ | $\frac{1}{6}$ |
|---|---|---|---|---|---|

**Step 4**   Compare your fraction models.

Which fraction is greater? Explain your answer.

Use the symbols < or > to write a sentence that tells which fraction is greater.

Who colored the greater part of his or her banner?

Compare $\frac{2}{5}$ and $\frac{3}{4}$.

**Step 5**   Shade 2 equal parts of a fraction strip that show fifths.

| $\frac{1}{5}$ | $\frac{1}{5}$ | $\frac{1}{5}$ | $\frac{1}{5}$ | $\frac{1}{5}$ |
|---|---|---|---|---|

What fraction does your model represent?

**Step 6**   Shade 3 equal parts of the fraction strip that show fourths.

| $\frac{1}{4}$ | $\frac{1}{4}$ | $\frac{1}{4}$ | $\frac{1}{4}$ |
|---|---|---|---|

What fraction does this model represent?

**Step 7**   Place your fraction models one below the other and compare them.

Which fraction is less? Explain your answer.

Use the symbols < or > to write a sentence that tells which fraction is less.

**Compare. Write < or >. Use your fraction strips.**

1. $\frac{1}{4}$ __?__ $\frac{3}{4}$

2. $\frac{1}{3}$ __?__ $\frac{1}{6}$

3. $\frac{1}{2}$ __?__ $\frac{5}{8}$

## Communicate

Math Journal

4. Explain in your Math Journal how you can compare fractions with the same denominators and with unlike denominators.

# Finding Part of a Number

Fractions help us find parts of a number.

Donna has 12 eggs in a carton. One half of them are brown. How many eggs are brown?

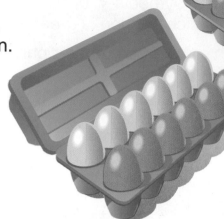

$\frac{1}{2}$ of 12 = __?__

To find $\frac{1}{2}$ of 12, divide by 2.

$\frac{1}{2}$ of 12 = __?__

$$12 \div 2 = 6 \qquad 2\overline{)12}^{\,6}$$

Six eggs are brown.

Remember: The denominator tells the total number of equal parts in a set.

## Study these examples.

$\frac{1}{5}$ of 15 = 3 ←

Think: 15 ÷ 5 = __?__

$\frac{1}{4}$ of 20 = 5 ←

Think: 20 ÷ 4 = __?__

## Find part of the number.

1. $\frac{1}{2}$ of 10 = __?__

2. 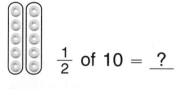 $\frac{1}{3}$ of 9 = __?__

3. $\frac{1}{3}$ of 6 = __?__

4. $\frac{1}{4}$ of 16 = __?__

5.  $\frac{1}{7}$ of 7 = __?__

6.  $\frac{1}{2}$ of 20 = __?__

**Find part of the number.**

7. $\frac{1}{3}$ of 6 = _?_

    6 ÷ 3 = _?_

8. $\frac{1}{4}$ of 8 = _?_

    8 ÷ 4 = _?_

9. $\frac{1}{2}$ of 4 = _?_

    4 ÷ 2 = _?_

10. $\frac{1}{3}$ of 12 = _?_

    12 ÷ 3 = _?_

11. $\frac{1}{4}$ of 16 = _?_

    16 ÷ 4 = _?_

12. $\frac{1}{2}$ of 10 = _?_

    10 ÷ 2 = _?_

13. $\frac{1}{3}$ of 12

14. $\frac{1}{2}$ of 8

15. $\frac{1}{3}$ of 15

16. $\frac{1}{2}$ of 20

17. $\frac{1}{4}$ of 32

18. $\frac{1}{3}$ of 27

19. $\frac{1}{4}$ of 20

20. $\frac{1}{5}$ of 25

21. $\frac{1}{10}$ of 40

22. $\frac{1}{7}$ of 21

23. $\frac{1}{5}$ of 45

24. $\frac{1}{6}$ of 18

## Critical Thinking

Draw and cut out 10 circles.

Divide the circles into 5 groups of 2.

**Use your circles to help you complete each.**

Communicate

25. If $\frac{1}{5}$ of 10 is 2, then $\frac{2}{5}$ of 10 is 4.

26. $\frac{3}{5}$ of 10 is _?_

27. $\frac{4}{5}$ of 10 is _?_

28. $\frac{5}{5}$ of 10 is _?_

29. Explain the pattern you see in exercises 25–28.

## PROBLEM SOLVING

30. Which is longer, $\frac{1}{4}$ of a foot or 4 inches?

31. Which is longer, $\frac{1}{3}$ of a yard or 5 feet?

32. Use patterning to determine which is longer, $\frac{2}{3}$ of a foot or 10 inches.

# Mixed Numbers

A **mixed number** is a number made up of a whole number and a fraction.

▶ Write the mixed number as: $1\frac{1}{2}$.

Read $1\frac{1}{2}$ as: one and one half.

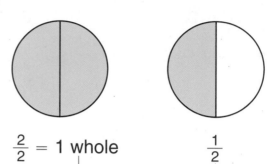

$\frac{2}{2} = 1$ whole      $\frac{1}{2}$

$1\frac{1}{2}$

▶ Write the mixed number as: $2\frac{3}{4}$.

Read $2\frac{3}{4}$ as: two and three fourths.

Think: 1 whole + 1 whole = 2 wholes

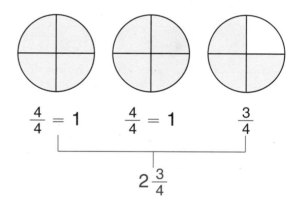

$\frac{4}{4} = 1$      $\frac{4}{4} = 1$      $\frac{3}{4}$

$2\frac{3}{4}$

## Write the mixed number for each.

1.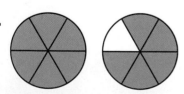

2.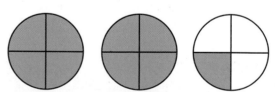

3.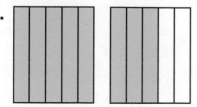

4.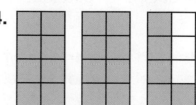

**Write the mixed number and the word name for each.**

5.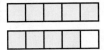

$1\frac{4}{5}$; one and four fifths

6.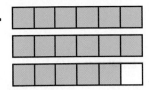

7.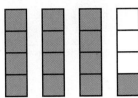

8.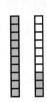

9.

10.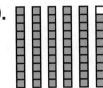

**Estimate about how many. Write the mixed number for each.**

11.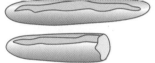

12.

13.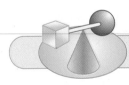

14.

## Challenge

**Solve. Use the pictograph.**

15. How many apples did each eat?

16. Who ate the most?

17. Who ate the least?

| Apples Eaten | | | | |
|---|---|---|---|---|
| Marsha | | | | |
| Janice | | | | |
| Jake | | | | |
| $\bigcirc$ = 1 apple. $\bigcirc$ = $\frac{1}{2}$ apple. | | | | |

# Adding Fractions

Ann, Tom, and Jody painted a fence. Ann painted $\frac{1}{8}$ of the fence and Tom painted $\frac{3}{8}$ of it. What fractional part of the fence did Ann and Tom paint together?

## Hands-On Understanding

**You Will Need:** 1 fraction strip that shows eighths, crayons

**Step 1**    Shade $\frac{1}{8}$ for the part of the fence that Ann painted.

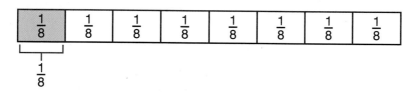

**Step 2**    Now shade $\frac{3}{8}$ more for the part of the fence that Tom painted.

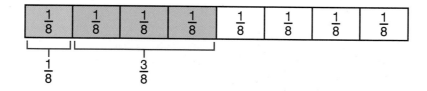

How many eighths are shaded in all?

Write this amount as a fraction.

Now write an addition sentence that tells the fractional part of the fence that Ann and Tom painted together.

| Step 3 | Locate the sum on the fraction table at the right. |
|---|---|

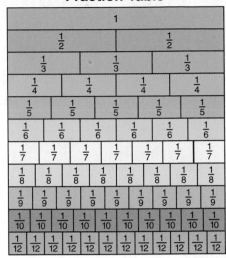

**Fraction Table**

Name all the equivalent fractions for the sum.

Which equivalent fraction uses the least number of parts of the whole?

$\frac{4}{8} = \frac{1}{2}$    $\frac{1}{2}$ is in **simplest form**.

**Use your fraction strips to model each exercise.
Then find the sum in simplest form.**

1. $\frac{1}{3} + \frac{1}{3}$

2. $\frac{3}{5} + \frac{1}{5}$

3. $\frac{2}{4} + \frac{1}{4}$

4. $\frac{3}{6} + \frac{2}{6}$

5. $\frac{3}{7} + \frac{3}{7}$

6. $\frac{3}{8} + \frac{2}{8}$

7. $\frac{6}{9} + \frac{2}{9}$

8. $\frac{3}{10} + \frac{5}{10}$

9. $\frac{7}{12} + \frac{2}{12}$

**Write an addition sentence to represent each model.**    Algebra

10.

11.

12.

## Communicate

Math Journal

13. Write a rule in your Math Journal to tell how to add fractions with the same denominator.

# Subtracting Fractions

Matthew ran $\frac{3}{10}$ of a mile. Charlie ran $\frac{1}{10}$ of a mile. How much farther did Matthew run than Charlie?

▶ To find how much farther, subtract: $\frac{3}{10} - \frac{1}{10} = \underline{\ ?\ }$

## Hands-On Understanding

**You Will Need:** 1 fraction strip that shows tenths, 6 fraction strips that show sixths, crayons, fraction table

**Step 1**   Shade 3 equal parts of a fraction strip that show tenths.

| $\frac{1}{10}$ | $\frac{1}{10}$ | $\frac{1}{10}$ | $\frac{1}{10}$ | $\frac{1}{10}$ | $\frac{1}{10}$ | $\frac{1}{10}$ | $\frac{1}{10}$ | $\frac{1}{10}$ | $\frac{1}{10}$ |

What fraction does your model represent?

**Step 2**   Now draw an X through 1 shaded part of your fraction strip.

| $\frac{1}{10}$ | $\frac{1}{10}$ | $\frac{1}{10}$ | $\frac{1}{10}$ | $\frac{1}{10}$ | $\frac{1}{10}$ | $\frac{1}{10}$ | $\frac{1}{10}$ | $\frac{1}{10}$ | $\frac{1}{10}$ |

How many shaded parts of your fraction strip have an X?

What fraction does this represent?

How many shaded tenths are left?

Write this amount as a fraction in simplest form.

Write a subtraction sentence to show how much farther Matthew ran than Charlie.

**Step 3** Use your fraction strips for sixths to find each difference.

$$\frac{6}{6} - \frac{1}{6} = \underline{\ ?\ }$$   $$\frac{5}{6} - \frac{1}{6} = \underline{\ ?\ }$$   $$\frac{4}{6} - \frac{1}{6} = \underline{\ ?\ }$$

$$\frac{3}{6} - \frac{1}{6} = \underline{\ ?\ }$$   $$\frac{2}{6} - \frac{1}{6} = \underline{\ ?\ }$$   $$\frac{1}{6} - \frac{1}{6} = \underline{\ ?\ }$$

What pattern do you notice?

**Use fraction strips to model each exercise.**
**Then find the difference in simplest form.**
Use a fraction table to help.

**1.** $\frac{7}{10} - \frac{4}{10} = \underline{\ ?\ }$   **2.** $\frac{6}{9} - \frac{2}{9} = \underline{\ ?\ }$   **3.** $\frac{4}{6} - \frac{3}{6} = \underline{\ ?\ }$

**4.** $\frac{7}{12} - \frac{2}{12} = \underline{\ ?\ }$   **5.** $\frac{6}{8} - \frac{5}{8} = \underline{\ ?\ }$   **6.** $\frac{3}{3} - \frac{1}{3} = \underline{\ ?\ }$

**Write a subtraction sentence to represent each model.**
**Write the difference in simplest form.**

Algebra

**7.**
| $\frac{1}{9}$ | $\frac{1}{9}$ | $\frac{1}{9}$ | $\frac{1}{9}$ | $\frac{1}{9}$ | $\frac{1}{9}$ | $\frac{1}{9}$ | $\frac{1}{9}$ | $\frac{1}{9}$ |

**8.**
| $\frac{1}{4}$ | $\frac{1}{4}$ | $\frac{1}{4}$ | $\frac{1}{4}$ |

**9.**
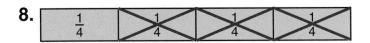

## Communicate

Math Journal

**10.** Write a rule in your Math Journal about how to subtract fractions with the same denominators.

**11.** Use fraction strips to model and explain why $\frac{3}{8} - \frac{1}{8}$ does not equal $\frac{2}{0}$.

385

## 12-9   Circle Graphs

A **circle graph** shows information as fractional parts of a circle.

**School Transportation**

This circle graph shows how the students in Ms. Ferrara's class come to school.

This line divides $\frac{1}{2}$ into two equal parts: $\frac{1}{4}$ and $\frac{1}{4}$.

This line divides the whole into two equal parts: $\frac{1}{2}$ and $\frac{1}{2}$.

Walk $\frac{1}{4}$   Car $\frac{1}{4}$

School Bus $\frac{1}{2}$

What part of the students walks to school **or** comes by car?

To find the part of the students that walks or comes by car, add the fractional parts from the graph.

Of all the students, $\frac{1}{4}$ walks and $\frac{1}{4}$ comes by car.

Add: $\frac{1}{4} + \frac{1}{4} = \frac{2}{4}$     $\frac{2}{4} = \frac{1}{2}$

So $\frac{1}{2}$ of the students walks to school or comes by car.

The circle graph also shows: $\frac{1}{2}$ of the students comes by bus.

---

**Use this circle graph to answer each problem.**

**Favorite Colors**

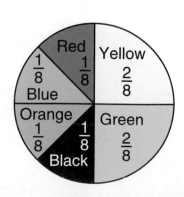

1. What fractional part of the class likes orange?

2. What fractional part of the class likes yellow?

3. What fractional part of the class likes red or blue?

4. What color is liked as much as green?

5. Do more students prefer red than green?

6. Do twice as many students prefer yellow than orange?

**Use this circle graph to answer each problem.**

7. What fractional part of the class likes fruit?

8. What fractional part of the class likes chicken?

9. What food is liked by most students?

10. What fractional part of the class likes spaghetti or cheese?

11. What fractional part tells how many more students like spaghetti than chicken?

12. What fractional part tells how many students do *not* favor chicken?

**Favorite Foods**

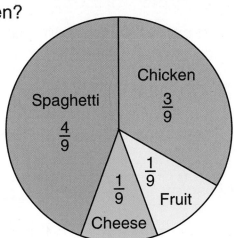

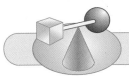

**Challenge**

Mia spends $20 on food, toys, and books at the fair. Jamie spends $16 on food, toys, and books.

**Use the information above and the circle graphs to solve each problem.**

13. How much money does Mia spend on food? toys? books?

14. How much money does Jamie spend on food? toys? books?

15. Mia and Jamie each spend $\frac{1}{4}$ of their money on food. Do they spend the same amount? Explain your answer.

**School Fair Spending Money**

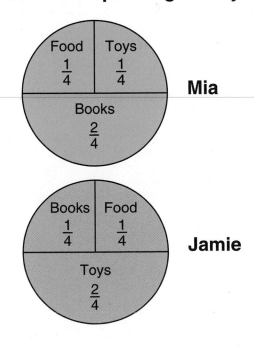

# TECHNOLOGY

## Fractions on a Calculator

You can use a fraction calculator to solve problems with fractions and mixed numbers.

▶ To enter the fraction $\frac{3}{4}$,

| | This key is used to show a fraction. |

Press these keys →

Display →

Display → $\boxed{3}$ $\boxed{3/}$ $\boxed{3/4}$

▶ To enter the mixed number $2\frac{5}{8}$,

Press these keys → $\boxed{2}$ $\boxed{\text{Unit}}$ $\boxed{5}$ $\boxed{/}$ $\boxed{8}$

Display → $\boxed{2}$ $\boxed{2\text{u}}$ $\boxed{2\text{u}5}$ $\boxed{2\text{u}5/}$ $\boxed{2\text{u}5/8}$

The **Unit** key is used to enter the whole number part of a mixed number.

▶ To add $\frac{1}{6} + \frac{3}{6}$ on a calculator,

Press these keys → $\boxed{1}$ $\boxed{/}$ $\boxed{6}$ $\boxed{+}$ $\boxed{3}$ $\boxed{/}$ $\boxed{6}$ $\boxed{=}$

Display → $\boxed{1}$ $\boxed{1/}$ $\boxed{1/6}$ $\boxed{1/6}$ $\boxed{3}$ $\boxed{3/}$ $\boxed{3/6}$ $\boxed{4/6}$

So $\frac{1}{6} + \frac{3}{6} = \frac{4}{6}$.

▶ To subtract $\frac{5}{8} - \frac{3}{8}$ on a calculator,

Press these keys →

Press these keys → $\boxed{5}$ $\boxed{/}$ $\boxed{8}$ $\boxed{-}$ $\boxed{3}$ $\boxed{/}$ $\boxed{8}$ $\boxed{=}$

So $\frac{5}{8} - \frac{3}{8} = \frac{2}{8}$.

Display → $\boxed{2/8}$

388

## Write the missing keys that will give each display.

**1.** [7] [?] [8] → | 7/8 |

**2.** [6] [?] [3] [/] [?] → | 6⌴3/5 |

**3.** [?] [?] [?] [?] [?] → | 4⌴1/9 |

**4.** [2] [/] [?] [+] [?] [/] [6] [=] → | 5/6 |

**5.** [?] [/] [1] [0] [−] [7] [/] [?] [?] [=] → | 2/10 |

## Match each set of keys with its display.

**6.** [5] [/] [1] [2]　　　　　　　　　　　　**a.** | 4/5 |

**7.** [5] [Unit] [1] [/] [2]　　　　　　　　　　**b.** | 5/12 |

**8.** [1] [/] [5] [+] [3] [/] [5] [=]　　　　　**c.** | 3/12 |

**9.** [5] [/] [1] [2] [−] [2] [/] [1] [2] [=]　　**d.** | 5 1/2 |

## Complete. Watch the + or −. Use a calculator.

**10.** $\frac{6}{7} - \frac{2}{7} = $ ?

**11.** $\frac{7}{10} + \frac{2}{10} = $ ?

**12.** $\frac{1}{6} + \frac{4}{6} = $ ?

**13.** $\frac{5}{9} - \frac{3}{9} = $ ?

**14.** $\frac{9}{15} - \frac{?}{15} = \frac{3}{15}$

**15.** $\frac{8}{11} + \frac{?}{11} = \frac{10}{11}$

**16.** $\frac{?}{5} - \frac{3}{5} = \frac{1}{5}$

**17.** $\frac{?}{16} + \frac{5}{16} = \frac{11}{16}$

**18.** $\frac{?}{8} + \frac{3}{8} = \frac{5}{?}$

**19.** $\frac{2}{3} - \frac{?}{3} = \frac{1}{?}$

**20.** $\frac{?}{?} + \frac{6}{12} = \frac{11}{12}$

**21.** $2\frac{5}{6} - \frac{1}{6} = $ ?

**22.** $9\frac{4}{11} + \frac{5}{11} = $ ?

**23.** $\frac{2}{4} + 27\frac{1}{4} = $ ?

**24.** $18\frac{7}{9} - \frac{4}{9} = $ ?

389

# 12-11 Problem Solving: Use a Drawing/Model

**Problem:** Kai cuts a large watermelon into 8 equal pieces. Each child eats one piece. They eat $\frac{3}{4}$ of the watermelon. How many children are there?

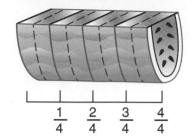

$\frac{1}{4}$   $\frac{2}{4}$   $\frac{3}{4}$   $\frac{4}{4}$

**1 IMAGINE**   Look at the drawing.
Think about the problem.

**2 NAME**   *Facts:*   8 pieces of watermelon
$\frac{3}{4}$ are eaten
Each child eats one piece.

*Question:*   How many children are there?

**3 THINK**   Look at the drawing.
It shows $\frac{3}{4}$ of 8 pieces.
To find the number of children, first
find $\frac{1}{4}$ of 8.

**4 COMPUTE**   $\frac{1}{4}$ of 8 = $\underline{\ ?\ }$
$8 \div 4 = 2$
Since $\frac{1}{4}$ of 8 = 2 pieces,
then $\frac{2}{4}$ of 8 = 4 pieces
and $\frac{3}{4}$ of 8 = 6 pieces.
Since each child eats 1 piece and 6 pieces
are eaten, there are 6 children.

**5 CHECK**   Use the drawing and count to find that
$\frac{3}{4}$ of 8 = 6.

Your answer checks.

**Use a drawing or a model to solve each problem.**

**1.** Kai finished $\frac{2}{3}$ of his school project.

Libby finished $\frac{4}{6}$ of her project.

Who has more of her or his project finished?

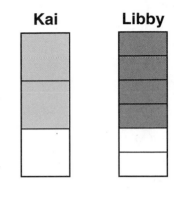
Kai    Libby

| IMAGINE | Look at the picture. Think about the problem. |
|---|---|

| NAME | *Facts:*   Kai has $\frac{2}{3}$ finished. |
|---|---|
| | Libby has $\frac{4}{6}$ finished. |
| | *Question:*  Who has more finished? |

| THINK | Use a model or fraction strips to compare. |
|---|---|
| | Shade $\frac{2}{3}$ of Kai's project. |
| | Shade $\frac{4}{6}$ of Libby's project. |

COMPUTE ———→ CHECK

**2.** Thirty-three children tried out for the soccer team. One third of the children were selected. How many children were selected for the team?

**3.** One half of the bran muffins were gone when Judy came home. If there were 16 muffins to begin with, how many were left?

**4.** Mike bought 16 cans of juice. He used $\frac{1}{8}$ of the cans at breakfast and $\frac{2}{8}$ of the cans for a party. Did Mike use more or less than $\frac{1}{2}$ of the cans?

**5.** What part of a farmer's field can hold corn if $\frac{1}{2}$ is planted with wheat, $\frac{1}{8}$ with oats, and $\frac{2}{8}$ with alfalfa?

**6.** A farmer's property is $\frac{1}{3}$ soil, $\frac{1}{4}$ water, and $\frac{1}{6}$ timber. Copy and shade the model to show these land forms.

**7.** Use the model for problem 6 to write and solve a problem.

**Farm Property**

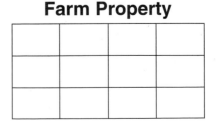

**Problem-Solving Applications**

**Solve each problem and explain the method you used.**

1. Mr. Fertal cut a loaf of bread into 8 equal parts. He gave 3 parts to a friend. Write a fraction for the part of the loaf Mr. Fertal kept.

2. Sue baked 2 cakes the same size. She cut one cake into tenths and the other into twelfths. Which cake had:
   **a.** more pieces?    **b.** larger pieces?

   Imagine

3. The Gomez family bought $\frac{2}{3}$ pound of cookies. The Shaw family bought $\frac{5}{9}$ pound of cookies. Which family bought more?

4. Ten fruit pies were sold. Five of them were apple. What fractional part was apple?

   Name

5. A pound cake is $4\frac{7}{8}$ in. high. An angel food cake is $5\frac{1}{4}$ in. high. Which cake is closest to 5 in. high?

   Think

6. Ms. DiFurio cut a pie into 8 pieces. She served 5 pieces. Does she have more or less than $\frac{1}{2}$ of the pie left?

   Compute

7. The baker sold 12 donuts. One third of them were jelly. The rest were plain. How many plain donuts did the baker sell?

   Check

8. Jo baked 1 tray of rolls. Drew baked 2 trays of rolls. Ed baked $\frac{2}{3}$ tray of rolls. How many trays of rolls did they bake in all?

392

**Choose a strategy from the list or use another strategy you know to solve each problem.**

Use these strategies:

Multi-Step Problem
Use a Graph
Use a Drawing or Model
Write a Number Sentence
Choose the Operation
Logical Reasoning

9. Tyrone baked a pie for $\frac{3}{4}$ of an hour. His sister baked a turnover for $\frac{1}{4}$ of an hour. How much longer did the pie take to bake?

10. Yvonne, Curt, and Joel each baked bread. Joel's bread rose less then Yvonne's. Joel's bread rose more than Curt's. If the tallest loaf rose $7\frac{1}{4}$ in., whose loaf was the tallest?

11. Amy used 2 cups of cream to make a pumpkin pie and $\frac{2}{3}$ cup of cream to make the topping. How much cream did she use altogether?

12. Ron gave one half of his 12 cookies away. He ate one half of what was left. How many cookies did Ron eat?

13. The baker made 6 cakes. He iced one third of them. How many cakes were iced? How many cakes were not iced?

14. Ms. Murphy cut a cake into twelve equal pieces. Each guest ate one piece. Altogether they ate $\frac{1}{3}$ of the cake. How many guests were there? What part of the cake was not eaten?

**The graph shows the results of a survey of 8 students. Use the graph for problems 15–16.**

15. Which kind of bread did the most students like? the fewest students like?

16. How many more students liked white bread than rye bread?

**Favorite Bread**

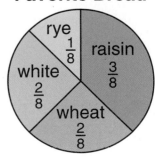

rye $\frac{1}{8}$ | raisin $\frac{3}{8}$ | white $\frac{2}{8}$ | wheat $\frac{2}{8}$

# Chapter Review and Practice

**Write the fraction for the shaded part of each figure.** *(See pp. 370–371.)*

1.
2.
3.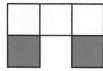
4.

**Write the equivalent fraction for each.** *(See pp. 372–373.)*

5. $\frac{1}{6} = \frac{?}{12}$
6. $\frac{3}{5} = \frac{?}{10}$
7. $\frac{1}{3} = \frac{?}{9}$
8. $\frac{4}{16} = \frac{?}{4}$

**Find part of the number.** *(See pp. 378–379.)*

9. $\frac{1}{3}$ of 9
10. $\frac{1}{5}$ of 25
11. $\frac{1}{7}$ of 42
12. $\frac{1}{10}$ of 90

**Write the mixed number and the word name for each.** *(See pp. 380–381.)*

13.
14.

## PROBLEM SOLVING

*(See pp. 370–371, 390–393.)*

15. A spinner has 6 equal parts. There are 2 red parts, 3 blue parts, and 1 green part. What part of the spinner is blue?

16. Nova strings $\frac{3}{8}$ of the beads and Des strings $\frac{5}{8}$ of the beads. Who strings more beads?

**Use this circle graph to answer each problem.** *(See pp. 382–387.)*

17. What fractional part of Earl's homework time is spent on math? on science and English?

18. What fractional part of Earl's homework time is *not* spent on math?

19. What fractional part tells how much more time Earl spends on math than history?

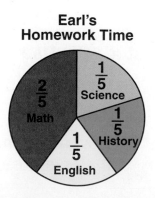

Earl's
Homework Time

(See *Still More Practice*, p. 453.)

# FINDING THE ORIGINAL NUMBER

Jaime received 10 prize baseball cards from José's collection. If this was $\frac{1}{5}$ of José's collection, how many cards did José have in his original collection?

$\frac{1}{5}$ of $\underline{\ ?\ }$ = 10

Draw a picture of José's original collection.

Think: Fifths means that you can separate the collection into 5 equal groups. There are 10 cards in each group.

To find how many cards,

add:  $10 + 10 + 10 + 10 + 10 = 50$

or

multiply:  $5 \times 10 = 50$

José had 50 baseball cards in his original collection.

## Find the original number.

1. $\frac{1}{2}$ of $\underline{\ ?\ }$ = 5
2. $\frac{1}{5}$ of $\underline{\ ?\ }$ = 7
3. $\frac{1}{4}$ of $\underline{\ ?\ }$ = 8

4. $\frac{1}{3}$ of $\underline{\ ?\ }$ = 6
5. $\frac{1}{8}$ of $\underline{\ ?\ }$ = 3
6. $\frac{1}{2}$ of $\underline{\ ?\ }$ = 10

7. $\frac{1}{7}$ of $\underline{\ ?\ }$ = 4
8. $\frac{1}{6}$ of $\underline{\ ?\ }$ = 5
9. $\frac{1}{3}$ of $\underline{\ ?\ }$ = 9

10. $\frac{1}{4}$ of $\underline{\ ?\ }$ = 3
11. $\frac{1}{2}$ of $\underline{\ ?\ }$ = 7
12. $\frac{1}{5}$ of $\underline{\ ?\ }$ = 4

# Check Your Mastery

## Performance Assessment

**Use fraction strips to solve.**

Dee folded this fraction strip in half. Then she colored 2 of the smaller sections red and 1 of them blue.

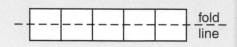
fold line

1. What fractional part of the strip is not colored?

2. Is more or less than $\frac{1}{2}$ of it not colored? Explain.

**Write the equivalent fraction for each.**

3. $\frac{1}{5} = \frac{?}{10}$    4. $\frac{5}{6} = \frac{?}{12}$    5. $\frac{2}{3} = \frac{?}{9}$    6. $\frac{6}{12} = \frac{?}{2}$

**Find part of the number.**

7. $\frac{1}{4}$ of 12    8. $\frac{1}{2}$ of 18    9. $\frac{1}{6}$ of 24    10. $\frac{1}{10}$ of 50

**Write the mixed number and the word name for each.**

11.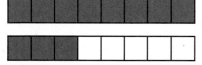

12.

**PROBLEM SOLVING**    *Use a strategy you have learned.*

13. Anita has 9 balloons. Seven are green and 2 are red. What fractional part is red?

14. Benvenido walks $\frac{5}{10}$ of a mile. Sam walks $\frac{3}{10}$ of a mile. Who walks the greater distance?

**Use this circle graph for problems 15–17.**

15. What fractional part of the students likes oranges best? pears or plums best?

16. What fractional part of the students likes oranges or pears best?

17. What fractional part tells how many more students like oranges than apples best?

**Favorite Fruits**

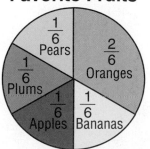

# 13
## Decimals

## Strategy for a Marathon

I will start
when the gun goes off.
I will run
for five miles.
Feeling good,
I will run
to the tenth mile.
At the tenth
I will say,
"Only three more
to the halfway."
At the halfway mark,
13.1 miles,
I will know
fifteen is in reach.
At fifteen miles
I will say,
"You've run twenty before,
keep going."
At twenty
I will say,
"Run home."

*Marnie Mueller*

**In this chapter you will:**

Relate fractions and decimals
Explore tenths and hundredths
Compare, order, add, and subtract decimals
Solve problems by finding a pattern

**Critical Thinking/Finding Together**

Thirteen and one-tenth is the halfway mark of the marathon. Write this number as a fraction. How long is the marathon?

## 13-1 Fractions and Decimals

You can write a fraction and a **decimal** to show parts of a whole.

 ## Hands-On Understanding

**You Will Need:** grid paper, ruler, 1 red and 1 green crayon

**Step 1**
Outline a 10-by-10 square on your grid paper.

How many columns does your square contain?

What fractional part of the whole square is each column?

**Step 2**
Shade 6 columns of your square red.

How many tenths are shaded red?

Write a fraction to show how many columns are shaded red.

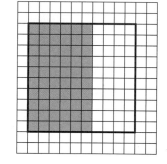

You can also write a decimal to represent how many columns are shaded red.

Numbers to the right of the ones digit represent **decimals**.

Write the decimal as: 0.6 ← Write a 6 after the decimal point to show 6 tenths.

Write a zero before the decimal point to show no ones.

decimal point

One place to the right of the ones place is the **tenths** place.

| ones. | tenths |
|-------|--------|
| 0. | 6 |

Read $\frac{6}{10}$ and 0.6 as six tenths.

$$\frac{6}{10} = 0.6$$

| Step 3 | Outline a 10-by-10 square on another sheet of grid paper. | 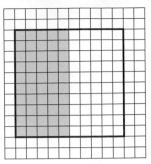 |
|---|---|---|
| Step 4 | Shade 5 out of the 10 columns green. | |

Write a fraction and a decimal to represent the number of columns shaded.

Write a fraction and a decimal to represent the number of columns *not* shaded.

## Communicate

Discuss ✓

1. How are fractions and decimals alike? How are they different?

**Write the decimal for the shaded part of each.**

2. 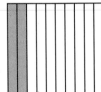    3. 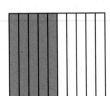    4. 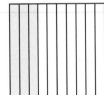    5.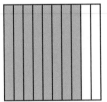

**Write as a decimal.**

6. $\frac{5}{10}$     7. $\frac{9}{10}$     8. $\frac{2}{10}$     9. $\frac{3}{10}$     10. $\frac{1}{10}$

11. four tenths     12. six tenths     13. eight tenths

14. seven tenths     15. three tenths     16. one tenth

**Write the word name for each.**

17. $\frac{1}{10}$     18. $\frac{10}{10}$     19. 0.5     20. 0.2     21. 0.9

**Complete this number line for fractions and decimals.**

22.

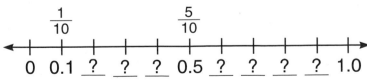

399

# Hundredths

Two places to the right of the ones place is the **hundredths** place.

| ones. | tenths | hundredths |
|-------|--------|------------|
|       |        |            |

## Hands-On Understanding

**You Will Need:** 2 10 × 10 grids, crayons

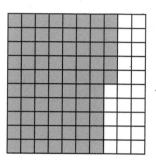

**Step 1**

Shade 75 out of 100 squares.

Write a fraction to show how many squares are shaded.

How many hundredths are shaded?

You can also write this amount as a decimal.

| ones. | tenths | hundredths |
|-------|--------|------------|
| 0.    | 7      | 5          |

Write the decimal as:  0.75

Read $\frac{75}{100}$ and 0.75 as seventy-five hundredths.

$\frac{75}{100} = 0.75$

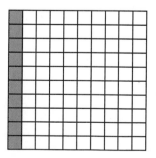

**Step 2**

Use another 10 × 10 grid to shade one column of 10 out of 100 squares.

Write a fraction and a decimal to represent the number of *columns* shaded.

Write a fraction and a decimal to represent the number of *squares* shaded.

Look at the decimals you wrote to represent your model.

What do you notice about 0.1 and 0.10?

How many hundredths are there in one tenth?

| **Step 3** | Now shade 9 more columns of the same model. |
|---|---|

How many columns are shaded now?

How many tenths are there in 1?

How many squares are shaded altogether?

How many hundredths are there in 1?

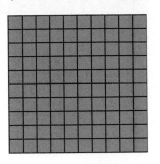

# Communicate

Discuss ✓

1. In 0.3 there are __?__ ones, __?__ tenths, and __?__ hundredths.

2. In 0.30 there are __?__ ones, __?__ tenths, and __?__ hundredths.

3. Are these numbers the same? Explain why or why not. Use models to prove your answer.

**Write as a decimal.**

4. $\frac{41}{100}$    5. $\frac{95}{100}$    6. $\frac{9}{100}$    7. $\frac{4}{100}$    8. $\frac{33}{100}$

9. thirty-two hundredths    10. nine hundredths    11. forty hundredths

**Write the fraction and word name for each.**

12. 0.43    13. 0.99    14. 0.04    15. 0.05    16. 0.90

17. 0.52    18. 0.87    19. 0.01    20. 0.07    21. 0.60

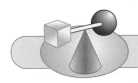

# Challenge

**Write each amount using a dollar sign and decimal point.**

22. forty cents                23. eight cents

24. two dollars              25. one dollar and two cents

26. $\frac{1}{2}$ of one dollar        27. $\frac{3}{4}$ of one dollar

# 13-3 Decimals Greater Than One

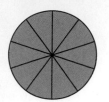 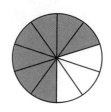 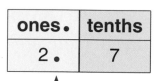

1 whole          1 whole          $\frac{7}{10}$ ——→ $2\frac{7}{10}$

Remember:
1 whole + 1 whole =
2 wholes

▶ The mixed number $2\frac{7}{10}$ can be
written as the decimal 2.7.

| ones. | tenths |
|-------|--------|
| 2. | 7 |

Write a 2 before
the decimal point
to show 2 ones.

Read $2\frac{7}{10}$ and 2.7 as two and seven tenths.

▶ The mixed number $2\frac{7}{10}$ and the decimal 2.7
name the same amount. You can locate fractions
and decimals on a number line.

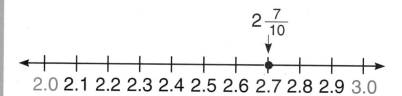

$2\frac{7}{10}$

2.0 2.1 2.2 2.3 2.4 2.5 2.6 2.7 2.8 2.9 3.0

$2\frac{7}{10}$ = 2.7

## Study these examples.

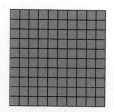 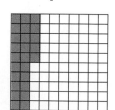     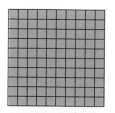 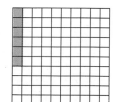

$1\frac{25}{100}$ = 1.25                    $1\frac{6}{100}$ = 1.06

one and twenty-five hundredths          one and six hundredths

402

**Write the mixed number and the decimal for each.**

1.

2.

**Write the decimal and word name for each.**

3. $2\frac{4}{10}$   4. $1\frac{8}{10}$   5. $6\frac{1}{10}$   6. $3\frac{9}{10}$   7. $4\frac{7}{10}$

8. $1\frac{2}{10}$   9. $2\frac{3}{10}$   10. $3\frac{7}{10}$   11. $4\frac{8}{10}$   12. $5\frac{1}{10}$

13. $2\frac{9}{10}$   14. $1\frac{5}{10}$   15. $4\frac{6}{10}$   16. $1\frac{7}{10}$   17. $2\frac{3}{10}$

18. $1\frac{34}{100}$   19. $4\frac{36}{100}$   20. $7\frac{28}{100}$   21. $5\frac{92}{100}$

22. $6\frac{20}{100}$   23. $9\frac{90}{100}$   24. $8\frac{8}{100}$   25. $2\frac{1}{100}$

**Copy and complete each number line.**

26.

2.0 2.1 _?_ _?_ _?_ 2.5 _?_ _?_ _?_ _?_ 3.0 _?_ _?_ _?_ _?_

27.

3.5 _?_ _?_ _?_ 3.9 _?_ _?_ _?_ _?_ 4.4 _?_ _?_ _?_ _?_ _?_ 5.0

**Write Yes or No. Explain your answer.**

28. Does $1\frac{4}{10} = 1.8$?   29. Does $2\frac{5}{100} = 2.5$?   30. Does $3\frac{1}{10} = 1.3$?

31. Does $4.7 = 4\frac{7}{10}$?   32. Does $6.3 = 3\frac{6}{10}$?   33. Does $1.93 = 1\frac{93}{100}$?

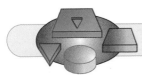 **Critical Thinking**   Algebra ✓

$3.58 = 3 + 0.5 + 0.08$

**Find the missing decimal.**

34. $1.35 = 1 + \underline{\ ?\ } + 0.05$   35. $4.91 = 4 + 0.9 + \underline{\ ?\ }$

36. $6.40 = 6 + 0.4 + \underline{\ ?\ }$   37. $5.02 = 5 + \underline{\ ?\ } + 0.02$

403

**Comparing and Ordering Decimals** Algebra

Marcia walks $\frac{5}{10}$ of a mile to school. Corey walks 0.7 of a mile to school. Who has the longer walk to school?

▶ You can use models to compare fractions and decimals.

Think: $\frac{5}{10} = 0.5$

0.5    ?    0.7

The model for 0.5 has fewer bars shaded than the model for 0.7.

So 0.5 < 0.7

Corey has the longer walk to school.

▶ You can use a number line to compare decimals.

Compare: 0.8 _?_ 0.3

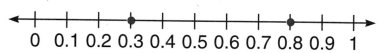

0    0.1 0.2 0.3 0.4 0.5 0.6 0.7 0.8 0.9    1

Think: 0.8 is to the right of 0.3.
      0.8 is greater than 0.3.

0.8 > 0.3

▶ You can use the place value of the digits to compare decimals.

Compare: 4.3 _?_ 4.9

• Look at the ones place.
  Compare the digits.   4 = 4

| ones . | tenths |
|--------|--------|
| 4.     | 3      |
| 4.     | 9      |

• Look at the tenths place.
  Compare the digits.   3 < 9

So 4.3 < 4.9

**Compare. Write < or >.**

**1.** 0.3 _?_ 0.4

**2.** 0.1 _?_ 0.9

**3.** $\frac{8}{10}$ _?_ 0.6

**4.** 0.5 _?_ $\frac{7}{10}$

**5.** 4.1 _?_ 4.5

**6.** 3.5 _?_ 3.2

**7.** 3.9 _?_ 1.5

**8.** 6.4 _?_ 2.4

**9.** 4.8 _?_ 7.9

---

### Ordering Decimals

Write in order from least to greatest: 1.1, 1.9, 1.7

To order decimals compare the place value of the digits.

Compare ones. ⟶ Compare tenths.

1.1, 1.9, 1.7          1.1, 1.9, 1.7

The ones digits are the same. Compare tenths.     $1 < 7$ and $7 < 9$

In order from least to greatest: 1.1, 1.7, 1.9

You can show order using a number line.

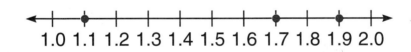

---

**Order from least to greatest. You may use a number line.**

**10.** 0.2, 0.8, 0.4

**11.** 1.4, 1.8, 1.2

**12.** 2.6, 2.3, 2.5

**13.** 7.0, 7.9, 7.5

**14.** 3.3, 3.0, 3.7

**15.** 4.7, 4.4, 4.1

 **Skills to Remember**      **Compute.**

**16.** $346 + $398

**17.** $411 − $183

**18.** $6.01 − $2.95

**19.** $7.90 + $2.89

**20.** $15.75 − $10.89

**21.** $53.30 + $39.92

**Adding and Subtracting Decimals**

On Monday Damon rode his bicycle 2.43 mi. The next day he rode 1.83 mi. How many miles did Damon ride altogether?

▶ To find how many miles altogether, add:  $2.43 + 1.83 = $ ?

2.43          +          1.83          =          ?

| Align the decimal points. | → | Add hundredths. | → | Add tenths. | → | Add ones. |
|---|---|---|---|---|---|---|

|  |  |  | $\overset{1}{ }$ | $\overset{1}{ }$ |  |
|---|---|---|---|---|---|
| 2.43 | 2.43 | 2.43 | 2.43 | 2.43 |
| + 1.83 | + 1.83 | + 1.83 | + 1.83 | + 1.83 |
|  | 6 | 26 | 4.26 | 4.26 |

**Remember:** Write the decimal point.

Damon rode 4.26 miles altogether.

On Monday Damon's friend Ron rode his bicycle 3.05 mi. How many more miles did Ron ride than Damon?

▶ To find how many more miles, subtract:  $3.05 - 2.43 = $ ?

| Align the decimal points. | → | Subtract hundredths. | → | Subtract tenths. | → | Subtract ones. |
|---|---|---|---|---|---|---|

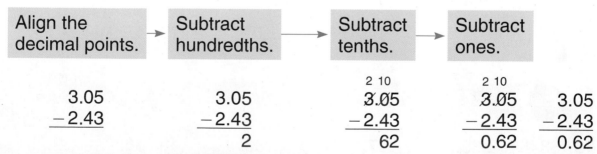

|  |  | $\overset{2\ 10}{ }$ | $\overset{2\ 10}{ }$ |  |
|---|---|---|---|---|
| 3.05 | 3.05 | 3.̸05 | 3.̸05 | 3.05 |
| − 2.43 | − 2.43 | − 2.43 | − 2.43 | − 2.43 |
|  | 2 | 62 | 0.62 | 0.62 |

Ron rode 0.62 more miles than Damon.

**Find the sum.**

| | | | | |
|---|---|---|---|---|
| 1.    0.2<br>    +0.5 | 2.    0.3<br>    +0.7 | 3.    0.12<br>    +0.24 | 4.    0.35<br>    +0.27 | 5.    2.5<br>    +2.2 |
| 6.    1.6<br>    +3.8 | 7.    9.2<br>    +1.6 | 8.    7.5<br>    +3.6 | 9.    3.64<br>    +2.21 | 10.    4.55<br>    +2.75 |
| 11.    4.28<br>    +3.72 | 12.    7.19<br>    +2.81 | 13.    16.04<br>    +22.75 | 14.    23.08<br>    +43.67 | 15.    65.32<br>    +24.77 |

**Find the difference.**

| | | | | |
|---|---|---|---|---|
| 16.    0.8<br>    −0.5 | 17.    1.4<br>    −0.7 | 18.    8.0<br>    −4.3 | 19.    5.5<br>    −3.7 | 20.    5.9<br>    −4.2 |
| 21.    4.55<br>    −3.05 | 22.    3.95<br>    −2.80 | 23.    4.80<br>    −3.05 | 24.    9.43<br>    −7.76 | 25.    6.87<br>    −4.29 |
| 26.    7.00<br>    −2.69 | 27.    8.27<br>    −3.19 | 28.    16.20<br>    −12.07 | 29.    63.27<br>    −15.08 | 30.    44.06<br>    −40.42 |

## Calculator Activity

**Use your calculator. Solve.**

Watch for addition and subtraction signs.

| | | | | |
|---|---|---|---|---|
| 31.    4.3<br>    2.6<br>    +8.8 | 32.    1.7<br>    6.2<br>    +3.1 | 33.    9.22<br>    6.84<br>    +1.73 | 34.    2.31<br>    4.79<br>    +8.27 | 35.    7.92<br>    3.95<br>    +2.19 |
| 36.    9.07<br>    −0.45 | 37.    8.41<br>    +0.59 | 38.    29.12<br>    − 6.08 | 39.    10.36<br>    + 7.04 | 40.    16.22<br>    − 4.53 |
| 41.    $5.09<br>    + 3.25 | 42.    $4.36<br>    − 1.94 | 43.    $12.02<br>    + 11.89 | 44.    $19.61<br>    − 10.02 | 45.    $9.99<br>    + 1.11 |

## 13-6 | Problem Solving: Find a Pattern

*Algebra* ✓

**Problem:** Ten boys were standing in line. Each time 3 boys sat down, one more boy came into the line. If this continues, how many boys will be sitting down when there are 2 boys standing in line?

**1 IMAGINE**  Create a mental picture.

**2 NAME**

*Facts:*  10 boys standing in line
3 boys sit and 1 boy comes into the line

*Question:*  How many boys will be sitting down when there are 2 boys standing in line?

**3 THINK**  Each time 3 boys sit down, one more boy comes into the line.

Subtract 3.
Add 1.

Start with 10. Subtract 3, then add 1 until the answer is 2.

Count the number of boys sitting down.

**4 COMPUTE**

| In line | Sitting | | One More | | |
|---|---|---|---|---|---|
| 10 − 3 = 7 | ⟶ | 7 + 1 = 8 in line |
| 8 − 3 = 5 | ⟶ | 5 + 1 = 6 in line |
| 6 − 3 = 3 | ⟶ | 3 + 1 = 4 in line |
| 4 − 3 = 1 | ⟶ | 1 + 1 = 2 in line |

Four groups of 3 boys are sitting.    $4 \times 3 = 12$
There are 12 boys sitting down.

**5 CHECK**  Act out the problem, or use a calculator to check your computations.

**Find a pattern to solve each problem.**

1. A computer solved 3 problems in 0.5 s, 6 problems in 1.0 s, 9 problems in 1.5 s, and so on. How long will it take to solve 21 problems?

| IMAGINE | Create a mental picture. |
|---|---|

| NAME | *Facts:* | 3 problems in 0.5 s |
|---|---|---|
| | | 6 problems in 1.0 s |
| | | 9 problems in 1.5 s |

*Question:* How many seconds will it take to solve 21 problems?

**THINK** Make a table and look for a pattern.

| Problems | 3 | 6 | 9 | 12 |
|---|---|---|---|---|
| Time | 0.5 | 1.0 | 1.5 | ? |

Count by 3.

Add 0.5.

**COMPUTE** ⟶ **CHECK**

2. If posts are marked each tenth of a mile on a speed-walking route, how many posts does Joel pass if he begins at one and five tenths of a mile and ends at two and two tenths?

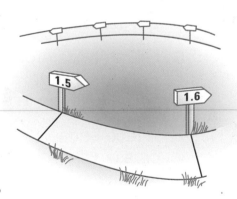

3. Wanda walks 1.2 mi on Sunday, 2.2 mi on Monday, 3.2 mi on Tuesday, and so on. How many miles will Wanda walk in all by Friday?

4. What numbers come next in Aileen's pattern?
2.9, 2.6, 3.6, 3.3, 4.3, 4.0, 5.0, _?_ , _?_

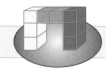

**Make Up Your Own**

5. Write your own problem modeled on problem 3 or 4. Have a classmate solve it.

**Problem-Solving Applications**

**Solve each problem and explain the method you used.**

1. April swam 0.6 km. Did she swim more or less than $\frac{1}{2}$ of a kilometer?

2. Each day Cheryl ran 2.5 mi, Kim ran 2.3 mi, and Boreta ran 2.7 mi. Which girl ran the farthest each day?

3. Glen's first jump was one hundredth less than the class record of 3.06 m. How far did Glen jump?

4. The track team was going to a meet 23.5 miles away. The bus broke down after going 18.3 miles. How far was the team from the meet?

5. Adrian won the race with a time of 58.3 seconds. Adrian beat Chino by 2.7 seconds. What was Chino's time?

6. Gil ran twice around the 2.9-km path. Sol ran 6.5 km. How much farther does Gil have to run to equal Sol's distance?

7. A race began 12.3 km from the bridge, crossed the bridge, and ended 13.4 km after the bridge. If the total distance was 26.5 km, how long was the bridge?

Imagine

Name

Think

Compute

Check

**Use the table for problems 8–10.**

8. Which relay runner had the slowest time?

9. Whose time was closest to 7.0 seconds?

10. Whose time was between 6.3 and 6.6 seconds?

**Relay Race**

| Runner | Time |
| --- | --- |
| Arman | 6.8 s |
| Roy | 6.2 s |
| Tina | 6.5 s |
| Fran | 6.9 s |

**Choose a strategy from the list or use another strategy you know to solve each problem.**

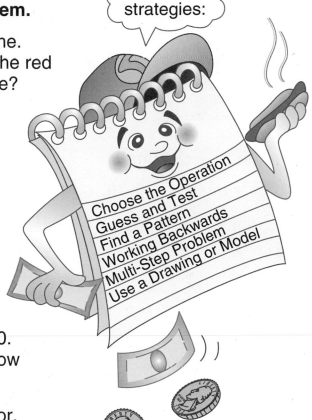

Use these strategies:

Choose the Operation
Guess and Test
Find a Pattern
Working Backwards
Multi-Step Problem
Use a Drawing or Model

11. There were 8 red flags at the finish line. Jo put 2 blue flags between each of the red flags. How many blue flags did Jo use?

12. Tim runs nine tenths of a mile each day. Ken runs 1.2 mi each day. How many miles will Ken have run when Tim has run 4.5 miles in all?

13. Ship School's team scored 89.15 points. Troy School's team scored 90.06 points. How many more points did the team from Troy School score?

14. Sonya and Brian have a total of $9.50. Sonya has $2.50 more than Brian. How much money does each one have?

15. Nick added 2.3 and 6.2 on a calculator. When he saw the sum 10.5 on the display, Nick knew he had not cleared the calculator. What number had not been cleared?

16. Jon and Rich divided some money evenly. After Jon spent $5.75 of his share at a ball game and $3.00 for lunch, he had $1.25 left. How much money did the boys have at first?

17. Carol lives 4.3 km north of the arena. Erin lives 7.2 km south of the arena. How far do the girls live from each other?

• Carol

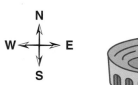

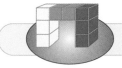

**Make Up Your Own**

• Erin

18. Write a problem using the information from problem 17.

411

**Write the fraction and the decimal for each.**    *(See pp. 398–401.)*

**1.**     **2.**     **3.**     **4.**

**Write as a decimal.**

**5.** one tenth    **6.** six tenths    **7.** four tenths    **8.** nine tenths

**9.** $\frac{3}{10}$    **10.** $\frac{5}{10}$    **11.** $2\frac{8}{10}$    **12.** $\frac{67}{100}$

**13.** $\frac{2}{100}$    **14.** $\frac{99}{100}$    **15.** $3\frac{3}{10}$    **16.** $7\frac{9}{100}$

**Write the word name for each.**    *(See pp. 398–403.)*

**17.** 0.2    **18.** 0.31    **19.** 9.7    **20.** 1.85

**Compare. Write < or >.**    *(See pp. 404–405.)*

**21.** 0.6 __?__ 0.8    **22.** 5.4 __?__ 5.2    **23.** $\frac{7}{10}$ __?__ 0.5

**Order from least to greatest.**

**24.** 1.8, 1.3, 1.9    **25.** 5.6, 5.2, 5.5

**Find the sum or difference.**    *(See pp. 406–407.)*

**26.**  0.3
      +0.6

**27.**  0.68
       +0.29

**28.**  4.37
       +1.28

**29.**  74.19
       +19.83

**30.**  8.7
       −2.4

**31.**  0.51
       −0.29

**32.**  6.07
       −1.98

**33.**  90.15
       −67.86

**PROBLEM SOLVING**

**34.** What numbers come next in Audra's pattern?

4.1, 3.4, 3.9, 3.2, 3.7, 3, __?__ , __?__

*(See Still More Practice, p. 454.)*

## EQUIVALENT FRACTIONS AND DECIMALS

### Name the fraction or decimal that does *not* belong.

**1.**
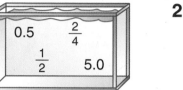
0.5    $\frac{2}{4}$
$\frac{1}{2}$    5.0

**2.**
$\frac{2}{3}$    $\frac{8}{12}$
0.09    $\frac{4}{6}$

**3.**
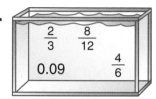
$\frac{1}{3}$    $\frac{3}{9}$
$\frac{2}{6}$    0.03

**4.**
0.01    $\frac{1}{4}$
$\frac{4}{16}$    $\frac{2}{8}$

**5.**
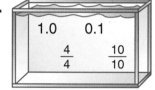
1.0    0.1
$\frac{4}{4}$    $\frac{10}{10}$

**6.**
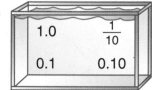
1.0    $\frac{1}{10}$
0.1    0.10

**7.**
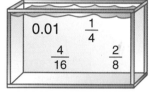
1.5    $1\frac{3}{4}$    $1\frac{1}{2}$

**8.**
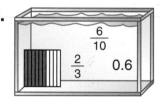
$\frac{6}{10}$
$\frac{2}{3}$    0.6

**9.**
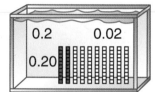
0.2    0.02
0.20

**10.**
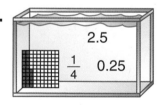
2.5
$\frac{1}{4}$    0.25

**11.**
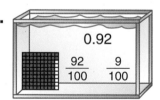
0.92
$\frac{92}{100}$    $\frac{9}{100}$

**12.**
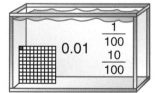
$\frac{1}{100}$
0.01    $\frac{10}{100}$

**13.** Make up your own equivalent fraction and decimal tank.

$\underline{\ ?\ }\ \ \frac{2}{4}\ \ \underline{\ ?\ }$

### Use models to find each sum as a fraction.

**14.** $0.6 + \frac{1}{10} = \frac{7}{\underline{10}}$    **15.** $\frac{6}{10} + 0.3$    **16.** $0.07 + \frac{20}{100}$    **17.** $\frac{4}{100} + 0.43$

Remember: $0.6 = \frac{6}{10}$

413

# Check Your Mastery

**Use the cards at the right.**

1. Make three different decimals by placing cards in the ones and tenths places.

2. Show each decimal you made on a number line. Then write them in order from least to greatest.

**Write the fraction and the decimal for each.**

3.     4.     5.     6.

**Write as a decimal.**

**7.** five tenths          **8.** nine tenths          **9.** four hundredths

**10.** $\frac{7}{10}$    **11.** $\frac{61}{100}$    **12.** $\frac{7}{100}$    **13.** $2\frac{7}{10}$    **14.** $9\frac{1}{100}$

**Write the word name for each.**

**15.** 0.7          **16.** 0.25          **17.** 1.4          **18.** 1.29

**Compare. Write < or >.**

**19.** 0.8 _?_ 0.5          **20.** 7.2 _?_ 7.6          **21.** 0.2 _?_ $\frac{1}{10}$

**Find the sum or difference.**

**22.**   0.4
      + 0.3

**23.**   0.59
      + 0.27

**24.**   1.96
      +4.59

**25.**   65.83
      +28.19

**26.**   5.9
      −2.3

**27.**   0.63
      − 0.49

**28.**   8.06
      −3.87

**29.**   70.21
      − 11.95

**PROBLEM SOLVING**   *Use a strategy you have learned.*

**30.** What numbers come next in Dave's pattern?

4.3, 5.1, 4.5, 5.3, 4.7, 5.5, _?_ , _?_

# Cumulative Test II

**Choose the best answer.**

**1.** What is the place of the underlined digit?

7<u>5</u>6, 849

    **a.** hundreds
    **b.** thousands
    **c.** ten thousands
    **d.** hundred thousands

**2.**
```
  6379
+ 1669
```

    **a.** 7438
    **b.** 7948
    **c.** 8048
    **d.** 8038

---

**3.** $80.00 − $8.03

    **a.** $71.97   **b.** $72.03
    **c.** $72.97   **d.** $71.03

**4.** $8 \times \underline{\ ?\ } = 40¢$

    **a.** 4¢   **b.** 20¢
    **c.** 32¢   **d.** 5¢

---

**5.** Which is a meaning of division?

    **a.** separating into unequal groups
    **b.** sharing equally
    **c.** joining together
    **d.** all of these

**6.** In a pictograph, each ☺ = 4 people. How many people are represented by ☺☺☺☺◖ ?

    **a.** 5
    **b.** 9
    **c.** 17
    **d.** 18

---

**7.** Which is greater than 5 meters?

    **a.** 50 mm   **b.** 58 dm
    **c.** 500 cm   **d.** none of these

**8.** $72 \div 8 = \underline{\ ?\ }$

    **a.** 7   **b.** 8
    **c.** 9   **d.** 6

---

**9.** Which is *not* a right angle?

    **a.** A   **b.** B
    **c.** C   **d.** all of these

**10.** Which is a rectangular prism?

    **a.** D   **b.** E
    **c.** F   **d.** none of these

---

**11.** Estimate.

$2\overline{)\$11.91}$

    **a.** $9.00   **b.** $6.00
    **c.** $24.00   **d.** $24.00

**12.**

$$\frac{2}{3} = \frac{?}{6}$$

    **a.** 1   **b.** 2
    **c.** 4   **d.** 6

**Add, subtract, multiply, or divide as indicated. Watch the signs.**

**13.**  $\begin{array}{r} 22 \\ \times\ 4 \\ \hline \end{array}$

**14.**  $\begin{array}{r} 48 \\ \times\ 5 \\ \hline \end{array}$

**15.** $2 \times \$343$

**16.** $9 \times 1721$

**17.** $56¢ \div 8$

**18.** $39 \div 3$

**19.** $6\overline{)50}$

**20.** $4\overline{)91}$

**21.** $\frac{2}{5} + \frac{1}{5}$

**22.** $\frac{5}{8} - \frac{2}{8}$

**23.** $\frac{11}{12} - \frac{5}{12}$

**24.** $\frac{7}{9} - \frac{4}{9}$

**25.**  $\begin{array}{r} 0.59 \\ -\ 0.05 \\ \hline \end{array}$

**26.**  $\begin{array}{r} 0.78 \\ +\ 0.52 \\ \hline \end{array}$

**27.**  $\begin{array}{r} 2.97 \\ +\ 4.69 \\ \hline \end{array}$

**28.**  $\begin{array}{r} 80.31 \\ -\ 12.96 \\ \hline \end{array}$

**29.** $\frac{1}{3}$ of 27

**30.** $\frac{1}{10}$ of 80

**31.** $\frac{1}{4}$ of 36

**32.** $\frac{1}{5}$ of 45

## PROBLEM SOLVING

**33.** Eric has 17 coins. Angela has 3 times as many. How many coins does Angela have?

**34.** A store clerk arranged 128 boxes in 8 equal rows. How many boxes are in each row?

**35.** Earl writes a number puzzle for Ruth to solve. The pattern is: add 7 hundredths then subtract 12 hundredths. Find the next number in the pattern: 2.58, 2.65, 2.53, <u>?</u>

**36.** Judy makes a box that is 4 cubes long, 2 cubes wide, and 3 cubes high. What is the volume of the box? Explain how you found your answer.

**37.** Kimiko drew the rectangle at the right.
  **a.** What is the perimeter of the rectangle?
  **b.** What is the area in square units?
  **c.** Explain how you found your answers in **a.** and **b.**

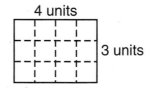

4 units

3 units

# For Rubric Scoring

**Listen for information on how your work will be scored.**

**38.** Which number sentences are *not* true? Explain how you know. Draw a picture to explain your reasoning.

$\frac{1}{2}$ is less than 0.5          0.9 is less than 1.0          1.5 is greater than 1.6

$\frac{1}{5}$ is less than $\frac{1}{4}$          $\frac{2}{6}$ is greater than $\frac{1}{3}$

# Moving On in Mathematics 14

## Marvelous Math

How fast does a New York taxi go?
What size is grandpa's attic?
How old is the oldest dinosaur?
The answer's in *Mathematics!*

How many seconds in an hour?
How many in a day?
What size are the planets in the sky?
How far to the Milky Way?

How fast does lightning travel?
How slow do feathers fall?
How many miles to Istanbul?
*Mathematics* knows it all!

*Rebecca Kai Dotlich*

**In this chapter you will:**

Learn about a million, divisibility rules, and the order of operations
Find missing digits and operations
Find common factors
Multiply and divide money amounts
Solve multi-step problems

**Critical Thinking/ Finding Together**

How many days old are you?
Don't forget about leap years.

417

# Moving On: Numeration

## 14-1 Place Value to a Million

More than 3,693,197 people own the book *The Cat in the Hat*.

To show this number, extend the *place-value chart*.

Read 3,693,197 as:

| | Millions Period | | | Thousands Period | | | Ones Period | | |
|---|---|---|---|---|---|---|---|---|---|
| | hundreds | tens | ones | hundreds | tens | ones | hundreds | tens | ones |
| | | | 3, | 6 | 9 | 3, | 1 | 9 | 7 |

three million,

six hundred ninety-three thousand,

one hundred ninety-seven

▶ To show 3,693,197 in *expanded form,*
write: 3,000,000 + 600,000 + 90,000 + 3000 + 100 + 90 + 7

▶ To show three million, six hundred ninety-three thousand, one hundred ninety-seven in *standard form,* write: 3,693,197

Study this example.

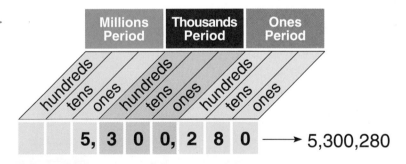

| | Millions Period | | | Thousands Period | | | Ones Period | | |
|---|---|---|---|---|---|---|---|---|---|
| | hundreds | tens | ones | hundreds | tens | ones | hundreds | tens | ones |
| | | | 5, | 3 | 0 | 0, | 2 | 8 | 0 |

5,300,280

Read 5,300,280 as:   five million, three hundred thousand, two hundred eighty

To show 5,300,280 in *expanded form,*
write:   5,000,000 + 300,000 + 200 + 80

Write the number in standard form.

**1.** 8 million

**2.** three million

**3.** 7,000,000 + 600,000

**4.** 4,000,000 + 30,000 + 2000

**5.** 5,000,000 + 200,000 + 20,000 + 4000 + 300 + 60 + 7

**6.** 1,000,000 + 700,000 + 90,000 + 2000 + 700 + 80 + 5

Complete the expanded form.

**7.** 9,000,000 = _?_ millions

**8.** 6,000,000 = _?_ millions

**9.** 2,080,020 = _?_ millions _?_ hundred thousands _?_ ten thousands _?_ thousands _?_ hundreds _?_ tens _?_ ones

**10.** 7,207,899 = _?_ millions _?_ hundred thousands _?_ ten thousands _?_ thousands _?_ hundreds _?_ tens _?_ ones

**11.** 5,142,000 = _?_ millions _?_ hundred thousand _?_ ten thousands _?_ thousands _?_ hundreds _?_ tens _?_ ones

Write the number in standard form.

**12.** nine million, three hundred seventy-eight thousand, fifty-two

**13.** six million, five hundred forty-four thousand, six hundred

**14.** two million, four hundred fifty thousand, two hundred sixty-one

In what place is the underlined digit? What is its value?

**15.** 4,5<u>7</u>9,673

**16.** <u>8</u>,700,346

**17.** 1,263,4<u>4</u>7

 **Challenge**

**18.** What number is 1,000,000 more than 2,450,000?

# 14-2 Add and Subtract Larger Numbers

*Kids' World* magazine sold 48,540 copies to children and 30,460 copies to adults. How many magazines did *Kids' World* sell in all? How many more magazines did *Kids' World* sell to children than to adults?

To find how many in all, add:  48,540 + 30,460

```
     1 1
   4 8,5 4 0
  +3 0,4 6 0
   7 9,0 0 0
```

| Remember: Regroup when necessary. |

To find how many more, subtract:  48,540 − 30,460

```
      4 14
   4 8,5̶ 4̶ 0
  −3 0,4 6 0
   1 8,0 8 0
```

*Kids' World* sold 79,000 magazines in all.

*Kids' World* sold 18,080 more magazines to children than to adults.

Study these examples.

```
     1 1
   6 1,2 5 4
  +2 7,9 8 3
   8 9,2 3 7
```

```
    8 10 6 15
  $6 9̶ 0̶ 7̶ 5̶
  −   5 4.6 7
  $6 3 6.0 8
```

```
      1 1
  $5 2 6.7 2
  +   3 5.4 0
  $5 6 2.1 2
```

```
    8 15 2 10
   2 9̶,5̶ 3̶ 0̶
  −1 8,7 2 5
   1 0,8 0 5
```

---

Add or subtract. Watch for + and −.

**1.**  43,257
+ 12,431

**2.**  76,989
− 40,743

**3.**  22,349
+ 50,158

**4.**  89,832
− 75,613

**5.**  21,878
+ 47,566

**6.**  18,358
−  4,962

**7.**  73,555
+  1,555

**8.**  69,700
− 58,245

**9.**  $364.21
−  143.68

**10.**  $491.49
+  461.58

**11.**  $209.87
+   94.22

**12.**  $308.50
−   89.45

Solve and check.

| 13. | 58,499<br>+ 30,275 | 14. | 48,225<br>− 23,550 | 15. | 12,750<br>+ 61,860 | 16. | 99,000<br>− 38,827 |
|---|---|---|---|---|---|---|---|
| 17. | $102.35<br>+ 304.60 | 18. | $779.98<br>− 104.89 | 19. | $199.95<br>+ 650.78 | 20. | $450.50<br>− 200.75 |

Align and solve.

**21.** 47,569 + 10,833    **22.** 86,532 − 75,732    **23.** 87,290 + 2809

**24.** $590.42 − $65.75    **25.** 5723 + 23,540    **26.** $671.01 + $98.99

**27.** $408.32 + $64.78    **28.** $200.15 − $86.35    **29.** $505.05 − $55.66

Problem Solving

**30.** Last year Comic Corner sold 34,902 comic books. This year the store sold 21,892 more comic books. How many comic books were sold this year?

**31.** A company prints 32,415 copies of one baseball card. The company gives away 1500 of these cards and sells the rest. How many baseball cards does the company sell?

**32.** A bookstore sold 38,065 adventure books and 25,109 mystery books. Each adventure book costs $6.98. Each mystery book costs $4.98. How many more adventure books than mystery books were sold?

## 14-3 Divisibility

A number is **divisible** by another number
when there is no remainder.

$$20 \div 2 = 10 \qquad 20 \div 5 = 4 \qquad 20 \div 10 = 2$$

Since there are no remainders, 20 is divisible
by 2, 5, and 10.

▶ To find which numbers are divisible by 2,
skip count by 2 starting at 0.

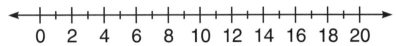

Look at the digits in the ones place.
What pattern do you see?

> Any number ending in
> 0, 2, 4, 6, or 8 is
> *divisible* by 2.

▶ To find which numbers are divisible by 5,
skip count by 5 starting at 0.

Look at the digits in the ones place.
What pattern do you see?

> Any number ending in
> 0 or 5 is *divisible* by 5.

▶ To find which numbers are divisible by 10,
skip count by 10 starting at 0.

Look at the digits in the ones place.
What pattern do you see?

> Any number ending
> in 0 is *divisible* by 10.

Is the number divisible by 2? Explain.

**1.** 8      **2.** 17      **3.** 31      **4.** 56      **5.** 100

**6.** 101      **7.** 589      **8.** 652      **9.** 4883      **10.** 7914

Is the number divisible by 5? Explain.

**11.** 19      **12.** 40      **13.** 85      **14.** 86      **15.** 100

**16.** 125      **17.** 137      **18.** 190      **19.** 1005      **20.** 4444

Is the number divisible by 10? Explain.

**21.** 16      **22.** 20      **23.** 40      **24.** 56      **25.** 99

**26.** 100      **27.** 140      **28.** 265      **29.** 3006      **30.** 4200

Copy and complete the table.

| | Divisible by | 45 | 50 | 72 | 91 | 102 | 225 | 4010 |
|---|---|---|---|---|---|---|---|---|
| **31.** | 2 | no | ? | ? | ? | yes | ? | ? |
| **32.** | 5 | yes | ? | no | ? | ? | ? | ? |
| **33.** | 10 | ? | yes | ? | ? | ? | ? | ? |

Problem Solving

**34.** I am a number between 42 and 48. I am divisible by 5. What number am I?

**35.** I am a number between 51 and 67. I am divisible by 10. What number am I?

**36.** I am a number that is divisible by 2. I have two digits. One digit is a 6 and one digit is a 9. What number am I?

**37.** I am a number that is divisible by 5. I have three digits. I am between 121 and 129. What number am I?

# Moving On: Algebra

## 14-4 Missing Digits

Rochelle's dog ripped her homework into pieces. Rochelle taped her homework together the best she could. Parts of her homework are missing. Help Rochelle find the numbers missing from her homework.

```
   2□
 +□3
 ─────
  1 0 2
```

Use Guess and Test to find each missing digit.

| Guess and test the top box. | Write the digit. Add. Guess and test. | Check your answer by adding. |
|---|---|---|
| $\downarrow$ 2□ <br> +□3 <br> ───── <br> 1 0 2 | 1 <br> 2 9 <br> +□3 <br> ───── <br> 1 0 2 | 1 <br> 2 9 <br> +7 3 <br> ───── <br> 1 0 2 |

Try 8.
```
    1
   2⃞8
 +□3
 ─────
      1   No.
```
There should be 2 ones.

Try 5.
```
    1
   2 9
 +5⃞3
 ─────
    8 2   No.
```
The sum should be 102.

Try 9.
```
    1
   2⃞9
 +□3
 ─────
      2   Yes!
```

Try 7.
```
    1
   2 9
 +7⃞3
 ─────
  1 0 2   Yes!
```

Study these examples.

? − 5 = 0
$\downarrow$

$4 - ? = 1 \rightarrow$
```
  4□8
 −□5□
 ─────
  1 0 2
```
$\leftarrow 8 - ? = 2$

```
  4 5 8
 −3 5 6
 ─────
  1 0 2
```

Will 1 work? No.
What about 0?
$\downarrow$
```
  1□7
 ×    6
 ─────
  6 4 2
```

```
      4
  1 0 7
 ×    6
 ─────
  6 4 2
```

Find the missing digits.

1.
```
  8□
+ □2
─────
  9 7
```

2.
```
  □5
+ 6□
─────
  7 5
```

3.
```
  7□
− □3
─────
  2 0
```

4.
```
  □1
− 4□
─────
  4 1
```

5.
```
  □43
+ 5□□
──────
  7 5 9
```

6.
```
  6□□
+ □11
──────
  7 3 8
```

7.
```
  9□5
− □2□
──────
  2 6 4
```

8.
```
  27□
− □□4
──────
    1 4
```

9.
```
  □2
+ 6□
─────
  8 1
```

10.
```
  4□
+ □8
─────
  1 2 6
```

11.
```
  □2
− 4□
─────
  1 5
```

12.
```
  8□
− □5
─────
    9
```

13.
```
  □1□
+ 4□9
──────
  9 4 2
```

14.
```
  3□7
+ □4□
──────
  6 7 2
```

15.
```
  2□0
− □2□
──────
  1 1 4
```

16.
```
  5□□
− □22
──────
  2 2 8
```

17.
```
  □38□
+ 1□05
───────
  3 5 8 5
```

18.
```
  69□3
+ □□1□
───────
  7 9 9 8
```

19.
```
  4□6□
− 38□4
───────
  1 1 5 1
```

20.
```
  □61□
− 5□□3
───────
  2 2 1 2
```

21.
```
  □92□
+ 1□□5
───────
  7 0 1 6
```

22.
```
  □6□3
+ 1□8□
───────
  9 8 3 2
```

23.
```
  □431
− 6□9□
───────
  2 1 3 6
```

24.
```
  385□
− □9□4
───────
  1 8 8 8
```

25.
```
  4□
×  3
─────
  1 2 6
```

26.
```
  7□
×  2
─────
  1 5 0
```

27.
```
  □1
×  7
─────
  4 9 7
```

28.
```
  □3
×  3
─────
  2 7 9
```

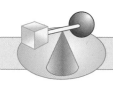

## Challenge

Use each digit only once.

29. How many ways can you find?

```
  □□□□□
+ □□□□□
─────────
  9 9 9 9 9
```

```
0 1 2 3 4
5 6 7 8 9
```

# 14-5 Order of Operations

Artie is playing a math game on his computer. He has to solve the problem, $3 \times 6 - 5$, to earn three points.

When a problem has more than one operation, solve using the order of operations.

$3 \times 6 - 5 = \underline{\ ?\ }$

> ### Order of Operations
>
> - Multiply or divide in order from left to right.
> - Add or subtract in order from left to right.

**Solve** $3 \times 6 - 5$

$3 \times 6 - 5$    First multiply 6 by 3.

$18 \quad - 5$    Then subtract 5 from 18.

$13$

**Solve** $16 - 8 + 3$

$16 - 8 + 3$    No multiplication or division. So work from left to right.

$8 \quad + 3$

$11$

**Study the example.**

$25 - 10 \div 2 \times 3$    First divide 10 by 2.

$25 - \quad 5 \quad \times 3$    Next multiply 3 by 5.

$25 - \quad\quad 15$    Then subtract 15 from 25.

$10$

Write the operation that should be done first.

**1.** $8 + 3 - 4$    **2.** $24 - 9 + 1$    **3.** $26 - 3 \times 8$    **4.** $12 + 4 \times 2$

**5.** $18 \div 9 + 4$    **6.** $5 - 2 \div 1$    **7.** $15 \div 3 \times 2$    **8.** $10 \times 4 \div 2$

**9.** $4 + 10 - 3 \times 3$            **10.** $6 \times 5 + 2 - 1$

**11.** $7 + 8 - 6 \div 2$            **12.** $9 - 2 + 8 \div 4$

Choose the correct solution.

**13.** $12 - 2 \times 3 + 1$    **a.** 31    **b.** 7    **c.** 4

**14.** $8 - 4 + 3 \times 5$    **a.** 5    **b.** 19    **c.** 35

**15.** $40 \div 4 + 6 - 2$    **a.** 14    **b.** 5    **c.** 2

**16.** $36 \div 9 - 3 + 6$    **a.** 12    **b.** 3    **c.** 7

Use the order of operations to solve.

**17.** $10 - 3 + 2$    **18.** $6 + 9 - 7$    **19.** $3 + 2 \times 8$    **20.** $5 \times 8 + 9$

**21.** $7 - 2 \times 3$    **22.** $9 \times 4 - 2$    **23.** $48 \div 6 + 2$    **24.** $8 + 8 \div 4$

**25.** $70 - 63 \div 9$    **26.** $25 - 20 \div 5$    **27.** $16 \div 2 \times 2$    **28.** $8 \times 10 \div 2$

**29.** $7 + 6 \times 6 - 5$            **30.** $6 \times 8 + 3 - 2$

**31.** $28 \div 4 + 6 - 2$            **32.** $7 + 18 \div 3 + 2$

 **Calculator Activity**

You can use the parentheses keys on a calculator to solve problems. Press the keys in this order.

$3 \times (4 + 5)$             The answer is 27.

Use a calculator to solve.

**33.** $(6 - 3) \times 2$    **34.** $(3 + 5) \times 2$    **35.** $40 \div (10 - 5)$    **36.** $16 \div (2 + 6)$

427

# Moving On: Algebra

## 14-6 Missing Operation

Jane and Vinny are playing a number game. Jane says, "I am thinking of the numbers 8 and 7. The answer is 56. What operation did I use?"

▶ Use Guess and Test to find the missing operation.

$$8 \bigcirc 7 = 56$$

**Look at the answer.**

Think: 56 is greater than both 8 and 7. So test addition or multiplication.

Test: $8 + 7 = 56$ not true
$8 \times 7 = 56$ true

Jane used multiplication to find the answer.

▶ Find the missing operation.

$$10 \bigcirc 4 = 6$$

**Look at the answer.**

Think: 6 is not less than 4, but 6 is less than 10. So test subtraction or division.

Test: $10 - 4 = 6$ true
$10 \div 4 = 6$ not true

Subtraction completes the number sentence.

---

Write + or − to complete.

**1.** $3 \bigcirc 7 = 10$    **2.** $1 \bigcirc 5 = 6$    **3.** $8 \bigcirc 5 = 3$    **4.** $9 \bigcirc 9 = 18$

**5.** $7 \bigcirc 7 = 0$    **6.** $7 \bigcirc 5 = 2$    **7.** $6 \bigcirc 5 = 11$    **8.** $5 \bigcirc 2 = 3$

Write × or ÷ to complete.

**9.** 8 ◯ 4 = 32    **10.** 81 ◯ 9 = 9    **11.** 16 ◯ 2 = 8    **12.** 3 ◯ 6 = 18

**13.** 6 ◯ 8 = 48    **14.** 36 ◯ 4 = 9    **15.** 5 ◯ 5 = 1    **16.** 9 ◯ 2 = 18

Write +, −, ×, or ÷. Check your answer with a calculator.

**17.** 5 ◯ 8 = 13    **18.** 40 ◯ 8 = 5    **19.** 3 ◯ 3 = 9    **20.** 7 ◯ 3 = 4

**21.** 7 ◯ 6 = 42    **22.** 6 ◯ 4 = 24    **23.** 36 ◯ 6 = 6    **24.** 9 ◯ 6 = 15

**25.** 8 ◯ 2 = 6    **26.** 9 ◯ 9 = 0    **27.** 16 ◯ 4 = 4    **28.** 7 ◯ 2 = 9

**29.** 8 ◯ 2 = 4    **30.** 54 ◯ 6 = 9    **31.** 8 ◯ 7 = 15    **32.** 8 ◯ 8 = 1

**33.** 21 ◯ 1 = 21    **34.** 23 ◯ 3 = 20    **35.** 2 ◯ 12 = 24    **36.** 3 ◯ 10 = 30

**37.** 55 ◯ 5 = 11    **38.** 80 ◯ 8 = 10    **39.** 80 ◯ 8 = 72    **40.** 80 ◯ 8 = 88

**41.** 6 ◯ 15 = 90    **42.** 9 ◯ 25 = 34    **43.** 3 ◯ 28 = 84    **44.** 100 ◯ 4 = 25

**45.** 65 ◯ 7 = 58    **46.** 90 ◯ 7 = 97    **47.** 96 ◯ 8 = 12    **48.** 3 ◯ 78 = 234

## Critical Thinking

Communicate ✓

Which of these number sentences has more than one answer? Why?

**49.** 3 ◯ 2 = 6    **50.** 0 ◯ 1 = 0    **51.** 8 ◯ 1 = 8    **52.** 2 ◯ 1 = 2

Problem Solving

**53.** Write a number sentence that can be solved using more than one operation.

**54.** Write a number sentence that can be solved using addition or multiplication.

## 14-7 Factors

Two or more numbers that are multiplied to give a product are called **factors**.

$$1 \times 12 = 12 \qquad 2 \times 6 = 12$$

factors                        factors

▶ You can use multiplication sentences to find all the factors of a number.

Find all the factors of 12.

$$1 \times 12 = 12$$
$$2 \times 6 = 12$$
$$3 \times 4 = 12$$

The factors of 12 are
1, 2, 3, 4, 6, and 12.

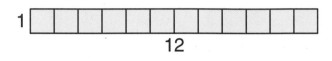

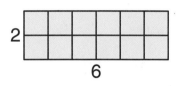

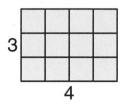

▶ **Common factors** are numbers that are factors of two or more products.

Find all the common factors of 12 and 18.

First list the factors of each number.

$$1 \times 12 = 12 \qquad 1 \times 18 = 18$$
$$2 \times 6 = 12 \qquad 2 \times 9 = 18$$
$$3 \times 4 = 12 \qquad 3 \times 6 = 18$$

Then list the common factors.

1, 2, 3, 6

The common factors of 12 and 18 are 1, 2, 3, and 6.

Find the missing factor.

**1a.** $\underline{\ ?\ } \times 16 = 16$     **b.** $\underline{\ ?\ } \times 8 = 16$     **c.** $\underline{\ ?\ } \times 4 = 16$

**2a.** $\underline{\ ?\ } \times 24 = 24$     **b.** $\underline{\ ?\ } \times 6 = 24$     **c.** $\underline{\ ?\ } \times 8 = 24$

**3a.** $1 \times \underline{\ ?\ } = 20$     **b.** $4 \times \underline{\ ?\ } = 20$     **c.** $2 \times \underline{\ ?\ } = 20$

List all the factors of each number.
You may use multiplication sentences.

**4.** 10     **5.** 9     **6.** 15     **7.** 21     **8.** 14

**9.** 27     **10.** 25     **11.** 35     **12.** 40     **13.** 45

List all the common factors of each set of numbers.

**14.** 2 and 6     **15.** 4 and 8     **16.** 5 and 10

**17.** 9 and 15     **18.** 6 and 15     **19.** 8 and 12

**20.** 6 and 12     **21.** 12 and 16     **22.** 12 and 24

**23.** 15 and 25     **24.** 20 and 4     **25.** 3 and 21

## Share Your Thinking

**26.** Look at the common factors for exercises 14–25. What number is a common factor in every set of numbers? Why?

**27.** How many factors can you list to make these number sentences true?

$$0 = \underline{\ ?\ } \times \underline{\ ?\ }$$

$$1 = \underline{\ ?\ } \times \underline{\ ?\ }$$

## 14-8 Multiplying Money

Rita makes stuffed animals. She sells fluffy bears for $3.89 each. How much will 6 bears cost?

▶ First estimate the product:

6 × $4.00 = $24.00

▶ Then to find the cost of 6 bears, multiply:  6 × $3.89 = ?

<table>
<tr><td>

5 5
$3.89
×    6
2 3 3 4

</td><td>

Follow the same rules for multiplying whole numbers.

</td></tr>
</table>

<table>
<tr><td>

5 5
$3.89
×    6
$2 3.3 4

</td><td>

Bring down the decimal point. Write the $ sign.

</td></tr>
</table>

Six bears will cost $23.34.

Think:  $23.34 is close to $24.00. The answer is reasonable.

Study these examples.

| 3 | 7 4 | 1 1 |
|---|---|---|
| $8.05 | $2.96 | $9.35 |
| ×    7 | ×    8 | ×    3 |
| $56.35 | $23.68 | $28.05 |

Estimate. Then multiply.

1. $2.24
   × 2

2. $2.12
   × 3

3. $1.90
   × 5

4. $4.50
   × 4

5. $8.83
   × 8

6. $6.49
   × 3

7. $5.02
   × 6

8. $3.05
   × 7

Multiply.

9. $.23
   × 3

10. $.34
    × 2

11. $2.18
    × 4

12. $3.17
    × 7

13. $5.05
    × 9

14. $7.00
    × 5

15. $9.20
    × 1

16. $8.01
    × 8

17. $6.27
    × 6

18. $3.95
    × 3

19. $5.59
    × 0

20. $2.89
    × 1

21. $8.10
    × 5

22. $9.03
    × 3

23. $4.21
    × 7

24. $7.52
    × 4

25. $6.42
    × 9

26. $5.74
    × 6

27. $8.44
    × 6

28. $1.75
    × 8

29. 7 × $.30

30. 6 × $9.90

31. 4 × $5.10

32. 2 × $7.66

Problem Solving

33. Ed makes felt hats for Rita's stuffed animals. Each hat costs $2.59. How much will 8 hats cost?

34. Rita sells stuffed penguins for $6.27 each. How much will 7 penguins cost?

35. Rita sells stuffed pigs for $4.59 each and stuffed penguins for $6.27 each. Karen has $30. Does she have enough money to buy 2 stuffed pigs and 3 stuffed penguins?

433

## 14-9 Dividing Money

A store sells 6 snack boxes of graham crackers for $2.70. How much does one box of graham crackers cost?

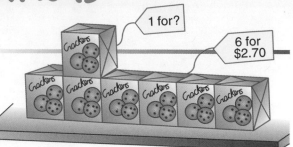

1 for?

6 for $2.70

▶ To find how much one box of crackers costs,
   divide:  $2.70 ÷ 6 = _?_

| Follow the same rules for dividing whole numbers. | Bring up the decimal point. Write the $ sign. |
|---|---|

```
      4 5
6)$2.7 0
 -2 4
    3 0
  - 3 0
      0
```

```
   $ .4 5
6)$2↑7 0
 -2 4
    3 0
  - 3 0
      0
```

One box of graham crackers costs $.45.

Matt bought 4 bags of bagel chips for $8.
How much did one bag of bagel chips cost?

▶ To find how much one bag cost,
   divide:  $8 ÷ 4 = _?_

| Write a decimal point and two zeros in the dividend before dividing. | Follow the same rules for dividing whole numbers. | Bring up the decimal point. Write the $ sign. |
|---|---|---|

```
4)$8.0 0
```

```
      2 0 0
4)$8.0 0
 -8
   0 0
 -   0
     0 0
   -   0
       0
```

```
     $2.0 0
4)$8↑0 0
 -8
   0 0
 -   0
     0 0
   -   0
       0
```

It cost $2.00 for one bag of bagel chips.

Copy and complete.

$$\begin{array}{r} \$.2\phantom{0} \\ 1.\ 2\overline{)\$.48} \end{array}$$
$$\begin{array}{r} \$.2\phantom{0} \\ 2.\ 3\overline{)\$.66} \end{array}$$
$$\begin{array}{r} \$.\phantom{00} \\ 3.\ 4\overline{)\$.80} \end{array}$$
$$\begin{array}{r} \$.\phantom{00} \\ 4.\ 3\overline{)\$.60} \end{array}$$

$$\begin{array}{r} \$\ .8\phantom{0} \\ 5.\ 8\overline{)\$6.72} \end{array}$$
$$\begin{array}{r} \$\ .8\phantom{0} \\ 6.\ 9\overline{)\$7.47} \end{array}$$
$$\begin{array}{r} \$3.\phantom{00} \\ 7.\ 3\overline{)\$9.63} \end{array}$$
$$\begin{array}{r} \$3.\phantom{00} \\ 8.\ 2\overline{)\$6.48} \end{array}$$

$$\begin{array}{r} \$\ .\phantom{00} \\ 9.\ 5\overline{)\$7.20} \end{array}$$
$$\begin{array}{r} \$\ .\phantom{00} \\ 10.\ 6\overline{)\$9.30} \end{array}$$
$$\begin{array}{r} \$\ .\phantom{00} \\ 11.\ 6\overline{)\$6.36} \end{array}$$
$$\begin{array}{r} \$\ .\phantom{00} \\ 12.\ 7\overline{)\$7.28} \end{array}$$

Divide and check.

**13.** $4\overline{)\$.20}$  **14.** $5\overline{)\$.35}$  **15.** $2\overline{)\$.92}$  **16.** $6\overline{)\$.96}$

**17.** $3\overline{)\$9.42}$  **18.** $2\overline{)\$4.62}$  **19.** $3\overline{)\$7.47}$  **20.** $2\overline{)\$9.14}$

**21.** $8\overline{)\$3.52}$  **22.** $5\overline{)\$1.80}$  **23.** $4\overline{)\$8}$  **24.** $3\overline{)\$45}$

**25.** $\$.84 \div 4$  **26.** $\$.78 \div 6$  **27.** $\$.80 \div 8$  **28.** $\$.90 \div 9$

**29.** $\$7.56 \div 7$  **30.** $\$3.21 \div 3$  **31.** $\$6 \div 4$  **32.** $\$8 \div 5$

Problem Solving

**33.** How much is 1 box of raisins?

**34.** How much is 1 jar of honey?

**35.** How much is 1 granola bar?

**36.** How much is 1 can of juice?

**37.** How much are 2 ice cream bars?

| Snack Shack Sale |
| --- |
| 8 boxes of raisins for $5.20 |
| 3 granola bars for $3.45 |
| 6 cans of juice for $3 |
| 2 jars of honey for $8.48 |
| 8 ice cream bars for $4 |

**38.** Mel buys 4 fruit roll-ups for $3.56. How much will 1 roll-up cost?

**39.** Paula buys 2 pounds of apples for $1.18. How much will 3 pounds cost?

435

# Moving On: Problem Solving

## 14-10 Strategy: More Multi-Step

**Problem:** In making a sign for a holiday sale, Mr. Thrifty wants to use one of these words: GREAT, HUGE, GIANT, or LARGE. After looking at the price of each letter, which word will he use if he wants to spend the least amount of money?

| Price Per Letter | |
|---|---|
| A ... 6¢ | L .... 4¢ |
| E ... 8¢ | N .... 6¢ |
| G ... 9¢ | R ... 11¢ |
| H ... 6¢ | T .... 4¢ |
| I ... 2¢ | U ... 11¢ |

**1 IMAGINE** Place yourself in the problem.

**2 NAME**

*Facts:* the words — GREAT, HUGE, GIANT, LARGE
the price of each letter

*Question:* Which word will cost the least amount of money?

**3 THINK** Find the price of each letter on the chart. Add the price of each letter in each word. Order the cost of the words.

**4 COMPUTE**

G   R   E   A   T
9¢ + 11¢ + 8¢ + 6¢ + 4¢ = 38¢
H   U   G   E
6¢ + 11¢ + 9¢ + 8¢       = 34¢
G   I   A   N   T
9¢ + 2¢ + 6¢ + 6¢ + 4¢ = 27¢
L   A   R   G   E
4¢ + 6¢ + 11¢ + 9¢ + 8¢ = 38¢

34¢ < 38¢ and 27¢ < 34¢,
so GIANT will cost the least amount of money.

**5 CHECK** Use a calculator to check the cost of each word.

Use the Multi-Step Problem strategy to solve each problem.

**1.** Hector doubled $1.89 and then found 2 nickels. Rosa doubled $2.06 and then lost a quarter. Who had more money?

| IMAGINE | Create a mental picture. |
|---|---|

| NAME | *Facts:* | Hector doubled $1.89 and found 2 nickels. Rosa doubled $2.06 and lost a quarter. |
|---|---|---|
| | *Question:* | Who had more money? |

| THINK | To double $1.89, multiply by 2. Then add 10¢. To double $2.06, multiply by 2. Then subtract 25¢. |
|---|---|

**COMPUTE** ────→ **CHECK**

**2.** A store owner bought 8 kites for $52.72. Then she sold each kite for $6.89. Did the store owner gain or lose money? How much did she gain or lose?

**3.** The weight of 10 feet of copper wire is 12 ounces. If Jeff's science project requires three pieces of 8-foot copper wire and two pieces of 18-foot copper wire, how many ounces must Jeff buy?

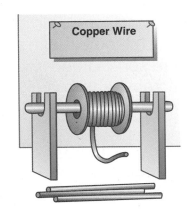
Copper Wire

**4.** The width of a rectangle is 24 m. If the length of the rectangle is 2 m more than the width, what is the perimeter?

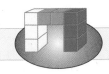

## Make Up Your Own

**5.** Write a multi-step problem. Have a classmate solve it.

# Moving On: Problem Solving

## 14-11 Applications

Solve each problem and explain the method you used.

1. Mr. Farmer's silo had 23,200 ears of corn. During the winter his pigs ate 12,375 ears. How many ears of corn were left?

2. A large poultry farm produced one million, nine hundred thousand eggs. How many more eggs are needed to reach a goal of 2,000,000?

Imagine

3. A beekeeper placed 17,500 hives in cherry groves and 13,800 hives in almond groves. How many hives did the beekeeper place in all?

Name

4. Jen's orchard produced 985 red apples and 1126 yellow apples. What color apples can she divide evenly between 5 stores? between 2 stores?

Think

5. Old MacDonald divided his 48 acres of farmland equally among his 3 children. One son planted wheat on 9 of his acres and soybeans on the rest. How many acres of soybeans did he plant?

Compute

6. Lauren spent $3.73 at a produce stand. She paid with a $5.00 bill. She received a dollar bill, a quarter, and 2 nickels as change. Was this correct? Explain.

Check

7. Each basket of strawberries costs $1.59. How much will 9 baskets cost?

8. Arlene spent $73.28 for 4 bushels of pineapples. What is the cost of 1 bushel?

438

Choose a strategy from the list or use another strategy you know to solve each problem.

Use these strategies:

Multi-Step Problem
Find a Pattern
Use a Drawing or Model
Write a Number Sentence
Interpret the Remainder
Guess and Test

9. Spot and King are two farm dogs. Spot is 4 years old. King is 12 years old. In how many years will King be twice Spot's age?

10. A farmer had 21 bales of hay. A week later she had 5 bales left. If she put the same number of bales into each of 8 stalls, how many bales would be in each stall?

11. A rancher sold half of her sheep. After buying 12 more, she had 21 sheep. How many sheep did she have at first?

12. What is the sum of the following numbers: 1000, 2000, 3000, 4000, 5000, 6000, 7000, 8000, and 9000?

13. How many cartons will a farmer need for shipping 330 cantaloupes if exactly 8 cantaloupes are packed into each full carton?

14. Carly has 1.8 kg of apples and 2.3 kg of berries. How many more kilograms of berries than apples does Carly have?

15. There are 10 rows of 6 trees in the orchard. Every fifth tree is pruned. How many trees still need to be pruned?

16. A gardener bought a plant that was 35 cm tall. Each time he cut it back 2 cm, it grew another 5 cm. If he cut it back four times, how tall would the plant be?

439

# Chapter Review and Practice

## Write the number in standard form.
(See pp. 418–419.)

**1.** $6{,}000{,}000 + 800{,}000 + 40{,}000 + 7000 + 500 + 30 + 5$

**2.** two million, three hundred thousand, twelve

## Add or subtract. Watch for + and −.
(See pp. 420–421.)

**3.**
$$\begin{array}{r} 16{,}526 \\ +\,15{,}198 \\ \hline \end{array}$$

**4.**
$$\begin{array}{r} 92{,}613 \\ -\,28{,}298 \\ \hline \end{array}$$

**5.**
$$\begin{array}{r} 46{,}227 \\ +\ \ 8{,}465 \\ \hline \end{array}$$

**6.**
$$\begin{array}{r} \$109.05 \\ -\ \ \ \ 79.86 \\ \hline \end{array}$$

## Find the missing digits.
(See pp. 424–425.)

**7.**
$$\begin{array}{r} 9\,\square \\ +\,\square\,3 \\ \hline 1\,0\,7 \end{array}$$

**8.**
$$\begin{array}{r} \square\,5\,\square \\ -\,2\,\square\,2 \\ \hline 6\,3\,4 \end{array}$$

**9.**
$$\begin{array}{r} \square\,9\,2 \\ -\,3\,\square\,\square \\ \hline 2\,4\,6 \end{array}$$

**10.**
$$\begin{array}{r} \square\,8 \\ \times\ \ \ 6 \\ \hline 2\,8\,8 \end{array}$$

## Compute.
(See pp. 426–427, 432–435.)

**11.** $9 \times 5 + 8 - 3$   **12.** $32 \div 4 + 5 - 3$   **13.** $7 \times 6 \div 2$   **14.** $50 - 32 \div 8$

**15.**
$$\begin{array}{r} \$6.28 \\ \times\ \ \ \ \ 6 \\ \hline \end{array}$$

**16.**
$$\begin{array}{r} \$4.09 \\ \times\ \ \ \ \ 7 \\ \hline \end{array}$$

**17.** $2\overline{)\$8.24}$   **18.** $6\overline{)\$6.36}$

## Write +, −, × or ÷ to complete.
(See pp. 428–429.)

**19.** $60 \ \underline{?} \ 8 = 52$   **20.** $84 \ \underline{?} \ 6 = 14$   **21.** $4 \ \underline{?} \ 70 = 280$

**22.** $9 \ \underline{?} \ 44 = 53$   **23.** $210 \ \underline{?} \ 7 = 30$   **24.** $3 \ \underline{?} \ 18 = 54$

## List all the common factors.
(See pp. 430–431.)

**25.** 3 and 12   **26.** 4 and 32   **27.** 9 and 15

## PROBLEM SOLVING
(See pp. 422–423.)

**28.** I am a number between 51 and 61. I am divisible by 5 and 10. What number am I?

(See *Still More Practice*, p. 454.)

## ALWAYS, SOMETIMES, NEVER

The sentences below are called **mathematical statements**. A statement is either *always* true, *sometimes* true, or *never* true.

Be careful when testing each statement. Try several examples before you answer.

**Write *always*, *sometimes*, or *never* for each statement. Give an example to support your answer.**

1. The sum of two odd numbers is an even number.

2. The product of two even numbers is an even number.

3. If you multiply an even number by 5, the ones digit in the product is 0.

4. The sum of an odd number and an even number is an even number.

5. When factors are doubled, their product is doubled.

6. When addends are doubled, their sum is doubled.

7. The product of two odd numbers is an odd number.

8. Rectangles have 4 sides and 4 right angles.

9. Curved figures are circles.

10. Pentagons have seven sides.

11. Rectangles are squares.

12. A circle has 4 edges.

**Performance Assessment**

Use these cards to complete 16 ◯ 8 ◯ 2.

1. Which two cards give an answer greater than 100?

2. Which two cards give an answer between 10 and 30?

**Write the number in standard form.**

3. 7,000,000 + 100,000 + 30,000 + 6000 + 400 + 20 + 1

4. nine million, four hundred sixty thousand, one hundred five

**Add or subtract. Watch for + and −.**

| 5. | 15,456 | 6. | 69,708 | 7. | 89,146 | 8. | $316.00 |
|---|---|---|---|---|---|---|---|
| | +18,338 | | −24,289 | | + 3,697 | | −    56.78 |

**Find the missing digits.**

| 9. | □5 | 10. | 2□6 | 11. | 3□□ | 12. | 3□ |
|---|---|---|---|---|---|---|---|
| | +5□ | | +15□ | | −□29 | | ×  4 |
| | 128 | | 373 | | 267 | | 140 |

**Compute.**

13. 7 + 20 − 9 ÷ 3          14. 7 × 3 − 4 + 6          15. 40 − 5 × 6

16. 4 × $5.07        17. 9 × $2.78        18. $6.28 ÷ 2        19. $8.16 ÷ 4

**List all the common factors.**

20. 5 and 30                21. 6 and 42                22. 7 and 28

**PROBLEM SOLVING**    *Use a strategy you have learned.*

23. I am a number between 61 and 79. I am divisible by 2, 5, and 10. What number am I?

## Practice 1-1

Write the number in standard form.

1. 200 + 60 + 3

2. 4 thousands 5 hundreds 0 tens 2 ones

3. 3 hundred thousands

In what place is the underlined digit?

**4a.** 7159      **b.** 404,712      **c.** 23,862

What is the value of the underlined digit?

**5a.** 193,215      **b.** 7966      **c.** 104,638

Compare. Write < or >.

**6a.** 43 _?_ 147      **b.** 213 _?_ 231

**7a.** 1976 _?_ 1979      **b.** $5.88 _?_ $5.86

Write in order from least to greatest.

**8a.** 299, 295, 305      **b.** 5631, 6501, 5650

Round to the nearest ten and nearest hundred.

**9a.** 639      **b.** 112      **c.** 435

Round to the nearest dollar.

**10a.** $7.73      **b.** $4.25      **c.** $6.56

Write the missing numbers.

11. 26, 29, 32, _?_ , _?_ , 41, _?_

Write the amount.

12. 2 quarters, 4 dimes, 3 pennies

13. 5 one-dollar bills, 5 dimes, 2 nickels

14. Cal has 1 one-dollar bill, 3 quarters, and 1 dime. How much money does Cal have?

15. Claire has $4.86. Frank has $4.68. Who has more money?

16. Carol buys a ball for 59¢. She gives the cashier 3 quarters. Name the change.

## Practice 1-2

Write the number in expanded form.

1. Two hundred fifty-four

2. One thousand, three hundred ten

**3a.** 523,014      **b.** 709,205

Compare. Write < or >.

**4a.** 7215 _?_ 8031      **b.** 9001 _?_ 5999

**5a.** 6135 _?_ 6253      **b.** 4432 _?_ 4142

**6a.** 5209 _?_ 5215      **b.** 3702 _?_ 3701

Round to the nearest thousand.

**7a.** 2402      **b.** 5764

**8a.** 3515      **b.** 4029

Write the amount.

9. 1 five-dollar bill, 2 dimes, 4 pennies

10. 2 ten-dollar bills, 1 half-dollar, 2 nickels

11. 2 ten-dollar bills, 1 five-dollar bill, 3 quarters, 7 pennies

# Chapter 2

## Practice 2-1

Regroup for addition.

**1.** 17 ones = _?_ ten _?_ ones

**2.** 14 tens = _?_ hundred _?_ tens

**3.** 13 hundreds = _?_ thousand _?_ hundreds

Estimate. Then add.

**4a.**  55    **b.**  44    **c.**  32
   +17       +26       +75

**5a.**  82    **b.**  604   **c.**  158
   +47      +126     +326

**6a.**  308   **b.**  605   **c.**  432
  + 93     + 97     + 69

**7a.**  757   **b.**  $1.46  **c.**  $4.07
  +153     + 3.93    + 3.86

**8.** Jeanna has 62 baseball cards. Bud has 45 baseball cards. How many baseball cards do they have in all?

**9.** Peggy has 28 wide-tip markers and 7 fine-tip markers. How many markers does she have in all?

**10.** Reggie read one book of 263 pages and another book of 284 pages. Find the total number of pages he read.

**11.** Ben spent $4.73 on paper and $3.84 on tape. How much did he spend in all?

**12.** Amy spent $6.09 in one store and $2.91 in another store. How much did she spend altogether?

**13.** Stacey has 36 tapes. Jeff has 17 more than Stacey. How many tapes does Jeff have?

## Practice 2-2

Estimate. Then add.

**1a.**  398   **b.**  546   **c.**  $3.95
  +267     +186     + 2.89

**2a.**  $2.99  **b.**  4379  **c.**  5063
  + 3.99    +2100    +1235

**3a.**  7082  **b.**  5242  **c.**  6919
  +2181    +1837    +2082

**4a.**  4557  **b.**  $19.86  **c.**  $27.59
  +4846    + 23.75    + 47.98

**5a.**  34    **b.**  32    **c.**  302
   94       85       193
  +99      +18      + 45

**6a.**  562   **b.**  $2.16  **c.**  $5.21
   308      3.75     3.96
  + 61     + 2.30    + 4.19

**7.** The Reilly family traveled 452 miles one day and 189 miles the next day. How many miles did they travel in all?

**8.** A chef prepared 285 lunches and 145 dinners. How many meals did she prepare altogether?

**9.** Mike put 125 roses and 385 tulips on the truck. How many flowers in all did he put on the truck?

**10.** Tricia spent $4.36. Anil spent $2.89 more than Tricia. How much money did Anil spend?

**11.** A poster costs $24.75. The frame costs $19.95. What is the total cost?

**12.** A pizza costs $8.99. A bottle of juice costs $2.25. Estimate the total cost.

## Practice 3-1

Regroup for subtraction.

**1.** 4 tens 6 ones = ? tens ? ones

**2.** 7 hundreds 9 tens = ? hundreds ? tens

**3.** 4 dollars 7 dimes = ? dollars ? dimes

Estimate. Then subtract.

**4a.** 35 − 24   **b.** 49 − 30   **c.** 67 − 38

**5a.** 48 − 19   **b.** 75¢ − 28¢   **c.** 87¢ − 69¢

**6a.** 486 − 124   **b.** 597 − 280   **c.** $6.58 − 4.16

**7a.** $9.89 − 4.17   **b.** 783 − 692   **c.** 539 − 347

**8.** Liz made 34 banners. She sold 19 of them. How many banners are left?

**9.** The Westwood School has 473 students. There are 292 boys. How many girls are there?

**10.** A dairy farmer has 325 cows. He has 175 brown cows. How many cows are not brown?

**11.** Ramon walked 33 miles the first week. He walked 22 miles the second week. How many more miles did he walk the first week?

**12.** Fran bought a bat for $8.79. She bought a ball for $2.99. How much more did she pay for the bat?

## Practice 3-2

Estimate. Then subtract.

**1a.** 718 − 549   **b.** 846 − 679   **c.** 932 − 368

**2a.** $4.36 − 2.47   **b.** $5.67 − 2.89   **c.** $7.58 − 4.29

**3a.** 500 − 449   **b.** 800 − 376   **c.** $9.00 − 4.53

**4a.** 3641 − 2530   **b.** 8975 − 7243   **c.** $47.65 − 36.04

**5a.** 6840 − 5673   **b.** 7936 − 4879   **c.** 6348 − 4769

Align and subtract. Check by adding.

**6a.** 36 − 29   **b.** 84 − 58

**7a.** $4.00 − $2.79   **b.** 523 − 395

**8.** Jude's book has 208 pages. He has read 162 pages. How many more pages are left to read?

**9.** The scouts are collecting cans. They have collected 275. Their goal is 500 cans. How many more cans do they need?

**10.** A living room set costs $800. Mrs. Borelli has saved $735. How much more money does she need?

**11.** Amy bought a book for $3.75. How much change did she receive from $5.00?

**12.** On Thursday the town recycled 3250 pounds of newspaper. On Friday 1189 pounds were recycled. How many fewer pounds were recycled on Friday?

# Chapter 4

## Practice 4-1

Write a multiplication sentence for each.

**1a.** $2 + 2 + 2 + 2$     **b.** $5 + 5 + 5 + 5 + 5$

Multiply.

**2a.** $\begin{array}{r} 2 \\ \times 1 \\ \hline \end{array}$     **b.** $\begin{array}{r} 4 \\ \times 0 \\ \hline \end{array}$     **c.** $\begin{array}{r} 5 \\ \times 5 \\ \hline \end{array}$

**3a.** $\begin{array}{r} 3 \\ \times 7 \\ \hline \end{array}$     **b.** $\begin{array}{r} 4 \\ \times 9 \\ \hline \end{array}$     **c.** $\begin{array}{r} 2 \\ \times 0 \\ \hline \end{array}$

Find the product.

**4a.** $3 \times 1$     **b.** $4 \times 2$     **c.** $6 \times 3$

**5a.** $5 \times 4$     **b.** $6 \times 2$     **c.** $5 \times 0$

**6a.** $8 \times 4$     **b.** $3 \times 5$     **c.** $6 \times 4$

**7a.** $7 \times 3$     **b.** $4 \times 5$     **c.** $2 \times 4$

**8a.** $9 \times 5$     **b.** $7 \times 2$     **c.** $4 \times 3$

**9a.** $6 \times 5¢$     **b.** $7 \times 4¢$     **c.** $9 \times 3¢$

Find the missing factor.

**10a.** $\underline{?} \times 4 = 24$     **b.** $7 \times \underline{?} = 21$

**11a.** $\underline{?} \times 5 = 40$     **b.** $32 = \underline{?} \times 4$

**12.** There are 3 rows of stamps. Each row has 4 stamps. How many stamps are there?

**13.** There are 5 children. Each child has 3 tapes. How many tapes are there?

**14.** There are 4 bags of apples. There are 3 apples in each bag. How many apples are there?

**15.** Pepe worked 4 hours every day for 5 days. How many hours did he work in all?

**16.** There are 4 children playing in the snow. Each child has a pair of gloves. How many gloves are there?

**17.** A jogger walks 2 miles a day for 5 days. How many miles does the jogger walk?

**18.** Tony reads 2 stories each day. How many stories does he read in 4 days?

# Chapter 5

## Practice 5-1

Find the quotient.

**1a.** $4 \div 2$     **b.** $3 \div 3$     **c.** $8 \div 4$

**2a.** $18 \div 3$     **b.** $12 \div 4$     **c.** $3 \div 1$

**3a.** $0 \div 3$     **b.** $8 \div 2$     **c.** $24 \div 3$

Divide.

**4a.** $2\overline{)12}$     **b.** $3\overline{)27}$     **c.** $3\overline{)9}$

**5a.** $4\overline{)16}$     **b.** $3\overline{)15}$     **c.** $2\overline{)10}$

**6a.** $2\overline{)18}$     **b.** $4\overline{)20}$     **c.** $4\overline{)4}$

**7.** There are 18 desks in 3 rows. At most, how many desks are in each row?

**8.** Sixteen oranges are shared equally among 4 children. How many oranges does each child get?

**9.** Zelda has 15 stickers. How many groups of 3 stickers can she make?

**10.** Joe has 12 pieces of art paper. It takes 2 pieces of paper to make a plane. How many planes can Joe make?

## Practice 5-2

Divide.

**1a.** 25 ÷ 5    **b.** 20 ÷ 4    **c.** 10 ÷ 5

**2a.** 9 ÷ 3    **b.** 6 ÷ 3    **c.** 2 ÷ 2

**3a.** 35 ÷ 5    **b.** 28 ÷ 4    **c.** 12 ÷ 2

**4a.** 12 ÷ 3    **b.** 0 ÷ 5    **c.** 18 ÷ 2

**5a.** 20¢ ÷ 5    **b.** 27¢ ÷ 3    **c.** 14¢ ÷ 2

**6a.** 30¢ ÷ 5    **b.** 10¢ ÷ 2    **c.** 32¢ ÷ 4

**7a.** $3\overline{)24}$    **b.** $5\overline{)45}$    **c.** $3\overline{)15}$

**8a.** $2\overline{)16}$    **b.** $3\overline{)18}$    **c.** $5\overline{)40}$

**9a.** $4\overline{)36¢}$    **b.** $5\overline{)5¢}$    **c.** $4\overline{)24¢}$

**10a.** $4\overline{)16¢}$    **b.** $3\overline{)21¢}$    **c.** $5\overline{)15¢}$

**11.** Donna and Dawn will share 18¢ equally. How much will each girl receive?

**12.** Each bird feeder is made from 2 feet of wood. How many bird feeders can be made from 8 feet of wood?

**13.** Charlie has a string 15 feet long. He must cut it into 5 pieces. If all the pieces are the same length, how long is each piece?

**14.** Five children have 30 minutes to work on the computer. Each child uses it for the same amount of time. How long does each child use the computer?

**15.** Bob shares 8 apples equally with Ted. How many apples does each boy get?

---

**Chapter 6**

## Practice 6-1

The tally chart shows the number of absences in each grade last month.

| Grade | Tally | Total |
|-------|-------|-------|
| 1 | ЖНТ ||| | ? |
| 2 | ЖНТ ЖНТ | | ? |
| 3 | ЖНТ ЖНТ ЖНТ | ? |
| 4 | ЖНТ ЖНТ | ? |

**1.** Copy and complete the tally chart.

**2.** Use the chart to make a bar graph.

**3.** Which grade had the most absences? the fewest absences?

**4.** Which grades had more than 10 absences?

**5.** Which grade had exactly 11 absences?

Make a tree diagram and a list.

**6.** Frances has a pair of blue and a pair of black jeans. She has a blue, a yellow, and a white sweater. How many different outfits can she make?

Use the spinner to find the probability:

**7.** of landing on red.

**8.** of landing on blue.

**9.** of landing on yellow.

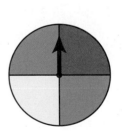

**10.** Is it certain or impossible that the spinner will land on green?

**11.** Write *equally likely* or *not equally likely* for the spinner landing on each color.

## Practice 7-1

Which unit is used to measure each: in., ft, yd, or mi?

**1.** length of a marker

**2.** height of a doorway

**3.** length of a soccer field

Compare. Write $<$, $=$, or $>$.

**4a.** 5 in. _?_ 5 yd    **b.** 8 ft _?_ 8 yd

**5a.** 1 ft _?_ 14 in.    **b.** 3 ft _?_ 3 mi

Use a ruler to measure each line segment to the nearest inch.

**6.**

**7.**

Draw a line for each length.

**8a.** 2 in.    **b.** $3\frac{1}{2}$ in.    **c.** $1\frac{3}{4}$ in.

Which unit is used to measure each: c, pt, qt, or gal?

**9.** juice in a small glass

**10.** house paint in a large can

Which unit is used to measure each: oz or lb?

**11.** tape recorder

**12.** pencil

Compare. Write $<$, $=$, or $>$.

**13a.** 3 qt _?_ 1 gal    **b.** 20 oz _?_ 1 lb

---

## Practice 7-2

Which unit is used to measure each: cm, dm, m, or km?

**1.** length of a pencil

**2.** length of a bridge

**3.** height of a television

Compare. Write $<$, $=$, or $>$.

**4a.** 200 m _?_ 200 cm   **b.** 3 dm _?_ 3 m

**5a.** 210 cm _?_ 2 m    **b.** 22 dm _?_ 2 m

**6a.** 1 km _?_ 1 dm    **b.** 1500 m _?_ 1 km

Use a centimeter ruler to measure each line segment to the nearest centimeter.

**7.**

**8.**

Draw a line for each length.

**9a.** 2 cm    **b.** 5 cm    **c.** 9 cm

Which unit is used to measure each: mL or L?

**10.** milk in a spoon

**11.** oil in a large jug

Which unit is used to measure each: g or kg?

**12a.** baby    **b.** cookie

Compare. Write $<$, $=$, or $>$.

**13a.** 3 mL _?_ 3 L    **b.** 2000 g _?_ 2 kg

## Practice 7-3

Write the letter of the most reasonable temperature.

1. classroom     **a.** 30°F    **b.** 70°F

2. ice cube     **a.** 30°F    **b.** 70°F

3. ice cream     **a.** 1°C    **b.** 50°C

4. boiling water   **a.** 10°C   **b.** 100°C

Write each time in standard form.

**5a.**      **b.** half past four

**6a.**      **b.** thirty-two minutes after seven

What time will it be in 15 minutes if it is now:

**7a.** 1:00?    **b.** 4:10?     **c.** 7:35?

**8.** How many days are in 1 week?

**9.** How many months are in 1 year?

**10.** How many days are in 1 year?

**11.** How many days are there in June?

---

Chapter 8

## Practice 8-1

Find the product.

**1a.** 6 × 6    **b.** 2 × 7    **c.** 8 × 6

**2a.** 4 × 6    **b.** 6 × 7    **c.** 5 × 6

**3a.** 4 × 7    **b.** 0 × 6    **c.** 8 × 7

**4a.**   7     **b.**   7     **c.**   6
    ×1         ×0         ×9

Find the quotient.

**5a.** 0 ÷ 6    **b.** 35 ÷ 7    **c.** 24 ÷ 6

**6a.** 7 ÷ 7    **b.** 54 ÷ 6    **c.** 14 ÷ 7

**7a.** 30 ÷ 6    **b.** 49 ÷ 7    **c.** 21 ÷ 7

**8a.** 7)56    **b.** 6)18    **c.** 7)28

**9a.** 6)48    **b.** 7)63    **c.** 6)6

Write the complete fact family for each.

**10a.** 4, 7, 28      **b.** 5, 6, 30

Multiply.

**11a.** 3 × 2 × 4     **b.** 2 × 1 × 5

**12.** Jo puts 42 books in 6 boxes. There are the same number of books in each box. How many books are in each box?

**13.** Louis worked 7 days a week for 8 weeks. How many days did he work?

**14.** Seven friends share 35 pencils equally. How many pencils does each friend get?

**15.** Sarah bought 8 packages of nails. There are 7 nails in each package. How many nails did Sarah buy?

## Practice 8-2

Find the product.

**1a.** 9 × 9    **b.** 4 × 9    **c.** 2 × 8

**2a.** 1 × 8    **b.** 8 × 9    **c.** 7 × 9

**3a.** 5 × 9    **b.** 7 × 8    **c.** 5 × 8

**4a.**    8    **b.**    9    **c.**    8
      ×4        ×3        ×0

Find the quotient.

**5a.** 9 ÷ 9    **b.** 72 ÷ 8    **c.** 81 ÷ 9

**6a.** 64 ÷ 8    **b.** 27 ÷ 9    **c.** 40 ÷ 8

**7a.** 36 ÷ 9    **b.** 0 ÷ 8    **c.** 72 ÷ 9

**8a.** 8)48    **b.** 8)24    **c.** 9)54

**9a.** 8)56    **b.** 9)45    **c.** 8)16

Write the rule. Then complete the pattern.

**10a.** 2, 4, 8, _?_     **b.** 24, 6, 8, 2, 4, _?_

Write the complete fact family for each.

**11a.** 4, 8, 32     **b.** 6, 9, 54

**12.** There are 36 cable cars running on 9 tracks. The same number of cars are on each track. How many cars are on each track?

**13.** Sam has 56 rolls of paper. He stacks the rolls in piles of 8. How many piles does he make?

**14.** Juan has 5 boxes of crayons. There are 8 crayons in each box. How many crayons does Juan have?

**15.** Find the product of 3, 2, and 5.

**16.** The computer club has 9 members. The art club has 3 times as many members. How many members are there in the art club?

---

**Chapter 9**

## Practice 9-1

Draw these figures on grid paper.

**1a.** square      **b.** triangle

**2a.** hexagon      **b.** rectangle

**3a.** circle      **b.** pentagon

**4a.** octagon      **b.** parallelogram

Name each: line, line segment, ray, or none of these.

**5a.**       **b.** ⟷

**6a.** ⌣      **b.** •—•

**7.** Draw two lines that are parallel.

Find two congruent figures.

**8a.**    **b.** ◯   **c.**    **d.** ▱

**9a.** ⬡   **b.**    **c.**    **d.** ▭

Find two similar figures.

**10a.** ☆   **b.** ◇   **c.** ☆   **d.** ♣

**11a.** ▢   **b.**    **c.** ▽   **d.** △

**12.** Name a figure that has 4 sides of the same length and 4 right angles.

## Practice 9-2

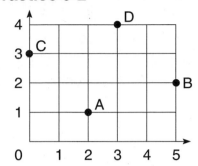

**1.** Write the ordered pair for D.

**2.** Write the letter for (0, 3).

Is each a line of symmetry?
Write *Yes* or *No*.

**3a.**   **b.**   **c.**

Write *slide, flip,* or *turn* for each.

**4a.**   **b.**

Write the name of each space figure.

**5a.**    **b.**

Find the perimeter.

**6a.** 2 in.
2 in. ⬜ 2 in.
2 in.

**b.** 4 cm △ 4 cm
4 cm

Write how many square units.

**7a.** [grid]   **b.** [grid]

Find the volume.

**8a.** [figure]   **b.** [figure]

---

### Chapter 10

## Practice 10-1

Estimate. First round, then use front-end estimation.

**1a.** 43
× 2

**b.** 26
× 4

**c.** $32
× 3

**2a.** 69
× 3

**b.** 52
× 4

**c.** $45
× 7

**3a.** 86
× 9

**b.** 34
× 4

**c.** $29
× 8

Find the product.

**4a.** 3 × 12   **b.** 2 × 21   **c.** 4 × $22

**5a.** 4 × 17   **b.** 2 × 28   **c.** 3 × $38

**6.** Mario has 4 packs of pencils. Each pack contains 20 pencils. How many pencils does Mario have?

**7.** A box contains 38 books. Each box has the same number of books. How many books are in 7 boxes?

**8.** If 9 students each read 21 books, how many books do they read altogether?

## Practice 10-2

Estimate. Then multiply.

**1a.** 123    **b.** 431    **c.** 2342
    $\times$ 3      $\times$ 2       $\times$ 2

**2a.** 114    **b.** 351    **c.** $1113
    $\times$ 7      $\times$ 2       $\times$ 6

**3a.** 486    **b.** 195    **c.** $287
    $\times$ 2      $\times$ 3       $\times$ 4

**4a.** 1187    **b.** 2159    **c.** 2387
    $\times$ 5      $\times$ 4      $\times$ 2

**5.** A tape recorder costs $125. Estimate the cost of 2 recorders.

**6.** Barbara saves $2145 a year. How much does she save in 3 years?

**7.** There are 250 sheets of computer paper in each package. Kippy has 4 packages of paper. How many sheets of computer paper does he have?

**8.** Jon has 4 bags of balloons. Each bag holds 125 balloons. How many balloons does he have?

---

### Chapter 11

## Practice 11-1

Complete.

**1.** $11 \div 3 = \underline{\ ?\ }$ remainder $\underline{\ ?\ }$

**2.** $19 \div 2 = \underline{\ ?\ }$ remainder $\underline{\ ?\ }$

**3.** $27 \div 3 = \underline{\ ?\ }$ remainder $\underline{\ ?\ }$

**4.** $33 \div 5 = \underline{\ ?\ }$ remainder $\underline{\ ?\ }$

Find the quotient and the remainder.

**5a.** $4\overline{)26}$    **b.** $5\overline{)20}$    **c.** $6\overline{)33}$

**6a.** $7\overline{)22}$    **b.** $8\overline{)65}$    **c.** $9\overline{)31}$

**7a.** $5\overline{)28}$    **b.** $3\overline{)14}$    **c.** $8\overline{)35}$

Divide and check.

**8a.** $42 \div 5$      **b.** $35 \div 8$

**9a.** $26 \div 9$      **b.** $34 \div 7$

**10a.** $51 \div 6$      **b.** $22 \div 4$

Find the quotient.

**11a.** $3\overline{)90}$    **b.** $3\overline{)900}$    **c.** $4\overline{)80}$

**12a.** $4\overline{)800}$    **b.** $2\overline{)60}$    **c.** $6\overline{)600}$

**13.** Sally has 29 stickers to share equally among 4 people. How many stickers will each person get? How many stickers will be left over?

**14.** Five shirts will fill each box. Tony has 24 shirts to pack. At most, how many boxes will he fill? How many shirts will be left over?

**15.** Louise has 14 postcards. She wants to send the same number of postcards to each of 6 people. How many postcards will each person get?

**16.** How many 6s are in 34? What is the remainder?

**17.** Len has 44 seashells to put into bags with 8 seashells in each. How many bags of 8 will he fill?

**18.** How many groups of 5 pipe cleaners can be made from a total of 27 pipe cleaners? How many will be left over?

## Practice 11-2

Divide and check.

**1a.** 4)72    **b.** 5)65    **c.** 9)99

**2a.** 8)92    **b.** 7)89    **c.** 5)77

**3a.** 4)53    **b.** 6)88    **c.** 9)95

**4a.** 7)79    **b.** 8)95    **c.** 5)63

Estimate the quotient.

**5a.** 3)31    **b.** 5)28    **c.** 6)58

**6a.** 5)$5.10    **b.** 3)$6.14    **c.** 4)$16.21

**7.** Estimate the cost of each pencil if 5 pencils cost $1.09.

**8.** Alex separates 105 books into 5 equal piles. How many books are there in each pile?

**9.** Two friends share 17 marbles equally. How many marbles does each get? How many marbles are left over?

---

**Chapter 12**

## Practice 12-1

Write the fraction for the shaded part.

**1a.**     **b.**     **c.**

**2a.**     **b.**     **c.**

Write the equivalent fraction.

**3a.**      **b.**

$\frac{2}{4} = \frac{?}{8}$      $\frac{2}{3} = \frac{?}{9}$

Compare. Write < or >.
Use your fraction strips.

**4a.** $\frac{3}{4}$ __?__ $\frac{1}{2}$     **b.** $\frac{2}{3}$ __?__ $\frac{5}{6}$

Estimate the fraction for the part of the set that is shaded. Write *less than* $\frac{1}{2}$ or *more than* $\frac{1}{2}$.

**5.**

Find part of the number.

**6a.** $\frac{1}{2}$ of 18 = __?__    **b.** $\frac{1}{4}$ of 16 = __?__

Write the mixed number.

**7a.**     **b.**

Use fraction strips to model each. Then find the sum or difference.

**8a.** $\frac{1}{6} + \frac{4}{6} =$ __?__    **b.** $\frac{3}{8} + \frac{4}{8} =$ __?__

**9a.** $\frac{2}{5} + \frac{2}{5} =$ __?__    **b.** $\frac{1}{3} + \frac{1}{3} =$ __?__

**10a.** $\frac{7}{8} - \frac{3}{8} =$ __?__    **b.** $\frac{3}{4} - \frac{1}{4} =$ __?__

**11a.** $\frac{9}{10} - \frac{3}{10} =$ __?__    **b.** $\frac{5}{7} - \frac{2}{7} =$ __?__

**12.** Of 18 bikes, $\frac{1}{3}$ have a racing stripe. How many have a racing stripe?

Use this circle graph to answer each question.

**13.** What fractional part of the students likes plums?

**14.** What fractional part of the students likes apples or pears?

**Favorite Fruits**

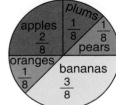

# Chapter 13

## Practice 13-1

Write the decimal.

**1a.** $\frac{3}{10}$     **b.** $\frac{5}{10}$     **c.** $\frac{2}{10}$

**2a.** four tenths     **b.** eight tenths

**3a.** $1\frac{2}{10}$     **b.** $3\frac{1}{10}$     **c.** $4\frac{7}{10}$

**4.** two and nine tenths

**5a.** $\frac{23}{100}$     **b.** $7\frac{41}{100}$     **c.** $\frac{6}{100}$

**6.** thirty-nine hundredths

Write the word name for each.

**7a.** $\frac{7}{10}$     **b.** 0.8     **c.** 0.1

**8a.** $\frac{4}{100}$     **b.** 0.05     **c.** 8.2

**9a.** $1\frac{9}{10}$     **b.** $7\frac{32}{100}$     **c.** 8.43

Compare. Write $<$ or $>$.

**10a.** 0.9 _?_ 0.6     **b.** $\frac{3}{10}$ _?_ 0.4

Order from least to greatest.

**11a.** 0.6, 0.9, 0.2     **b.** 0.7, 0.1, 0.4

**12a.** 1.9, 1.3, 1.5     **b.** 7.6, 7.8, 7.5

Find the sum.

**13a.** $\begin{array}{r} 0.3 \\ +0.4 \end{array}$    **b.** $\begin{array}{r} 0.36 \\ +0.22 \end{array}$    **c.** $\begin{array}{r} 6.84 \\ +2.05 \end{array}$

**14a.** $\begin{array}{r} 0.5 \\ +0.9 \end{array}$    **b.** $\begin{array}{r} 0.58 \\ +0.57 \end{array}$    **c.** $\begin{array}{r} 1.7 \\ +9.3 \end{array}$

**15a.** $\begin{array}{r} 8.95 \\ +4.33 \end{array}$    **b.** $\begin{array}{r} 6.32 \\ +9.84 \end{array}$    **c.** $\begin{array}{r} 52.95 \\ +13.45 \end{array}$

Find the difference.

**16a.** $\begin{array}{r} 0.8 \\ -0.3 \end{array}$    **b.** $\begin{array}{r} 0.75 \\ -0.42 \end{array}$    **c.** $\begin{array}{r} 9.66 \\ -3.04 \end{array}$

**17a.** $\begin{array}{r} 2.04 \\ -1.05 \end{array}$    **b.** $\begin{array}{r} 8.23 \\ -4.75 \end{array}$    **c.** $\begin{array}{r} 73.00 \\ -51.42 \end{array}$

**18a.** $8.01 - 2.55$     **b.** $6.20 - 3.71$

# Chapter 14

## Practice 14-1

Write the number in standard form.

**1a.** 3 million     **b.** 5 million

**2.** one million, two hundred thousand, ten

Add or subtract. Watch for $+$ and $-$.

**3a.** $\begin{array}{r} 49,128 \\ +16,493 \end{array}$    **b.** $\begin{array}{r} 81,423 \\ -\ \ 3,987 \end{array}$

**4a.** $\begin{array}{r} 37,982 \\ +29,416 \end{array}$    **b.** $\begin{array}{r} 59,465 \\ -21,846 \end{array}$

Complete.

**5a.** $\begin{array}{r} 6\square5 \\ +\square7\square \\ \hline 999 \end{array}$    **b.** $\begin{array}{r} \square9\square \\ -7\square6 \\ \hline 197 \end{array}$

Compute.

**6a.** $16 - 3 \times 4$     **b.** $6 - 2 \times 2$

**7a.** $12 \div 2 \times 3$     **b.** $15 \div 3 \times 5$

**8a.** $\begin{array}{r} \$4.36 \\ \times\ \ \ 4 \end{array}$    **b.** $\begin{array}{r} \$7.06 \\ \times\ \ \ 5 \end{array}$

**9a.** $3\overline{)\$6.39}$     **b.** $4\overline{)\$8.36}$

Write $+$, $-$, $\times$, or $\div$ to complete.

**10a.** $9 \bigcirc 9 = 81$     **b.** $9 \bigcirc 3 = 3$

**11a.** $5 \bigcirc 4 = 9$     **b.** $18 \bigcirc 9 = 9$

List all the common factors.

**12a.** 4 and 12     **b.** 5 and 13

## TEST 1

Compute.
**1a.** $7 + \underline{\ ?\ } = 15$    **b.** $13 = 6 + \underline{\ ?\ }$
**2.** $10 - \underline{\ ?\ } = 9$    **3.** $6 = 12 - \underline{\ ?\ }$
**4.** $8¢ + \underline{\ ?\ } = 10¢$    **5.** $9¢ - \underline{\ ?\ } = 7¢$

Compare. Write $<$, $=$, or $>$.
**6.** $9 - 3 \ \underline{?}\ 8 - 2$
**7.** $5 + 3 \ \underline{?}\ 9 - 2$
**8.** 5 thousand 2 $\underline{\ ?\ }$ 5007
**9.** 3 half-dollars, 2 nickels $\underline{\ ?\ }$ $1.70
**10.** 4 quarters, 5 dimes $\underline{\ ?\ }$ $1.50
**11.** How many odd numbers are between 163 and 175?

**12.** Beth bought 17 muffins and Amy bought 8 muffins. How many more muffins does Amy need to have as many as Beth?
**13.** Gil had 5 stamps from Brazil and double that number from Chile. How many stamps did he have in all?
**14.** Leo's plant is taller than Alan's. Leo's plant is shorter than Ryan's. Who has the shortest plant?
**15.** Carly had 20 pennies. With each step she took, she dropped 3 pennies. She has 8 pennies left. How many steps did she take?

## TEST 2

Compute.
**1a.** $300 + 40 + 8$    **b.** $4000 + 50 + 2$
**2.** $127 - 30$    **3.** $2000 - 90$
**4.** $19.55 + $.45$    **5.** $10.00 - $1.75$

Compare. Write $<$, $=$, or $>$.
**6.** $153 + 50 \ \underline{?}\ 450 - 200$
**7.** $17,378 \ \underline{?}\ 17,830$
**8.** 7 dimes $\underline{\ ?\ }$ 14 nickels
**9.** 3 half-dollars $\underline{\ ?\ }$ 7 dimes

Order from greatest to least.
**10.** 265, 2650, 2615, 26
**11.** 101, 11, 110, 1101

**12.** John added $29 + 36$ on his calculator. When 100 came up on the display, he knew he forgot to clear the calculator. What was the original amount on the calculator?
**13.** Fill in the missing addend.
$32 + 19 + \underline{\ ?\ } = 116$
**14.** Lucia gave 32 stickers to a friend. Then she had 14 stickers left. How many stickers did she have in the beginning?
**15.** A book and a pen cost $2.00 in all. The book cost $1.40 more than the pen. What did the pen cost?

## TEST 3

Compute.
**1.** 5 tens + 13 ones
**2.** $265 + 374$
**3.** 24 hundreds $= \underline{\ ?\ }$ thousands $\underline{\ ?\ }$ hundreds
**4.** $6425 + 1796$
**5.** 2 quarters, 1 dime, 3 nickels

Compare. Write $<$ or $>$.
**6.** $4 \times 2 \ \underline{?}\ 3 \times 4$
**7.** $2 \times 9 \ \underline{?}\ 4 \times 4$
**8.** 1 dollar, 1 quarter $\underline{\ ?\ }$ 6 quarters
**9.** Round $8.32 to the nearest dollar.
**10.** Round $2.38 to the nearest dime.

**11.** Nan has 12 children behind her and 11 girls and 13 boys ahead of her. What place is Nan in line?
**12.** Akemi has 67 marbles in one bag and 42 in another. Round the numbers to tell about how many marbles he has in all.
**13.** Lucio gave 10 stickers to a friend. He had 14 left. How many did he have at first?
**14.** Make this statement true.
$24 - \underline{\ ?\ } = 16 + \underline{\ ?\ }$
**15.** Of 1245 tickets, 956, 175, and 36 were sold on 3 days. How many tickets are left?

455

## TEST 4

Compute.
1. 75 − 56
2. 723 − 259
3. 8 × 2 + 3
4. 2 + 4 × 4
5. $7.00 − $2.34
6. $8.37 − $3.99

Compare. Write < or >.
7. 5934 + 325 _?_ 352 + 5903
8. 1875 − 250 _?_ 1485 + 165

Estimate. Round to the nearest thousand.
9. 6251 + 2798
10. 9329 − 6980
11. How many groups of 5 can you make with 35 pennies? 50 pennies?

12. Tom's pencil case holds 7 pencils. Julie's case holds twice as many. How many pencils can they hold in all?
13. After bagging 18 rolls, Mac counted 14 left. How many were there at first?
14. Lan stacks 7 rows of cans in a pattern for a store display. She puts 31 cans on the bottom row, 27 on the next row, and 23 on the next. How many cans will be on the top row?
15. Shannon has 75¢ in nickels and dimes. If there are 10 coins, how many of each coin does she have?

## TEST 5

Compute.
1. 3 × _?_ = 15
   _?_ ÷ 3 = 5
2. 4 × 7 = _?_
   28 ÷ _?_ = 7
3. 27 ÷ 3 − 3 = _?_
4. 30 ÷ 5 − 6 = _?_
5. (32 ÷ 8) + (6 × 5) = _?_
6. 3 × 4 × 2 = _?_
7. 20¢ × 4 = _?_

Use +, −, ×, or ÷ to make the sentence true.
8. 9 ___ 4 = 40 ___ 8
9. 56 ___ 8 = 6 ___ 1

Compare. Use < or >.
10. 6 × 9 _?_ 8 × 7
11. 72 ÷ 9 _?_ 25 − 16

12. Ernie wants to buy a train set for $18.95 and a model car for $7.89. Is $28.00 enough?
13. On Monday the grass is 2 cm high. By Tuesday the grass is 4 cm high. Each day the grass doubles its height from the day before. How high will the grass be on Friday?
14. Mary bought 5 packages of hair bows, each containing 1 dozen bows. If she gave away 15 bows, how many bows does she have left?
15. Rick and Spot are pet dogs. Rick is 3 years old and Spot is 12. How old will each dog be when Spot is twice as old as Rick?

## TEST 6

Use your ruler to draw a line for each length.
1. $7\frac{1}{2}$ in.
2. $4\frac{1}{4}$ in.
3. $8\frac{3}{4}$ in.

Compare. Use < or >.
4. 3 pt _?_ 2 qt
5. 2 gal _?_ 6 qt
6. 12 oz _?_ 1 lb
7. 2 m _?_ 1 km

Compute.
8. 10 kL = _?_ L
9. 20 kg = _?_ g
10. 40 km = _?_ cm

11. How many ways can Jessie, Than, Jake and Katie stand in line?
12. Would you use a pictograph or a bar graph to compare the heights of 6 people?

Use the graph to answer the questions.

**Favorite Flower**

| Violet | ⚇⚇⚇ |
|--------|------|
| Tulip | ⚇⚇⚇⚇⚇ |
| Rose | ⚇⚇⚇⚇⚇⚇ |
| Daisy | ⚇⚇ |
| Key: Each ⚇ = 4 children. | |

13. How many children like roses?
14. How many more children like violets than like daisies?
15. How many children have named their favorite flowers?

**TEST 7**

Compute.
1. $6 \times 7 = \underline{\ ?\ }$
   $42 \div \underline{\ ?\ } = 7$
2. $8 \times 6 = \underline{\ ?\ }$
   $\underline{\ ?\ } \div 8 = 6$
3. $9 \div 1 - 3 \times 2$
4. $8 \div 8 + 6 \times 6$

Compare. Use $<$ or $>$.
5. $3 \times 1 \times 6\ \underline{\ ?\ }\ 2 \times 3 \times 7$
6. $2 \times 4 \times 4\ \underline{\ ?\ }\ 4 \times 1 \times 4$

Estimate. Round to the nearest hundred.
7. $6532 + 3793$
8. $9291 - 8583$

Choose the better answer.
9. The length of a nail: 4 cm or 4 m?
10. A lump of sugar: 1 kg or 1 g?

11. Friday at 9 A.M. the thermometer read 63°F. Saturday at 9 A.M. the temperature reads 8° higher. What is Saturday's temperature?
12. At what Celsius temperature could you go ice-skating?
13. It is 150 m to Louisa's house. If Sally has already gone 80 m, how much further does she have to go?
14. It takes Mr. Perez 2 hours to drive to the beach. If he leaves his home at 10:15 A.M., at what time will he reach the beach?
15. If today is Monday, January 1st, what day of the week will it be on January 22nd?

**TEST 8**

Compute.
1. $\$.72 \div 9 = \underline{\ ?\ }$
2. $\$5.60 \div 7 = \underline{\ ?\ }$
3. $2 \times 10 \times 3 = \underline{\ ?\ }$
4. $10 \times 6 - 5 = \underline{\ ?\ }$
5. $623 + 284 = \underline{\ ?\ }$
6. $723 - 259 = \underline{\ ?\ }$

Complete.
7. $8 \times 5 = 40$ is to $40 \div 5 = 8$
   as $7 \times 4 = 28$ is to $\underline{\ ?\ }$
8. $7 + 9 = 16$ is to $16 - 7 = 9$
   as $8 + 5 = 13$ is to $\underline{\ ?\ }$

Give the total money value.
9. 3 quarters, 5 dimes, 3 nickels
10. 9 quarters, 8 dimes, 2 nickels, 4 pennies

11. If I can do 60 push-ups in a half hour, how many can I do in 15 minutes?
12. Bob begins work at 11:00 P.M. and works 8 hours. What time does he finish?
13. Jim's boat is between 30 meters and 40 meters long. It is longer than $6 \times 6$. It is an even number. How long is Jim's boat?
14. Mr. Sanchez wishes to store 72 boxes on 6 shelves. If every shelf will have the same number of boxes, how many will he store on each shelf?
15. Tickets to the ball game are 2 for $5.00. How many people can go to the game for $40.00?

**TEST 9**

1. Make a line segment 4 cm long.
2. Which is a ray? ⟷  •⟶
3. Which is a polygon? ▭  ◯
4. Which is a right angle? ⌐  ∠
5. Which shapes are congruent?
   a. ▱  b. ◯  c. ☁  d. ▱
6. Which shapes are similar?
   a. ☆  b. ◇  c. ☆  d. ♣

Is each a line of symmetry?
7.
8.

Is the figure a slide, flip, or turn?
9.
10. ▱▱

11. Draw three different figures that have a perimeter of 24 cm. Name each figure and label each side.
12. If the perimeter of a square is 36 m, what is the length of each side.?
13. Which has the larger area, a rectangle that measures 4 m by 6 m, or a square which is 5 m on each side?
14. On a grid, plot the following points: (1,1), (5,1), (5,4), and (1,4). Connect them in order. What figure did you draw? What is the area?
15. Which has a greater volume: a building with 4 floors and 4 rooms on each floor or a building with 8 floors and 2 rooms on each floor?

**TEST 10**

Estimate by rounding.
1. $4 \times 26$    2. $8 \times 263$    3. $9 \times 175$

Compute.
4. $5 \times 716$    5. $6 \times 324$    6. $5 \times 209$
7. $9 \times \$8.32$    8. $6 \times \$3.59$

Compare. Use $<$ or $>$.
9. $24 \div 6 \underline{\ ?\ } 48 \div 8$    10. $4\overline{)72} \underline{\ ?\ } 3\overline{)57}$

11. Pete has \$23.87. About how many baseball cards can he buy if the cards sell 4 for \$2.00?

12. About how many books are in the 4 sections of a library if there are 265 books in each section?
13. There are 25 children in line for the roller coaster. If 2 children sit in each car, how many cars will be needed?
14. Pat pastes 4 large beans and 3 small beans on each picture. When she has used 28 large beans, how many small beans has she used?
15. What is the sum of 15 and the product of 6 and 7?

---

**TEST 11**

Compute.
1. $\frac{5}{8} = \frac{?}{40}$    2. $\frac{3}{5} = \frac{?}{25}$    3. $\frac{8}{9} = \frac{?}{27}$

Compare. Use $<$ or $>$.
4.     5.

$\quad \frac{2}{3} \underline{\ ?\ } \frac{1}{3}$    $\quad \frac{1}{4} \underline{\ ?\ } \frac{1}{2}$

6. $\frac{5}{8} + \frac{1}{8} \underline{\ ?\ } \frac{1}{4} + \frac{3}{4}$    7. $\frac{7}{9} - \frac{2}{9} \underline{\ ?\ } \frac{2}{3} - \frac{1}{3}$
8. $\frac{1}{4}$ of $40 \underline{\ ?\ } \frac{1}{2}$ of $10$    9. $\frac{1}{3}$ of $27 \underline{\ ?\ } \frac{1}{6}$ of $18$

Write the mixed number.
10. 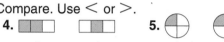 ?

11. Sam can type 64 words per minute. At this rate how many words can he type in $\frac{1}{2}$ an hour?
12. Anna made 18 sandwiches. She took $\frac{2}{3}$ of them to class. How many did she not take?
13. The recipe calls for $2\frac{1}{3}$ c of flour. If Dana plans to make a double batch of cookies, how much flour will she need?
14. Frank bakes 12 cupcakes. Angela bakes 3 times as many as Frank. Together they sell 25 cupcakes, how many are left?
15. Thirty people visited the Liberty Bell the first hour, 35 in the second hour, and 40 in the third hour. At this rate, how many visited the Bell during the seventh hour?

---

**TEST 12**

Change each to a decimal.
1. $\frac{7}{10}, \frac{8}{10}, 3\frac{4}{10}, 8\frac{1}{10}, 9\frac{6}{10}$    2. $\frac{1}{100}, \frac{34}{100}, \frac{52}{100}, \frac{92}{100}, \frac{73}{100}$

Compare. Write $<$ or $>$.
3. $0.05 + 0.35 \underline{\ ?\ } 0.04 + 0.42$
4. $0.56 + 0.1 \underline{\ ?\ } 0.51 + 0.3$
5. $0.75 - 0.35 \underline{\ ?\ } 0.72 - 0.49$
6. $0.32 - 0.01 \underline{\ ?\ } 0.23 - 0.10$

Fill in missing digits.
7. $\quad 6\square 2 1$
$\quad +\ 2 9 \square 9$
$\quad \overline{\square 3 5 \square}$

8. $\quad \$2\square 3\square$
$\quad \times \qquad 5$
$\quad \overline{\$\square\square,1 8 0}$

9. $5\overline{)705}$    10. $8\overline{)\$131.36}$

11. Jim ran 8.2 miles. Darcy ran 6.9 miles. Write a number sentence to tell how much longer Jim's run was than Darcy's.
12. Choose the numbers that are divisible by 2, 5, and 10: 8, 30, 15, 50, 10, 36.
13. Thuy baked 3 batches of 12 cookies each. Nina baked 4 batches of 11 cookies each. Ken baked 96 cookies in all. How many more cookies did Ken bake than Thuy and Nina together?
14. Use $+, -, \times,$ or $\div$ to make this sentence true. $20 \underline{\ ?\ } 5 = 240 \underline{\ ?\ } 60$
15. The third grade students wanted to raise \$75.00. On Saturday they made \$48.75 and on Sunday, \$40.32. Did they reach their goal? How much more or less did they make?

**SET 1**

1. 9 + 1; 7 + 2; 4 + 2; 5 + 3; 8 + 0
2. 9 − 2; 8 − 3; 10 − 4; 7 − 5; 6 − 4
3. Add 3 to: 4, 14, 24, 34, 44, 54, 64, 74, 84, 94
4. Subtract 2 from: 7, 17, 27, 37, 47, 57, 67, 77, 87, 97
5. Double: 2, 4, 6, 8, 1, 3, 5
6. _?_ nickels = 1 dime

7. Brad had 12 cows. He sold 5 of them. How many are left?
8. April has 4 shells and Babs has 3 shells. How many shells do they have in all?
9. A box holds 9 plums and 2 pears. Another holds 2 plums and 9 pears. How much fruit is in each box?
10. What number added to 7 is 14?

**SET 2**

1. 12 − 5; 11 − 6; 13 − 6; 17 − 9; 15 − 8
2. 8 + 4; 9 + 3; 8 + 5; 7 + 6; 9 + 5
3. Add 5 to: 12, 22, 32, 42, 52, 62, 72, 82, 92, 102
4. Subtract 6 from: 7, 17, 27, 37, 47, 57, 67, 77, 87, 97
5. Give 4 facts: 12, 5, 7; 16, 9, 7; 13, 4, 9; 17, 8, 9; 15, 8, 7; 14, 6, 8
6. Ami collected 13 dolls. Sue has 8. How many more dolls does Ami have?

7. Lyn has 8 white and 7 red roses. How many roses does she have in all?
8. _?_ pennies = 1 quarter
9. Todd needs 12 pieces of wood. He has 9 pieces. How many more pieces does he need?
10. Carlos has 6 pencils. Leo has double the number of pencils. How many pencils does Leo have?

**SET 3**

1. Even or odd? 7, 34, 16, 51, 67, 58
2. Order: 45, 450, 145; 118, 89, 189; 676, 760, 67; 330, 303, 333; 191, 19, 119
3. _?_ tens _?_ ones: 82, 6, 35, 97
4. Find the sums for 14: 5 + _?_ ; 6 + _?_ ; 7 + _?_ ; 8 + _?_ ; 9 + _?_ ; 10 + _?_ ; 11 + _?_ ; 12 + _?_
5. Add 2 to: 3 + 1; 4 + 2; 5 + 3; 2 + 7; 1 + 4; 3 + 2; 5 + 4; 6 + 2; 7 + 3

6. Complete the pattern: 5, 105, 205, 305, _?_ , _?_ , _?_
7. Teresita is 3 years old. Dad is 30 years older. How old is Dad?
8. Carly is twelfth in line. How many children are in front of her?
9. Ann has 1 red, 4 white, and 5 blue cars. How many cars does she have in all?
10. Which number is larger: 500 + 60 + 1 or 500 + 10 + 6?

**SET 4**

1. 6 + 8; 7 + 7; 9 + 5; 5 + 9; 6 + 9; 6 + 7; 8 + 6; 2 + 5; 7 + 6; 7 + 8
2. Count by 2, by 5, and by 10: 10 – 40; 30 – 70; 110 – 140; 270 – 300; 475 – 600
3. Give the value: 25, 17, 91, 106, 234, 165, 290, 6
4. Round: 18, 24, 47, 62, 28, 83, 32, 45, 69, 51, 11, 74, 97
5. How many tens in: 14, 54, 72, 103, 37, 116, 81, 125, 200?

6. Kyle has 2 quarters, 1 dime, and 3 nickels. How much does he have?
7. Ann bought one item for 9¢ and one item for 5¢. How much did she spend?
8. Popcorn costs 99¢. If Tim has 9 dimes, can he buy the popcorn?
9. Give the standard form: 7000 + 600 + 30 + 2
10. A ball costs 25¢. I pay with $1.00. How much change will I get?

**SET 5**

1. Add 6¢ to: 22¢, 32¢, 42¢, 52¢, 62¢, 72¢, 82¢
2. Subtract 5¢ from: 21¢, 31¢, 41¢, 51¢, 61¢, 71¢, 81¢, 91¢
3. Count by 100 from: 230 to 930, 102 to 902, 16 to 1016
4. ? tens ? ones: 16 ones; 12 ones; 2 tens 17 ones; 3 tens 11 ones; 18 ones
5. >, <: 67 ? 76; 111 ? 101; 365 ? 395; 243 ? 23; 515 ? 551

6. Dot sold lemonade for 10¢ a glass. How much did she make if she sold 6 glasses?
7. How many quarters are in $3.00?
8. John has 63 white buttons and 26 colored ones. About how many buttons does he have?
9. Trudy has 8 tomatoes. Beth has 4. How many fewer tomatoes does Beth have than Trudy?
10. Write the standard numeral for: twenty-four thousand, four hundred.

**SET 6**

1. Estimate: 36 + 43; 68 + 14; 17 + 12; 28 + 33; 56 + 49
2. ? hundreds ? tens: 15 tens; 13 tens; 3 hundreds 12 tens; 20 tens; 5 hundreds 14 ones; 16 tens
3. Add 8 to: 18, 28, 38, 48, 58, 68, 78, 88, 98
4. Add: 4 + 5 + 5; 7 + 3 + 7; 5 + 6 + 4; 3 + 8 + 2; 2 + 9 + 1
5. Round to the nearest dollar: $1.23, $4.68, $1.87, $2.49, $3.41

6. $1.87 = ? dimes, ? pennies
7. Jan has 16 green, 24 blue, and 12 red marbles. How many marbles does she have in all?
8. About how many dollars: $1.58 + $3.24?
9. Mother baked 14 butter cookies and 16 spice cookies. How many did she bake in all?
10. Kim collected 389 pennies. Can she buy a toy that costs $3.09?

1. 8 + 6;  7 + 6;  9 + 3;  6 + 4;
   14 − 8; 13 − 7; 12 − 9; 10 − 6
2. Add 5 to: 15, 25, 45, 75, 95, 55, 85,
   5, 35, 65
3. Subtract 7 from: 9, 19, 29, 39, 49,
   79, 59, 69, 89, 99
4. ? hundreds ? tens: 21 tens;
   18 tens; 4 hundreds 2 tens;
   7 hundreds 3 tens; 25 tens
5. Add: 8 + 1 + 2; 3 + 6 + 7;
   5 + 5 + 5; 6 + 1 + 4; 7 + 7 + 3;
   9 + 7 + 1

6. To 19 books Tim added 21 more.
   How many books are there?
7. The number before 221 is ? .
8. We saw 12 crows and 8 robins.
   How many birds did we see?
9. On Monday 3300 tickets were sold
   and on Tuesday 6200. How many
   tickets were sold in all?
10. The library has 135 books on
    sports, 430 on history, and 210 on
    science. How many books are in
    the library?

1. Subtract 2 from: 10, 20, 30, 40, 50,
   60, 70, 80, 90, 100
2. Estimate: 26 − 13; 46 − 17;
   55 − 21; 61 − 37; 78 − 49;
   33 − 11
3. Regroup tens to ones:
   4 tens 5 ones; 1 ten 6 ones;
   2 tens 8 ones
4. 17 − 6; 12 − 9; 15 − 6; 13 − 9;
   15 − 6; 11 − 8; 14 − 8; 16 − 9
5. Round to the nearest dollar:
   $1.85, $2.10, $3.45, $4.76, $5.99

6. Bob has 28 roses. Tess has 14.
   How many fewer roses does
   Tess have?
7. The number before 300 is ? .
8. We had 24 balloons but 7 of them
   burst. How many were left?
9. One year's rainfall was 90 in. In
   May, 21 in. fell. How many inches
   fell the rest of the year?
10. There were 7500 people at a
    game. Only 5500 stayed to the
    end. How many people went
    home early?

1. Estimate: 180 − 120; 560 − 240;
   450 − 130; 490 − 380; 320 − 310
2. Which is greater: $45.00 or $4500;
   $3.90 or $390; $270 or $27.00;
   $.54 or $54.00; $6200 or $6.20?
3. Subtract 3 from: 21, 71, 51, 11, 41,
   91, 31, 61, 81, 101
4. Subtract 6 from: 10, 20, 30, 40, 50,
   60, 70, 80, 90, 100
5. Regroup thousands to hundreds:
   2 thousands 3 hundreds;
   6 thousands 4 hundreds;
   8 thousands 1 hundred

6. $4.30 = ? dimes
7. Dan had $7.60. He bought a
   toy for $3.00. How much did he
   have left?
8. Harry is 20 years old. Gail is 13.
   How much older is Harry?
9. Stacy is 46 inches tall. Gwen is
   8 inches shorter than Stacy. What
   is Gwen's height?
10. Together Bill and Jill have 48
    stickers. If Bill has 22, how many
    does Jill have?

1. Solve: 4 twos, 5 threes, 3 twos, 2 fives, 2 sixes
2. Multiply by 1 and by 0: 3, 5, 2, 1, 6, 8, 0, 7, 9
3. Double these numbers: 2, 4, 6, 8, 1, 3, 5, 7, 9
4. Multiply by 2, by 3, and by 4: 9, 2, 4, 6, 8, 5, 0, 3
5. Give the time a half hour later than: 11:00, 4:00, 2:30, 10:30, 12:15, 1:15, 3:45, 4:45

6. One cookie costs 8 cents. How much do 2 cookies cost?
7. Ron walks 3 mi a day. How many miles does he walk in 8 days?
8. Dot has 4 apples. Her brother has 3 times as many. How many apples does he have?
9. The product is 24. One factor is 4. What is the other factor?
10. Roy has 9 tickets. If Matilda gives him 7 more, how many tickets will he have in all?

1. Multiply by 4: 2, 3, 5, 1, 7, 8, 0, 9, 6, 4
2. Multiply by 3, by 4, and by 5: 0¢, 2¢, 4¢, 6¢, 8¢, 10¢, 9¢, 5¢, 7¢, 3¢
3. Use +, −, or ×: $8 \underline{\ ?\ } 2 = 10$; $4 \underline{\ ?\ } 2 = 8$; $11 \underline{\ ?\ } 3 = 8$; $3 \underline{\ ?\ } 2 = 6$; $4 \underline{\ ?\ } 9 = 13$
4. Add 7 to: 2, 12, 22, 42, 52, 72, 32, 62, 82, 92
5. $9 \times \underline{\ ?\ } = 27$; $4 \times \underline{\ ?\ } = 16$; $\underline{\ ?\ } \times 4 = 32$; $\underline{\ ?\ } \times 5 = 45$; $7 \times \underline{\ ?\ } = 14$; $6 \times \underline{\ ?\ } = 18$

6. Mom spent $5. Dad spent 3 times that much. How much did Dad spend?
7. There were 17 pencils. Sam took 8 of them. How many were left?
8. If 5 students are sitting in each of 8 rows, how many students are there in all?
9. The product is 0. One factor is 4. What is the other factor?
10. The product is 9. One factor is 9. What is the other factor?

1. Multiply by 2: 2, 4, 8, 1, 3, 5
2. Subtract 4 from: 28, 24, 20, 16, 12, 8, 4
3. Divide by 1 and by the number itself: 3, 6, 8, 4, 5, 2, 1, 9, 7
4. Divide by 2: 4, 8, 14, 16, 2, 10, 18, 6, 12
5. Divide by 3: 18, 12, 3, 9, 21, 27, 15, 6, 24
6. If one factor is 6 and the other 4, what is the product?

7. One class has 5 rows with 2 children in each row. Another class has 2 rows with 5 children in each. How many children are in each class?
8. Eve has 2 pens. Ed has 4 times as many. How many does Ed have?
9. Three girls have 7¢ each. What is the total amount?
10. The product is 14. What are the factors?

1. Divide by 4: 36, 16, 20, 4, 12, 24, 8, 28, 32
2. Divide by 5: 25, 5, 40, 35, 10, 45, 15, 20, 30
3. Subtract 3 from: 27, 24, 21, 18, 15, 12, 9, 6, 3
4. Multiply by 5 and by 4: 1, 3, 5, 7, 9, 2, 4, 6, 8, 10
5. Count by 2 and by 4: 8–20; 12–28; 4–16; 16–36; 12–32

6. Pencils come 2 to a box. How many boxes are needed for 10 pencils?
7. Write the missing numbers: 4, _?_, 12, _?_, 20, 24, _?_, _?_, 36
8. Jo buys 5 balloons at 8¢ each. How much does Jo spend?
9. There are 5 children in each row. How many children are in 6 rows?
10. Rex had 32¢. He spent 4¢. How much did he have left?

1. Multiply by 6, then add 1: 3, 6, 2, 9, 0, 1, 4, 7, 8
2. $3)\overline{9}$  $2)\overline{12}$  $4)\overline{8}$  $5)\overline{15}$
   $2)\overline{16}$  $3)\overline{27}$  $5)\overline{25}$  $4)\overline{20}$
3. Give the related fact: $8 \div 2 = 4$; $28 \div 4 = 7$; $14 \div 2 = 7$; $9 \div 3 = 3$
4. Subtract 8 from: 11, 13, 15, 17, 12, 14, 16, 10, 18, 19
5. $8 + 7$  $4 + 7$  $7 + 9$  $5 + 6$
   $6 + 9$  $8 + 6$  $5 + 6$  $9 + 8$
6. How many twos are there in 18?

7. A horse trainer has 17 black horses and 8 brown horses. How many horses does he have in all?
8. The dividend is 35. The divisor is 5. What is the quotient?
9. Kim read 36 pages of her book. Chip read only 25 pages. How many more pages had Kim read than Chip?
10. The quotient is 3. The divisor is 2. What is the dividend?

1. Add 3 to: 8, 18, 28, 38, 48, 58, 68, 78, 88, 98
2. Divide by 5: 5, 15, 25, 35, 45, 10, 20, 30, 40, 50
3. $21 \div 3$  $16 \div 4$  $2 \div 2$  $4 \div 1$
   $0 \div 3$  $27 \div 3$  $5 \div 5$  $5 \div 1$
4. Subtract 4 from: 11, 12, 13, 14, 10, 9, 8, 7, 6, 5
5. $8 \div \underline{?} = 4$  $12 \div \underline{?} = 3$
   $15 \div \underline{?} = 5$  $25 \div \underline{?} = 5$
   $36 \div \underline{?} = 9$  $24 \div \underline{?} = 8$

6. How many ways can a bird, a mouse, and a cat sit on a fence?
7. Key: Each ♂ = 2: ♂ ♂ ♂ ♀ How many people are there?
8. Each box = 5 points: ▯▯▯▯ How many points are there?
9. A jar is filled with 5 red marbles. What is the probability of choosing a yellow marble? a red marble?
10. Team tallies for each quarter are ЖII, ЖIII, III, Ж. What is the total team score?

1. Add 12 to: 12, 24, 36, 48, 60, 72, 84
2. Multiply by 2 and by 4: 3, 6, 9, 0, 4, 2, 5, 8, 1, 7
3. Multiply by 3, then add 2: 3, 6, 9, 1, 7, 4, 8, 5, 0, 2
4. Compare using $<, =, >$:
   4 qt ? 1 gal; 2 c ? 2 qt; 3 pt ? 1 qt
5. Count by 100: 0 to 800; 200 to 900; 100 to 1000; 150 to 750

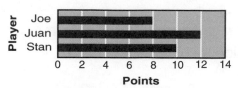

6. How many points did Juan score?
7. How many fewer points did Joe score than Stan?
8. How many more points did Juan score than Joe?
9. If Joe scores 6 more points, how many points will he have?
10. To make 20, how many points does Juan need?

1. Add 16 to: 32, 80, 64, 16, 48, 51
2. Which month is: third, first, 5th, seventh, 10th, twelfth?
3. ? $\times 8 = 64$   ? $\times 6 = 42$
   ? $\times 5 = 20$   ? $\times 7 = 14$
   ? $\times 9 = 36$   ? $\times 7 = 35$
4. $4\overline{)36}$  $4\overline{)28}$  $4\overline{)16}$  $4\overline{)8}$
   $4\overline{)20}$  $4\overline{)32}$  $4\overline{)12}$  $4\overline{)4}$
5. Add 9 to: 9, 8, 7, 6, 5, 4, 3, 2, 1

6. 2 lb = ? oz
7. Choose the customary unit of measure for the length of: scissors, walls, pens, rugs.
8. Sally has 1 meter of ribbon. How many centimeters is that?
9. John bought a pound of tea. How many ounces of tea did he buy?
10. 24 students went to the zoo. The trip was 15 mi each way. How many miles did they travel?

1. Divide by 3: 12, 18, 9, 6, 27, 21, 3
2. Multiply by 6 and by 7: 6, 4, 9, 1, 8, 2, 5, 7, 3
3. Add 7 to: 8, 18, 28, 38, 48, 58, 68, 78, 88, 98
4. 
   | 6 | 7 | 6 | 8 | 9 |
   |---|---|---|---|---|
   | 2 | 4 | 3 | 2 | 1 |
   | +3 | +0 | +3 | +2 | +3 |
5. Count by 1000: 0 to 7000; 1000 to 8000; 2000 to 9000; 100 to 9100

6. Name the months with 31 days.
7. How long does it take the big hand of the clock to go from one number to the next?
8. How long does it take the small hand of the clock to go from one number to the next?
9. ? g = 1 kg
10. ? min = 1 h

1. Multiply by 3 and by 6: 2, 4, 7, 8, 1, 3, 5, 6, 0, 9
2. Divide by 7: 21, 49, 28, 56, 35, 63, 14, 42, 7
3. $1 \times 8$   $8 \times 1$   $8 \overline{)8}$    $8 \div 1$
    $2 \times 8$   $8 \times 2$   $8 \overline{)16}$   $16 \div 2$
4. Multiply by 8 and by 9: 5, 3, 6, 7, 2, 1, 8, 4, 9
5. $6 \overline{)12}$   $6 \overline{)24}$   $6 \overline{)18}$   $6 \overline{)54}$
   $6 \overline{)6}$    $6 \overline{)30}$   $6 \overline{)42}$   $6 \overline{)36}$
6. There are 4 quarts in a gallon. How many quarts are in 5 gallons?

7. In a classroom 36 desks are arranged in rows of 6 desks each. How many rows are there?
8. The baby drank 6 oz of milk twice a day and 8 oz of juice once. How much did the baby drink in all?
9. The game started at 4:00 P.M. and ended 2 hours later. What time was the game over?
10. Phil's calculator is 3 in. wide. Bill's is twice as wide. How wide is Bill's calculator?

1. Add 8 to: 8, 18, 24, 38, 40, 58, 68, 78, 88, 98
2. Subtract 9 from: 16, 26, 36, 46, 56, 66, 76, 86, 96, 106
3. $5 \times 8$   $8 \times 5$   $8 \overline{)40}$   $40 \div 5$
   $6 \times 8$   $8 \times 6$   $8 \overline{)48}$   $48 \div 6$
4. $7 \times 8$   $8 \times 7$   $8 \overline{)56}$   $56 \div 7$
   $8 \times 8$         $8 \overline{)64}$   $64 \div 8$
5. $5 \times 9$   $6 \times 9$   $7 \times 9$   $8 \times 9$
   $9 \times 5$   $9 \times 6$   $9 \times 7$   $9 \times 8$

6. Mrs. Olsen uses 6 oranges a day. How many will she use in 6 days?
7. In the summertime would it be 85° or 32° Fahrenheit?
8. Mary rode 19 km on Sunday and 12 km on Monday. How many km did she ride in all?
9. Six boys caught 18 fish. Each caught the same number of fish. How many fish did each catch?
10. _?_ nickels = 1 quarter

1. Divide by 8: 64, 8, 24, 32, 48, 56, 40
2. Divide by 9: 18, 45, 63, 9, 54, 27, 81, 72, 36
3. $2 \times 4 \times 2$   $3 \times 3 \times 3$   $1 \times 4 \times 2$
   $3 \times 2 \times 1$   $4 \times 3 \times 0$   $1 \times 0 \times 3$
4. What is the time 2 h later? 12:00, 3:00, 11:00, 10:30, 1:30
5. Multiply by 5: 10¢, 20¢, 30¢, 40¢, 50¢, 60¢, 70¢, 80¢

6. Write the Roman numeral for 46.
7. Jean has 20 pennies. Sal has 4 times as many. How much does Sal have?
8. Each girl made 7 cards. How many cards did 8 girls make?
9. Fill in the missing numbers: 8, 16, _?_, 32, _?_, _?_, 56, _?_, 72
10. Dana's kitten is 35 days old. How many weeks is that?

1. Multiply by 6, by 7, and by 8:
   8, 6, 9, 0, 5, 3, 4, 2, 1

2. $6\overline{)30}$  $6\overline{)36}$  $6\overline{)42}$  $6\overline{)48}$
   $6\overline{)24}$  $6\overline{)18}$  $6\overline{)12}$  $6\overline{)54}$

3. Add 10 to: 9, 19, 29, 39, 49, 59,
   69, 79, 89, 99

4. Double: 10, 50, 40, 30, 20, 60, 90,
   70, 80

5. Name each space figure:

6. Five ribbons are needed to make a banner. Each ribbon is 20 cm long. How many cm must be bought in all?

7. At school 18 children are playing baseball with 9 on a team. How many teams are there?

8. The product of $7 \times 7$ is __?__ .

9. How many nines are in 81?

10. __?__ minutes = one half hour

1. $3 \times 1 \times 5$  $6 \times 2 \times 0$  $2 \times 5 \times 2$
   $4 \times 2 \times 2$  $3 \times 0 \times 3$  $1 \times 7 \times 2$

2. Subtract 7 from: 12, 13, 14, 15,
   16, 17, 18, 19

3.
   ```
    5   5   6   6   7
    3   4   2   3   2
   +2  +1  +2  +1  +1
   ```

4. Divide by 8 then add 1: 8, 64, 40,
   16, 72, 56, 32

5. Divide by 9: 36, 72, 63, 9, 18, 45,
   27, 54

6. A picture frame is 8 in. by 10 in. What is its perimeter?

7. Don made a floor plan of his bedroom. How many square units does the floor cover?

8. Tasha is 4 years old. Dad is 9 times her age. How old is Dad?

9. Name two space figures with 6 faces, 12 edges, and 8 corners.

10. Can a line of symmetry be drawn through the letter **L**?

1. Multiply by 2: 10, 20, 30, 40, 50,
   60, 70

2. Estimate: $18 \times 2$; $24 \times 3$; $13 \times 4$;
   $28 \times 2$; $16 \times 5$

3. Multiply by 4 then add 1: 1, 0, 5, 3,
   6, 8, 2, 7, 9, 4

4. Name the numerator and the denominator:
   $\frac{1}{3}, \frac{2}{5}, \frac{7}{8}, \frac{3}{7}, \frac{4}{10}, \frac{3}{4}, \frac{5}{6}$

5. How many cents: 3 nickels, 5 nickels, 2 nickels, 4 dimes, 7 dimes, 2 quarters?

6. Jon has 6 coins. Two are dimes. What part of the coins are dimes?

7. To change a quarter from heads to tails do you need to do a slide, a turn, or a flip?

8. Using cubic units, Sam built a 6-floor hotel with 4 rooms on each floor. Find the volume.

9. Jen is 13 years old. Jen is 4 years older than Ned. How old is Ned?

10. The square playground measures 9 yards on each side. What is its perimeter?

1. Estimate: 12 × 2; 28 × 3; 11 × 5; 39 × 2; 23 × 4; 17 × 3
2. Subtract 1 from: 10, 20, 30, 40, 50, 60, 70, 80, 90, 100
3. 5 × 6  7 × 6  6 × 5  6 × 7  6 × 6  8 × 6  6 × 9  6 × 8
4. Divide by 7: 49, 50, 7, 9, 63, 68, 14, 20, 42, 46
5. Multiply by 10 then add 3: 0, 3, 7, 5, 4, 9, 6, 2, 8, 1

6. To find the product, you  ?  .
7. How many right angles are in a square? a right triangle?
8. A recipe calls for 2 c of sugar. To double the recipe how much sugar would you use?
9. Frozen yogurt is 25¢ a scoop. How much will a triple scoop cost?
10. Andy drove 22 mi north and 33 mi west. How far did he go?

1. Divide by 2 and name the remainder: 3, 4, 5, 6, 7, 8, 9, 10, 11, 12
2. $\underline{\ ?\ } \times 9 = 36$   $\underline{\ ?\ } \times 9 = 27$
   $\underline{\ ?\ } \times 9 = 54$   $\underline{\ ?\ } \times 9 = 45$
3. Give other names for 1:
   $\frac{2}{?}, \frac{5}{?}, \frac{?}{6}, \frac{?}{3}, \frac{4}{?}, \frac{?}{10}$
4. Subtract 7 from: 9, 19, 29, 49, 59, 79, 39, 69, 89, 99
5. Add 4 to: 5, 15, 35, 25, 45, 75, 55, 65, 85, 95

6. What is the Roman numeral for 50?
7. Name the months with 30 days.
8. XVII =  ?
9. Ed packs 12 eggs in 1 carton. How many eggs will fill 3 cartons?
10. Last year Macavoy weighed 35 kg. This year he weighs 40 kg. How many kg did he gain?

1. 12 ÷ 2   15 ÷ 3   25 ÷ 5
   16 ÷ 4   18 ÷ 9   36 ÷ 6
   28 ÷ 7   81 ÷ 9   40 ÷ 8
2. Add 6 to: 7, 10, 8, 11, 21, 33, 9, 27, 44, 101
3. Divide by 3 and by 6: 6, 18, 24, 12, 30, 60
4. Divide by 3 and name the remainder: 16, 17, 19, 21, 22, 23, 25, 26
5. Divide by 5 and name the remainder: 12, 13, 14, 18, 19, 22, 23, 36

6. Six pencils fit in a case. If there are 27 pencils, how many cases can be completely filled? How many pencils are extra?
7. School is designated by the ordered pair (4, 3) on a graph. What does that mean?
8. There are 24 cashiers and 12 people helping to bag groceries. How many workers are there in all?
9. There are 5 horses with a shoe on each hoof. How many shoes are there?
10. Al wants to buy a $.39 horn. He has $.30. How much more does he need?

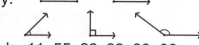
1. How many sides are in a triangle? a square? a rectangle?
2. Identify: ——  ↑ ——→
   ←——→  ∠  ∟  ∠
3. Divide by 11: 55, 22, 33, 99, 66, 44, 88, 77, 11, 110
4. Estimate: $8.12 ÷ 4; $5.85 ÷ 3; $2.24 ÷ 2; $4.90 ÷ 5
5. Give the related fact: 24 ÷ 4 = 6; 15 ÷ 3 = 5; 14 ÷ 7 = 2; 24 ÷ 8 = 3

6. Paul sold 20 chances for 5¢ each. How much money did he receive?
7. Our family drove 30 km a day. How many km did we drive in 5 days?
8. There are 54 peaches in 6 baskets. Each basket holds the same number. How many peaches are in one basket?
9. $9 \times 8$ is the same as $8 \times \underline{\ ?\ }$.
10. What is the remainder of $26 \div 7$?

1. $\frac{1}{3} = \frac{?}{6} = \frac{?}{9} = \frac{?}{12} = \frac{?}{15}$
2. Compare using < or >: $\frac{2}{2} \ ? \ \frac{1}{2}$, $\frac{1}{4} \ ? \ \frac{3}{4}$, $\frac{1}{5} \ ? \ \frac{3}{5}$, $\frac{7}{8} \ ? \ \frac{6}{8}$, $\frac{2}{7} \ ? \ \frac{5}{7}$
3. Divide by 4 and name the remainder: 9, 11, 15, 19, 23, 27, 31, 33
4. Multiply by 9: 4, 7, 2, 8, 1, 5, 9, 6, 3
5. $\frac{1}{2} = \frac{?}{4} = \frac{?}{6} = \frac{?}{8} = \frac{?}{10}$

6. Flat figures with straight sides are called $\underline{\ ?\ }$.
7. How many 9¢ treats can Dan buy with 36¢?
8. Ed has 9 rabbits. He puts a pair into each hutch. How many hutches does he fill? Are any rabbits left over?
9. Give the family of facts for: $30 \div 5 = 6$.
10. There are 66 books divided equally across 3 shelves. How many books are on each shelf?

1. $\frac{1}{6}$ of 42    $\frac{1}{6}$ of 48    $\frac{1}{6}$ of 54
   $\frac{1}{6}$ of 24    $\frac{1}{6}$ of 30    $\frac{1}{6}$ of 36
2. $\frac{1}{6} + \frac{4}{6} = \frac{?}{6}$    $\frac{3}{10} + \frac{4}{10} = \frac{?}{10}$
   $\frac{2}{5} + \frac{1}{5} = \frac{?}{5}$    $\frac{1}{4} + \frac{1}{4} = \frac{?}{4}$
3. $\frac{1}{5}$ of 10    $\frac{1}{5}$ of 20    $\frac{1}{5}$ of 45
   $\frac{1}{5}$ of 60    $\frac{1}{5}$ of 30    $\frac{1}{5}$ of 35
4. Use +, −, ×, or ÷: $5 \ ? \ 7 = 35$; $49 \ ? \ 7 = 7$; $56 \ ? \ 8 − 7$; $21 \ ? \ 8 = 29$; $16 \ ? \ 7 = 9$
5. $\frac{1}{4} = \frac{?}{8} = \frac{?}{12} = \frac{?}{16} = \frac{?}{20}$

6. Lines on a flat surface that never meet are called $\underline{\ ?\ }$.
7. Mary ate $\frac{1}{2}$ of a pizza. Sy ate $\frac{1}{4}$. Who ate more?
8. Ed colored 4 of 6 squares red. Is that more or less than half?
9. One sixth of 54 desks is $\underline{\ ?\ }$ desks.
10. First Mark ate $\frac{1}{4}$ of his apple. Then he ate another $\frac{1}{4}$. How much of the apple did he eat?

1. $\frac{3}{4} - \frac{1}{4} = \frac{?}{4}$  $\frac{4}{5} - \frac{2}{5} = \frac{?}{5}$
   $\frac{7}{8} - \frac{2}{8} = \frac{?}{8}$  $\frac{5}{7} - \frac{4}{7} = \frac{?}{7}$

2. Read: 0.34; 0.62; 0.73; 0.02; 0.40; 0.07; 0.96

3. Yes or No? $1.2 = 1\frac{2}{10}$; $2.1 = 2\frac{5}{10}$;
   $3.7 = 3\frac{7}{10}$; $4.5 = 4\frac{5}{10}$; $1.9 = 1\frac{6}{10}$

4. Read: 7.3; 6.8; 5.4; 2.9; 1.7; 3.5; 4.1

5. Add 1.1 to: 2.8, 6.4; 1.7; 8.6; 3.4; 5.3; 9.2; 4.6

6. Mother bought 1.1 yd of blue material and 2.5 yd of red. How much material did she buy?

7. Dan and Sal were riding bikes. Dan rode 2.3 mi. Sal rode 1.9 mi. Who rode farther?

8. There are 12 pieces of fruit in a basket. One third of the fruit are apples. How many are apples?

9. Which is greater: 0.08 or 0.8?

10. The sum of $.19 and $.04 is  ? .

1. 0.4, 1.4, 2.4,  ? ,  ? ,  ?

2.  ?  + 11 = 44;  ?  − 10 = 67;
    ?  × 2 = 40;  ?  ÷ 3 = 9;
   150 +  ?  = 250; 320 −  ?  = 20;
   5 ×  ?  = 55

3. Add 10,000 to: 7000; 80,000; 4000; 30,000; 6000; 50,000

4. Subtract 2.2 from: 4.6, 2.9, 6.3, 9.5, 7.4, 8.7, 5.5

5. How many equal parts? $\frac{7}{9}$; $\frac{1}{5}$; $\frac{2}{7}$; $\frac{4}{9}$; $\frac{5}{6}$

6. Tanya had $\frac{1}{2}$ c of sugar. She used $\frac{1}{4}$ c. How much is left?

7. Penguins live on a 6-mile-long island. Its width is $\frac{1}{3}$ of its length. How wide is the island?

8. Four boxes of computer paper weigh 36 kg. How much does each box weigh?

9. Than's zoo ticket costs $6.75. What is his change from $10?

10. Find the sum of 21,000 and 48,300.

1. (9 × 2) + 1  (4 × 3) + 3
   (7 × 5) + 2  (6 × 2) + 4

2. (8 ÷ 2) − 3  (12 ÷ 6) − 2
   (24 ÷ 3) − 5  (28 ÷ 4) − 7

3. Give the value: 3256, 1598, 24,000, 598, 134, 520

4. Multiply by 7: 10¢, 20¢, 60¢, 30¢, 50¢, 90¢, 80¢, 40¢, 70¢

5. Which are divisible by 2, by 5, and by 10? 20, 12, 15, 50, 35, 40

6. The stadium holds 62,000. If 42,000 are already seated, how many seats are empty?

7. Brian bought 2 T-shirts for $4.95 each. How much change did he receive from $10?

8. Give the standard form for: 100,000 + 30,000 + 4000 + 800 + 20 + 7

9. $.06 + $.07 + $.02 + $.03 =  ?

10. A pie has 8 pieces. Meg ate 1 piece and Ted ate 2. How many pieces are left?

# A

**actual distance** A real-world length.

**addend** A number that is added to another number or numbers.

$$3 + 4 = 7$$
addends

**addition** A joining operation on two or more numbers that gives a total.

**A.M.** Letters that indicate time from midnight to noon.

**angle** The figure formed by two rays that meet at a common endpoint.

**area** The number of square units needed to cover a flat surface.

**associative (grouping) property** Changing the grouping of the addends (or factors) does not change the sum (or product).

**average** A quotient derived by dividing a sum by the number of its addends.

# B

**bar graph** A graph that uses bars of different lengths to show data.

**benchmark** An object of known measure that can be used to estimate the measure of other objects.

# C

**calender** A system used to organize the days of the week into months and years.

**capacity** The amount, usually of liquid, that a container can hold.

**centimeter (cm)** A metric unit used to measure small objects; 1 cm = 10 dm.

**circle** A simple closed curve; all the points on the circle are the same distance from the center point.

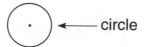
circle

**commutative (order) property** Changing the order of the addends (or factors) does not change the sum (or product).

**compatible numbers** Two numbers, one of which divides the other evenly.

**cone** A space or solid figure that has one circular base.

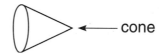
cone

**congruent figures** Figures that have the same size and the same shape.

**corner** The place where two or more lines meet.

**cube** A space or solid figure whose six faces are congruent squares.

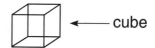
cube

**cup (c)** A customary unit of capacity; 2 c = 1 pt.

**customary system** The measurement system that uses inch, foot, yard, and mile; cup, pint, quart, and gallon; and ounce and pound.

**cylinder** A space or solid figure that has two circular bases.

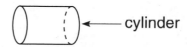
cylinder

## D

**data** Facts or information.

**decimal** A number in base ten that is written with a decimal point.

2.04 ◄—— decimal
↑
└—— decimal point

**decimeter (dm)** A metric unit of length equal to 10 cm.

**degree Celsius (°C)** A unit for measuring temperature. Water freezes at 0°C.

**degree Fahrenheit (°F)** A unit for measuring temperature. Water freezes at 32°F.

**denominator** The numeral below the bar in a fraction; it names the total number of equal parts.

$\frac{1}{2}$ ◄—— denominator

**difference** The answer in subtraction.

**digit** Any one of the numerals 0, 1, 2, 3, 4, 5, 6, 7, 8, or 9.

**dividend** The number to be divided.

24 ÷ 4          4)24
↑                    ↑
└——— dividend ———┘

**divisible** One number is divisible by another if it can be divided by that number and yield no remainder.

**division** An operation on two numbers that tells how many equal sets or how many are in each equal set.

**divisor** The number by which the dividend is divided.

36 ÷ 9          9)36
   └——— divisor ———┘

## E

**edge** The line segment where two faces of a space figure meet.

edge

**elapsed time** The amount of time between two given times.

**endpoint** The point at the end of a line segment or ray.

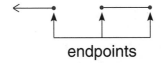
endpoints

**equation** (See **number sentence.**)

**equivalent fractions** Different fractions that name the same amount.

$\frac{1}{2} = \frac{2}{4}$

**estimate** An approximate answer; to find an answer that is close to an exact answer.

**even number** Any whole number that has 0, 2, 4, 6, or 8 in the ones place.

**expanded form** A way to write a number that shows the place value of each of its digits.

400 + 20 + 8 = 428

## F

**face** A flat surface of a space figure surrounded by line segments.

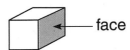
face

**factors** Two or more numbers that are multiplied to give a product.

5 ◄┐
×3 ◄┘ factors
15

3 × 5 = 15
↑    ↑
└——┘
factors

471

**family of facts** A set of related addition and subtraction facts or multiplication and division facts that use the same digits.

**flip** The movement of a figure over a line so that the figure faces in the opposite direction.

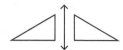

**foot (ft)** A customary unit used to measure length; 1 ft = 12 in.

**fraction** A number that names part of a whole or part of a set.

**front-end estimation** A way of estimating by using the front, or greatest, digits to find an approximate answer.

## G

**gallon (gal)** A customary unit of capacity; 1 gal = 4 qt.

**gram (g)** A metric unit used to weigh light objects; 1000 g = 1 kg.

**graph** A pictorial representation of data.

## H

**half gallon** A customary unit of capacity; 1half gallon = 2 qt.

**half hour** A unit of time; $\frac{1}{2}$ h = 30 min.

**hexagon** A polygon with six sides.

 hexagon

**hour (h)** A unit of time; 1 h = 60 min.

## I

**identity property (property of one)** The product of one and a number is that number.

**inch (in.)** A customary unit of length; 12 in. = 1 ft.

**intersecting lines** Lines that meet or cross.

## K

**key** A symbol that identifies the meaning of each picture in a pictograph.

**kilogram (kg)** A metric unit used to weigh heavy objects; 1 kg = 1000 g.

**kilometer (km)** A metric unit of length equal to 1000 meters; it is used to measure distance.

## L

**line** A straight set of points that goes on forever in both directions.

**line of symmetry** A line on which a figure can be folded so that the two halves exactly match.

**line segment** A part of a line that has two endpoints.

**liter (L)** A metric unit used to measure large amounts of liquid; 1 L = 1000 mL.

472

## M

**map distance**   A scaled length representing an actual length.

**mass**   The measure of the amount of matter an object contains.

**meter (m)**   A metric unit used to measure long lengths; 1 m = 100 cm; 1 m = 10 dm.

**metric system**   The measurement system that uses centimeter, decimeter, meter, and kilometer; milliliter and liter; and gram and kilogram.

**mile (mi)**   A customary unit used to measure distance; 1 mi = 5280 ft; 1 mi = 1760 yd.

**milliliter (mL)**   A metric unit of capacity. 1000 mL = 1 L

**minus (–)**   A minus symbol indicates subtraction. Read 10 – 8 as *10 minus 8*.

**minute (min)**   A unit of time; 60 min = 1 h.

**missing addend**   An unknown addend in addition.

$$\underline{\ ?\ } + 8 = 17$$

**missing factor**   An unknown factor in multiplication.

$$9 \times \underline{\ ?\ } = 27$$

**mixed number**   A number that is made up of a whole number and a fraction.

$$1\tfrac{1}{2} \longleftarrow \text{mixed number}$$

**multiplication**   A joining operation on two or more numbers to find a total of equal groups.

## N

**number line**   A line that is used to show the order of numbers.

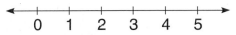

**number sentence**   An equation or inequality.

$$16 = 9 + 7 \qquad 28 < 52$$

**numerator**   The numeral above the bar in a fraction; it names the number of parts being considered.

$$\tfrac{3}{4} \longleftarrow \text{numerator}$$

## O

**odd number**   Any whole number that has 1, 3, 5, 7, or 9 in the ones place.

**ordered pair**   A pair of numbers that is used to locate a point on a graph.

**ordinal number**   A number that indicates order or position.

2nd (second)

5th (fifth)

**ounce (oz)**   A customary unit used to weigh small objects; 16 oz = 1 lb.

## P

**parallel lines**   Lines in the same plane that are always the same distance apart.

**parallelogram**   A quadrilateral that has opposite sides parallel and equal.

**pentagon**   A polygon with five sides.

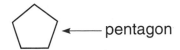

**perimeter** The distance around a figure.

**period** A group of three digits set off by commas in a whole number.

**pictograph** A graph that uses pictures or symbols to represent data.

**pint (pt)** A customary unit of capacity; 1 pt = 2 c.

**plane figure** A flat figure.

**plus (+)** A plus symbol indicates addition. Read 4 + 5 as *4 plus 5*.

**P.M.** Letters that indicate time from noon to midnight.

**point** An exact location.

**polygon** A simple closed flat figure made up of three or more line segments.

**pound (lb)** A customary unit used to weigh large objects; 1 lb = 16 oz.

**probability** The chance or likelihood of an event occurring.

**product** The answer in multiplication.

**pyramid** A space or solid figure that has a polygon for a base and has triangular faces that meet at a point.

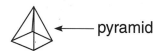
pyramid

## Q

**quadrilateral** Any four-sided polygon.

**quart (qt)** A customary unit of capacity; 1 qt = 2 pt.

**quarter hour** A unit of time; $\frac{1}{4}$ h = 15 min.

**quotient** The answer in division.

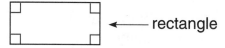

$$56 \div 8 = 7 \qquad 8\overline{)56}$$

quotient

## R

**ray** A part of a line with one endpoint that goes on forever in one direction.

**rectangle** A quadrilateral with four right angles whose opposite sides are parallel.

rectangle

**rectangular prism** A space or solid figure whose six faces are rectangles.

**regrouping** Trading one from a place for ten from the next lower place, or ten from a place for one from the next higher place.

**remainder** The number left over after dividing.

$$3\overline{)23} \quad 7\,R\,2$$
$$-\underline{21}$$
$$2 \quad\leftarrow\text{remainder}$$

**right angle** An angle that forms a square corner.

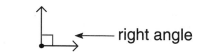
right angle

**right triangle** A triangle that has one right angle.

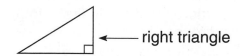
right triangle

**Roman numerals** Symbols for numbers used by the Romans.

**rounding**   Writing a number to the nearest ten or ten cents, hundred or dollar, and so on.

## S

**scale**   The ratio of the length in a drawing to the actual length.

**side**   A line segment that forms part of a polygon.

**similar figures**   Figures that have the same shape. They may or may not be the same size.

**skip counting**   Counting by a whole number other than one.

**slide**   The movement of a figure along a line.

**space (or solid) figure**   A figure that is not flat but that has volume.

**sphere**   A space or solid figure that is shaped like a ball.

sphere

**square**   A quadrilateral that has four congruent sides and four right angles.

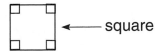
square

**square number**   A product that can be represented by a square array.

**standard form**   The usual way of using digits to write a number.

**subtraction**   A separating operation on two numbers that results in a difference.

**sum**   The answer in addition.

## T

**tally**   A count made by using tally marks.

**temperature**   The measurement of how cool or warm something is.

**thermometer**   An instrument used to measure temperature.

**trading**   (See **regrouping**.)

**triangle**   A polygon with three sides.

**turn**   The movement of a figure around a point.

## V

**volume**   The number of cubic units needed to fill a space figure.

## W

**weight**   The heaviness of an object.

**whole number**   Any of the numbers such as 0, 1, 2, 3, 4, . . .

## Y

**yard (yd)**   A customary unit used to measure length; 1 yd = 36 in.; 1 yd = 3 ft.

## Z

**zero (identity) property of addition**   The sum of zero and a number is that number.

**zero property of multiplication**   The product of zero and a number is zero.

## Mathematical Symbols

| | | | | | | |
|---|---|---|---|---|---|---|
| = | is equal to | . | decimal point | $\overleftrightarrow{AB}$ | line $AB$ |
| ≠ | is not equal to | ° | degree | $\overline{AB}$ | line segment $AB$ |
| < | is less than | + | plus | $\overrightarrow{AB}$ | ray $AB$ |
| > | is greater than | − | minus | ∠$ABC$ | angle $ABC$ |
| $ | dollars | × | times | ‖ | is parallel to |
| ¢ | cents | ÷ | divided by | ⊥ | is perpendicular to |
| | | | | (3, 4) | ordered pair |

## Table of Measures

### Metric Units

**Length**

100 centimeters (cm) = 1 meter (m)
  10 centimeters     = 1 decimeter (dm)
  10 decimeters     = 1 meter
1000 meters        = 1 kilometer (km)

**Capacity**

1000 milliliters (mL) = 1 liter (L)

**Mass**

1000 grams (g)     = 1 kilogram (kg)

### Customary Units

**Length**

 12 inches (in.) = 1 foot (ft)
  3 feet      = 1 yard (yd)
 36 inches   = 1 yard
5280 feet     = 1 mile (mi)
1760 yards   = 1 mile

**Capacity**

 8 fluid ounces (fl oz) = 1 cup (c)
 2 cups         = 1 pint (pt)
 2 pints        = 1 quart (qt)
 4 quarts      = 1 gallon (gal)

**Weight**

 16 ounces (oz)    = 1 pound (lb)
2000 pounds      = 1 ton (T)

**Money**

| | | |
|---|---|---|
| 1 nickel | = | 5¢ or $ .05 |
| 1 dime | = | 10¢ or $ .10 |
| 1 quarter | = | 25¢ or $ .25 |
| 1 half dollar | = | 50¢ or $ .50 |
| 1 dollar | = | 100¢ or $1.00 |
| 2 nickels | = | 1 dime |
| 10 dimes | = | 1 dollar |
| 4 quarters | = | 1 dollar |
| 2 half dollars | = | 1 dollar |

**Time**

| | |
|---|---|
| 60 seconds (s) | = 1 minute (min) |
| 60 minutes | = 1 hour (h) |
| 24 hours | = 1 day (d) |
| 7 days | = 1 week (wk) |
| 12 months (mo) | = 1 year (yr) |
| 52 weeks | = 1 year |
| 365 days | = 1 year |
| 366 days | = 1 leap year |